AN INTRODUCTION TO RADIO ASTRONOMY

Third Edition

Written by two prominent figures in radio astronomy this well-established, graduate-level textbook is a thorough and up-to-date introduction to radio telescopes and techniques. It is an invaluable overview for students and researchers turning to radio astronomy for the first time.

The first half of the book describes how radio telescopes work – from basic antennas and single-aperture dishes through to full aperture-synthesis arrays. It includes reference material on the fundamentals of astrophysics and observing techniques. The second half of the book reviews radio observations of our Galaxy, stars, pulsars, radio galaxies, quasars and the cosmic microwave background.

This third edition describes the applications of fundamental techniques to newly developing radio telescopes, including ATA, LOFAR, MWA, SKA and ALMA, which all require an understanding of aspects specific to radio astronomy. Two entirely new chapters now cover cosmology, from the fundamental concepts to the most recent results of WMAP.

BERNARD F. BURKE is William A. M. Burden Professor of Astrophysics, Emeritus, in the Department of Physics, Massachusetts Institute of Technology. He was the co-discoverer of radio noise from Jupiter, and he was later involved in the development of very-long-baseline interferometry. He has been a Visiting Professor at the University of Leiden and the University of Manchester, is a member of the National Academy of Science, and is on the governing board of the National Science Foundation.

F. GRAHAM-SMITH is an Emeritus Professor at the Jodrell Bank Observatory, University of Manchester. He has been Director of the Royal Greenwich Observatory and President of the Royal Astronomical Society, and was the 13th Astronomer Royal. He is a Fellow of the Royal Society, and researches in many fields of radio astronomy, particularly pulsars.

AN INTRODUCTION TO RADIO ASTRONOMY

Third Edition

BERNARD F. BURKE

Massachusetts Institute of Technology

F. GRAHAM-SMITH

Jodrell Bank Observatory, University of Manchester

CAMBRIDGE
UNIVERSITY PRESS

The Edinburgh Building, Cambridge CB2 8RU, UK

Published in the United States of America by Cambridge University Press, New York

Cambridge University Press is part of the University of Cambridge.

It furthers the University's mission by disseminating knowledge in the pursuit of education, learning and research at the highest international levels of excellence.

www.cambridge.org
Information on this title: www.cambridge.org/9781107672604

© B. Burke and F. Graham-Smith 2010

First published 1996
Second edition 2002
Third edition 2010
First paperback edition 2013

A catalogue record for this publication is available from the British Library

ISBN 978-0-521-87808-1 Hardback
ISBN 978-1-107-67260-4 Paperback

Contents

Preface

Astronomy makes use of more than 20 decades of the electromagnetic spectrum, from radio to gamma rays. The observing techniques vary so much over this enormous range that there are distinct disciplines of gamma-ray, X-ray, ultraviolet, optical, infrared, millimetre and radio astronomy, often concentrated in individual observatories. Modern astrophysics depends on a synthesis of observations from the whole wavelength range, and the concentration on radio in this text needs some rationale. Apart from the history of the subject, which developed from radio communications rather than as a deliberate extension of conventional astronomy, there are two outstanding characteristics that call for a special exposition. First, the astrophysics: long-wavelength radio waves are most often observed as a continuum in which the interaction with matter follows classical electrodynamics. High-energy electrons are involved; they are created in a variety of circumstances, and their radiation as they circulate in magnetic fields gives evidence of new phenomena, often showing a close link to the phenomena observed in X-rays and gamma-rays. At the shorter wavelengths the low quantum energy gives access to spectral lines from atomic and molecular species at comparatively low temperatures. Second, the techniques: radio astronomy takes account of the phase as well as the intensity of incoming radio waves, allowing the development of interferometers of astonishingly high angular resolution and sensitivity.

The third edition of this *Introduction* was stimulated by recent remarkable advances both in techniques and astrophysics. Without question the most important advance has been in the observations of the cosmic microwave background by the WMAP satellite. We present the results of the 5-year data reductions, which give a large number of fundamental cosmological constants with unprecedented accuracy. Also, a new generation of radio telescopes, with dramatically improved performance, is under construction, most of them such large enterprises that they necessarily involve international collaboration. The techniques follow well-established principles, but the advent of massive computer power and broad-band fibre-optic communications has only recently brought these schemes within the range of possibility. At the same time, the success of existing telescopes has shown what can be achieved by the new telescopes in several astrophysical domains, such as pulsars and black-hole physics, and particularly in addressing fundamental cosmology.

We aim therefore to extend our exposition of the fundamentals of radio astronomy in two directions, cosmology and technology. The cosmological discoveries of WMAP demonstrate new directions for CMB measurements, with polarization having particularly strong potential. The techniques of aperture synthesis have developed to allow the use of very large collecting areas. With the ever-advancing technology of digital circuits and wide-band, low-noise amplifiers, the attendant increased sensitivity and high angular resolution, including wide field coverage, open new areas of astrophysical research. These instruments will demand the efforts of a large work force, and they will provide material for a large new body of observers and astrophysicists. Our aim is to provide a basic introduction for this expanding community.

The plan of the book is twofold: we hope that the scope and impact of radio-astronomy observations will be demonstrated in the astrophysical discussion, and at the same time we intend to give a brief but comprehensive treatment of the elegant technologies that have developed. The breadth of the subject matter necessarily limits the length of the treatment for each subject; we have tried, therefore, to provide recent, comprehensive references to the extent that they are available. Cosmology, and especially the study of the cosmic microwave background, has been transformed in the last decade; here we have attempted a basic exposition as well as a presentation of the astounding conclusions from recent observations.

In addition to the astronomy graduate student and those professionally committed to radio astronomy, there is a wider audience for whom this book is intended: the interested astronomers from outside the field who want to be informed of the principal ideas current in radio astronomy, and may even be thinking of carrying out radio observations that would complement other work in progress. Even though we have mainly kept our discussions within the boundary of radio astronomy for the sake of convenience, everyone is aware that the boundaries between disciplines have dwindled in importance. Radio observations would have been a baffling puzzle if the optical identifications of sources had not been made, and both radio and X-ray astronomers have long been aware of their kinship, since both study high-energy phenomena, though at the opposite ends of the spectrum. The techniques vary, but the astronomer of the future should have access to the entire electromagnetic spectrum.

The text of this third edition has been extensively rewritten, especially in the important technical areas of interferometry and aperture synthesis and in most areas of astrophysics and cosmology. Keeping up with such a rapidly moving subject is impossible, but we have taken the advice of many colleagues, and have attempted to keep to the original objectives. We hope we have succeeded in providing an introduction that is useful both to the observer and to the astrophysicist; perhaps it will appeal most to those who, like ourselves, enjoy membership of both categories.

Bernard F. Burke
F. Graham-Smith

1

Introduction

1.1 The role of radio observations in astronomy

The data give for the coordinates of the region from which the disturbance comes, a right ascension of 18 hours and declination of $-10°$.

(Karl G. Jansky 1933)

Jansky's discovery of radio emission from the Milky Way is now seen as the birth of the new science of radio astronomy. Most astronomers remained unaware of this momentous event for at least the next decade, and its full significance became apparent only with the major discoveries in the 1950s and 1960s of the 21-cm hydrogen line, the quasars, the pulsars and the cosmic microwave background. These are now fully assimilated into astronomy, and radio is now regarded as one among the several tools available to astronomers in their pursuit of the astrophysics of our Galaxy, or of neutron stars, black holes or cosmology. Nevertheless, radio astronomy has its own distinctive character, not least in its techniques and the particular expertise which they demand. It also has several fields of application in which it is uniquely useful: there is no other way of exploring the cosmic microwave background, it allows spectroscopic investigation of molecular clouds in the Milky Way and it reveals a previously unseen Universe through the synchrotron emission of high-energy particles in stars, galaxies and quasars.

The history of this development is outlined at the end of this book in Appendix 3. Our purpose in the main text is to set out those parts of astrophysics and observational techniques which are particularly appropriate in radio astronomy, so that the subject may be fully available to all astronomers and to physicists with a wide range of backgrounds. There is, throughout all radio astronomy, a close relation with other wavebands, and we shall acknowledge, for example, the necessity of using optically measured redshifts for distances of quasars and the components of gravitational lenses, and the need to bring together the X-ray and radio observations to obtain a coherent picture of neutron stars in our Galaxy.

The Milky Way, our Galaxy, which is the origin of the radio noise first observed by Jansky, is a complex assembly of stars of widely varying ages, embedded in an *interstellar*

1

medium (ISM) of ionized and neutral gas, itself displaying a great diversity and complexity throughout the electromagnetic spectrum. Optical astronomy primarily addresses the surfaces of the stars, or nearby gas ionized by those stars, where the temperatures bring thermal radiation naturally into the visible range. X-ray astronomy deals with much hotter regions, such as the million-degree ionized gas which is found in such diverse places as the solar corona and the centres of clusters of galaxies. Infrared astronomy studies relatively cool regions, where thermal radiation from the dust component of the ISM is a prominent feature; warmer regions are also studied, where the thermal radiation from star-forming regions is also strong. Radio astronomy, using much longer wavelengths, addresses a broad range of both thermal and non-thermal phenomena, including the thermal radiation from the 21-cm line of neutral hydrogen in the ISM, and the thermal radiation from a wide variety of molecular lines, coming from dense, extremely cold gas concentrations that are found within the ISM. The radio noise discovered by Jansky belongs to an entirely different, non-thermal regime; it comes from charged particles, usually electrons, with very high energies, moving at relativistic velocities in the magnetic fields of the ISM. This regime of *synchrotron* radiation also accounts for the intense radio emission from quasars and other interesting objects in the Universe. Synchrotron radiation, although it can generate radiation up to X-ray and beyond, is a particularly prominent long-wavelength phenomenon, giving radio astronomy a unique role in the investigation of some of the most energetic objects in the Universe.

Radio is uniquely suited to observations of the cosmic microwave background (CMB), which is a black body whose 3 K temperature has a maximum emissivity at wavelengths of order 1–2 mm. Radio observations of the CMB have created a virtually new subject of precision cosmology.

The methods of the radio astronomer often appear to be quite different from those of the optical astronomer, but there is the same aim in all of astronomy, from the radio to the X-ray and gamma-ray domains. Nature presents us with a distribution of brightness on the sky, and it is the task of the astronomer to deduce, from this brightness distribution of electromagnetic radiation, what the radiating sources are, and what physical processes are acting. The distinguishing feature of a radio telescope is that the radiation energy gathered by the instrument is not measured immediately, a process known as *detection* in radio terminology. Instead, the radiation is amplified and manipulated coherently, preserving its wave-like character, before it is finally detected. The instrumental goals of the radio astronomer – obtaining a larger collecting area, greater angular resolution and more sensitive detectors – are otherwise the same as they are for all the astronomical disciplines.

To illustrate the relation between radio and other astronomies, the energy flux of electromagnetic radiation arriving at the Earth's surface from the cosmos is plotted in Figure 1.1. The wavelength scale runs from the standard radio broadcast band to the X-ray region, and the atmospheric windows are indicated schematically. The optical window is seen to be relatively narrow, blocked at the ultraviolet end by ozone absorption and at still shorter wavelengths by oxygen and nitrogen, while at the infrared end the principal absorbing

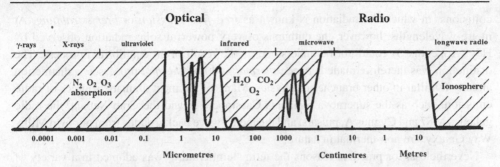

Figure 1.1. The electromagnetic spectrum, showing the wavelength ranges of the 'windows'. The radio range is limited by the ionosphere at wavelengths greater than a few metres. Atmospheric absorption becomes significant in the sub-millimetre range.

agents are water vapour and carbon dioxide. The short-wavelength blockage is so complete that ultraviolet and X-ray work must be carried out above the atmosphere. There are occasional windows in the atmosphere at infrared wavelengths, allowing observations to be made from high, dry mountain sites, but for the most part the observations must be taken from airplanes, balloons or satellites, depending on the particular spectral region. It is easy to see that there is a great stretch of spectrum at the radio end that has relatively little trouble with the atmosphere. Before Jansky's discovery, however, there was no reason to expect much of interest in the radio spectrum; if stars were the principal sources of radiation, very little radio emission could be expected. The maximum radiation from even the coolest of the known stars falls at visible or infrared wavelengths, and their contribution to the radio end of the spectrum is almost negligible. The slow response to Jansky's discovery is understandable both in terms of technical difficulty and lack of expectation.

The radio window is often divided into bands by frequency and wavelength: HF (below 30 MHz), VHF (30–300 MHz), UHF (300–1000 MHz), microwaves (1000–30 000 MHz), millimetre-wave and sub-millimetre-wave. Certain microwave bands have acquired particular names: L-band ($\approx$20 cm), S-band ($\approx$10 cm), C-band($\approx$6 cm), X-band ($\approx$3 cm), Ku-band (sometimes called U-band, $\approx$2 cm) and K-band ($\approx$1 cm). The names of the microwave bands are rooted in history, like the s, p, d states of atoms; they are commonly met with in practice.

1.2 Thermal and non-thermal processes

During the pioneering stage of radio astronomy, as a wide range of celestial objects turned out to be detectable in the radio spectrum, two broad classes of emitter became clearly distinguished. At centimetric wavelengths the radio emission from the Sun, as observed by Southworth (see Appendix 3), could be understood as a thermal process, with an associated temperature. The term *temperature* implies that there is some approximation to an equilibrium, or quasi-equilibrium, condition in the emitting medium; in this case the medium is the ionized solar atmosphere. The mechanism of generation is electron–ion collisions, in which the radiation is known as *free–free emission* or *bremsstrahlung*. At metre wavelengths, however, the outbursts of very powerful solar radiation observed by Hey (Appendix 3) could not be understood as the result of an equilibrium process. The distinction was therefore made between *thermal* and *non-thermal* processes, a distinction already familiar in other branches of physics. Many of the most dramatic sources of radio emission, such as the supernova remnants Cassiopeia A and the Crab Nebula, the radio galaxies M87 and Cygnus A, pulsars, and the metre-wave backgrounds from our own Milky Way Galaxy, are non-thermal in nature.

Nevertheless, for practical reasons the term 'temperature' was adopted in a variety of contexts, following practices that had been used widely in physics research during the 1940s. Within a system, individual components can exhibit different temperatures. In a plasma excited by a strong radio-frequency field, for example, the electron and ion components of the gas may each show velocity distributions that can be approximated by Maxwell–Boltzmann distributions, but at quite different temperatures. Each component is in a state of approximate thermal equilibrium, but the systems are weakly coupled and derive their excitation from different energy sources. One can speak, therefore, of two values of *kinetic temperature*, the *electron temperature* and the *ion temperature*.

A two-state system such as the ground state of the hydrogen atom, in which the proton and electron spins can be either parallel or antiparallel, can be used as a simple and illuminating example of how the temperature concept can be generalized. Given an ensemble of identical two-state systems at temperature T_s, the mean relative population of the two states, $\langle n_2/n_1 \rangle$, is given by the Boltzmann distribution

$$\langle n_2/n_1 \rangle = \exp[-\epsilon/(kT_s)], \qquad (1.1)$$

where the energy separation ϵ corresponds to a photon energy $h\nu$. If the two states are degenerate, the statistical weights g_1 and g_2 must be applied. The relationship can be inverted, however: for any given average ratio of populations, there is a corresponding value of the temperature, defined by the Boltzmann equation. This defines the *state temperature*, T_s.

The state temperature need not be positive. When $\langle n_2/n_1 \rangle$ is greater than 1, Equation (1.1) requires a negative state temperature, and this is precisely the condition for *maser* or *laser*[1] action to occur. A resonant photon beam traversing a medium at a positive state temperature suffers absorption, whereas it will be amplified if the state temperature is negative. Naturally occurring masers are common in astrophysics, particularly in star-forming regions and in the atmospheres of red giants. These are treated in Chapters 7 and 9. The population inversion is maintained by a pumping mechanism, which can be either radiative or collisional; the action may be to fill the upper state faster than the lower state, or to populate both states, but with the lower state being drained of population more rapidly.

Atomic and molecular systems almost always have a large number of bound states, and one can associate a state temperature with each pair of states. If the system is in a state

---

[1] These acronyms stand for **M**icrowave (and **L**ight) **A**mplification by the **S**timulated **E**mission of **R**adiation.

of thermal equilibrium, all these temperatures will be the same. The case of blackbody radiation, which is an example of a system with a continuum of energy states, is treated in Chapter 2 and the idea of *brightness temperature* is introduced. In brief, this assigns to an emitter of radiation at frequency ν the temperature that it would have to have if it were a black body. This need not correspond to a physical temperature, and the powerful non-thermal emitters exhibit brightness temperatures that can exceed 10^{12} K. Conversely, a thermal source of radiation need not have a brightness temperature that is constant over a wide spectrum and equal to the physical temperature; the variation of brightness temperature with frequency will be determined by the equation of radiative transfer, introduced in Chapter 7.

The temperature concept is also extended to the practice of receiver measurement. An ideal amplifier should add as little noise as possible to the system, although the laws of quantum mechanics prevent an amplifier from being entirely noise-free. The total excess noise is described as the *system noise temperature*; the definition arises from the properties of a resistor as a noise generator. Every physical resistance generates noise, because of thermal fluctuations in the sea of conduction electrons, and the noise power per unit bandwidth that can be extracted from the resistor is proportional to its temperature. The excess noise observed with any radio-astronomy receiver can be described by stating what temperature a resistive load would have to have, when connected to the input, to generate the observed noise. This turns out to be an entirely practical way to describe the system, because the faint continuum radio signals that one deals with are most conveniently calibrated by using as a reference the continuum noise generated by a hot (or cold) resistive load.

1.3 Radiation processes and radio observations

The use of thermodynamic concepts has more than a formal value. General properties of radiation processes and general theorems about antennas and receivers can be deduced from thermodynamic considerations. In a blackbody enclosure, where the radiation is in equilibrium with all matter in the enclosure, there is no need to specify any details of emission or absorption processes. The best practical example in radio astronomy is the CMB radiation, which at wavelengths shorter than 20 cm becomes the predominant source of the sky brightness (except for a strip about 3 degrees wide along the Galactic plane caused by thermal radiation from the interstellar medium). The CMB is specified completely by the temperature 2.74 K, from which the intensity over a wide range of wavelengths can be calculated. No radiation process need be invoked in the calculation; the radiation was originally in equilibrium with matter in an early stage of cosmic evolution, and has preserved its blackbody spectrum in the subsequent expansion and cooling.

The sky at long radio wavelengths is, however, very much brighter than is expected from the cosmic background alone; its brightness temperature at wavelength 10 m (30 MHz) exceeds 100 000 K. This radiation originates from high-energy electrons circulating in the magnetic field that permeates the interstellar medium in the Galaxy, which radiate predominantly at long wavelengths, with a spectrum that is completely different from that of a black body. This brings in an immediate cross-disciplinary contact with the study of the very energetic cosmic rays, a connection that might have been thought unlikely because of the low energies of radio photons. The relevance of radio observations to high-energy phenomena has continued, since the radio and X-ray observations of active galactic nuclei and quasars have close relationships to one another. One might note that there is a complementarity with optical observations as well; regions of high X-ray and radio luminosity tend to be faint optically, since the effective temperatures are so extreme that the matter is often highly ionized. Nevertheless, the optical observations are essential to understanding what types of object are the sources of emission, and to put the radio observations into a physical and astrophysical context.

Observations do not take place as an abstract process, and the diligent observer will have a knowledge of the characteristics of the instrument that is being used to take the data. With this familiarity, advantage can be taken of new and unexpected uses of an instrument, and caution can be exercised in interpreting data that may contain instrumentally induced flaws. The basic properties of radio telescopes are summarized in Chapter 4, followed by expositions of interferometry and aperture synthesis in Chapters 5 and 6. Both single-aperture telescopes and synthesis arrays are in intensive use, and the language of Fourier transforms is appropriate to both kinds of instrument. It will be obvious that Fourier-transform methods have wide applications to nearly all fields of science and technology, including radiation processes, antenna theory and, especially, aperture-synthesis interferometry. Most readers will be familiar, to a greater or lesser extent, with the Fourier transform; as an aid to the memory, Appendix 1 summarizes its basic properties and applications.

A careful observer will always be aware that the statistical significance of a result must be evaluated. In radio astronomy, one is nearly always looking for signals in the presence of noise, and Chapter 3 gives an exposition of the properties of random noise. Here, too, Fourier methods are essential, both in radiometry and spectroscopy.

The propagation of radio waves through stellar atmospheres, the interstellar medium and the terrestrial atmosphere differs in some important ways from the more familiar optical case. In Chapter 7 we deal with radiative transfer, leading to a brief exposition of maser action, and with the various effects of refraction in the ionized stellar medium. These effects are part of the tools of radio astronomy, giving access to such diverse quantities as the dynamics of gas motions close to an active galactic nucleus and the configuration of the magnetic field of our Galaxy.

Chapters 8–13 show how these various radio techniques have provided new insights into the astrophysics of stellar atmospheres, neutron stars, galaxies and quasars. Observations of the CMB, which is the subject of Chapters 14 and 15, have transformed our understanding of the Universe, and have given access to some of the most fundamental aspects of cosmology. Chapter 16 deals with the population of radio galaxies and quasars within the Universe. Finally, in Chapter 17, we note that radio astronomy is entering a new era in which very large and sensitive radio telescopes are being built as international projects, covering the whole of the radio range from metre to sub-millimetre wavelengths.

2

The nature of the radio signal

All telescopes – radio, optical and X-ray – couple the electromagnetic radiation from sources in the Universe to the astronomer's measuring devices. Spacecraft can explore the solar system directly, but otherwise the Universe is accessible to us only by observing the distribution of electromagnetic radiation across the sky, including its variation with time, frequency and state of polarization. For the radio astronomer, the incoming radiation can be treated as a superposition of classical electromagnetic waves, whereas for the optical or X-ray astronomer, the radiation is arriving as photons, discrete quanta of energy. Infrared astronomy is between these extreme regimes; the 'far' infrared is close to millimetric-wavelength radio in techniques, while the 'near' infrared is regarded as an extension of the optical regime. All astronomical observing starts with the telescope intercepting the incoming electromagnetic radiation. The received radiation goes to a radiometer, followed by a detection apparatus, which may be integral with the radiometer. The principal difference between radio astronomy and astronomy at other wavelengths is the use of low-noise amplifiers prior to signal detection and the consequent possibilities of using signal-processing techniques. The laws of quantum mechanics limit the use of amplifiers at shorter wavelengths in most cases.

This chapter is concerned with the properties of electromagnetic radiation, with an emphasis on fields rather than photons. Radio telescopes are treated in a generic sense, linking engineering and astronomical aspects. One difference in terminology should be noted, because radio telescopes are sometimes called radio antennas. Here, the terms radio telescope and radio antenna will be used interchangeably; in general usage, the term antenna is used when the angle of reception is large, as it is for television antennas. When the angle of reception is small, as it is for steerable paraboloids (which are, indeed, analogous to optical telescopes), they are always called radio telescopes.

2.1 Flux density: the jansky

In this section, monochromatic operation at wavelength λ and corresponding frequency ν is usually assumed. In the real world the astronomer's instrument measures a flux S that is the rate at which energy E crosses area A perpendicular to the direction of propagation:

$$S = (dE/dt)/A. \tag{2.1}$$

The incoming flux is distributed over a finite receiving band, and it is usually a function of frequency. For this reason, we introduce the concept of a *spectrum* described by the flux per unit bandwidth S_ν, defined as the *flux density* or *specific flux*. The observed flux is equal to the flux density integrated over the receiving bandwidth:

$$S = \int S_\nu \, d\nu. \tag{2.2}$$

In later chapters, quantities will be introduced that are functions of more than one variable, but these variables play the role of frequencies. Such a function $Q(\nu_1, \nu_2, \ldots)$ can still be called a spectrum, albeit a multivariate spectrum.

An antenna can be treated either as a receiving device, gathering the incoming radiation field and conducting the electrical signals to the output terminals, or as a transmitting system, launching electromagnetic waves outwards. The two cases are equivalent because of time reversibility: the solutions of Maxwell's equations are valid when time is reversed. As a transmitter the antenna produces a beam of radiation whose solid angle is determined by the size of the aperture: the larger the aperture, the narrower is the beam and the greater is the maximum power flux at the centre of the beam. The concept of the *power gain* of an antenna, which arises in transmission, is therefore closely related to that of *effective area*, which applies to reception.

When a telescope is used in the receiving mode, it is natural to think of it as a receiving area, intercepting a power flux S and yielding a received power P_{rec}. The *effective area* A_{eff} is directionally dependent, and is a function of direction $\hat{\mathbf{a}}$, measured with respect to the antenna axis (generally the direction $\hat{\mathbf{a}}_0$ of maximum response), so that

$$P_{rec} = A_{eff} S. \tag{2.3}$$

The range of directions over which the effective area is large is the *antenna beamwidth*; from the laws of diffraction the beamwidth of an antenna with characteristic size d is of order λ/d.

As a transmitter, the same antenna would have a power gain $G(\hat{\mathbf{a}})$ in a direction $\hat{\mathbf{a}}$ that is the ratio of the power flux $S(\hat{\mathbf{a}})$ that would be measured at some large distance and the power flux from a hypothetical isotropic radiator measured at the same distance. For a transmitter power P_{tr}

$$S(\hat{\mathbf{a}}) = \frac{G(\hat{\mathbf{a}})\mathbf{P}_{tr}}{4\pi r^2}. \tag{2.4}$$

Conservation of energy requires that the integral of $S(\hat{\mathbf{a}})$ over the whole sky, i.e. 4π solid angle, must equal P_{tr}, so that

$$\int_0^{4\pi} G(\hat{\mathbf{a}}) d\Omega = 4\pi. \tag{2.5}$$

Most antennas concentrate the greater part of the radiation into a single principal beam of effective solid angle Ω_0. For order-of-magnitude calculations, therefore, the radiation outside the principal beam is neglected, and the gain in the principal beam can be approximated by

$$G = 4\pi/\Omega_0. \tag{2.6}$$

For a symmetrical aperture having an area of order D^2 the principal beam has a width of order λ/D and solid angle λ^2/D^2. It follows that the effective area A_{eff} is proportional to the gain:

$$A_{\text{eff}} = \lambda^2 G/(4\pi). \tag{2.7}$$

This presentation is only illustrative, but the relation is exact, as shown in Section 2.2 below, and holds for both radio and optical telescopes.

The detailed telescope power gain over all angles is known, from radio engineering, as its *polar diagram*. The polar diagram, in addition to having a principal beam, exhibits the complete diffraction pattern, and the response outside the principal beam is referred to as the *sidelobe response*. In optical terminology, the power gain response is usually called the *point-spread function*.

Astronomical sources, both radio and optical, have a finite angular extent, even though they may appear as point sources, and therefore their appearance is characterized by a flux per unit solid angle, called the *brightness*, $B(\theta, \phi)$. Furthermore, the radiated power is a function of frequency, and is designated the *specific brightness* B_ν, which is the flux per unit solid angle per frequency interval:

$$dB_\nu \equiv (dS/d\Omega)d\nu = (dE/d\Omega)dA\, d\nu\, dt. \tag{2.8}$$

Thus the total flux is given by the integral over solid angle and frequency band

$$S = \int\int B_\nu(\theta, \phi)d\Omega\, d\nu. \tag{2.9}$$

The radiative flux per unit solid angle per unit bandwidth is generally designated the *specific intensity* I_ν.

We reserve the term brightness B_ν for the specific intensity at the emitting surface, while I_ν is used for the local value along a ray path, since it may be changed by emission and absorption processes (Chapter 7). In free space, I_ν, or B_ν, is invariant; consequently in the absence of absorption the surface brightness is independent of the distance of the emitting object.

In the radio context mks units are generally used. The *specific power*, or *power density* P_ν, has units of W Hz^{-1}, and the *specific flux*, or *flux density*, S_ν, has units of $\text{W m}^{-2}\text{ Hz}^{-1}$. A convenient unit of flux density has been designated the *jansky* (Jy):

$$1\,\text{Jy} = 10^{-26}\,\text{W m}^{-2}\,\text{Hz}^{-1}.$$

The jansky is also widely used in the infrared spectrum; for example the Spitzer infrared satellite observatory has been calibrated over a wide band of wavelengths in terms of

Table 2.1. *Flux densities of a source with visual magnitude $m_v = 0$*

Band name	Wavelength (μm)	S_v (Jy)
V (visual)	0.556	3540
J	1.215	1630
H	1.654	1050
K	2.179	655
L	3.547	276
M	4.769	160
N	10.472	35

janskys (Werner 2004). The commonly encountered infrared bands J, H, K, L, M and N are determined by the Earth's atmosphere, which has numerous absorbing bands of water and carbon dioxide that block observations between the bands. The wavelength equivalents of these bands are shown in Table 2.1, which also shows the flux density in these bands for a star with visual magnitude $m_v = 0$.

The magnitude scale of optical astronomy was established in ancient times, and is now defined as a logarithmic scale on which a magnitude difference of 5 corresponds to a factor of 100 in flux ratio. The scale is fixed by measurements of the bright star Vega, an A0 star with a spectrum close to that of a black body. Separate calibrations are necessary for the individual photometric bands conventionally used in optical photometry.[1] The flux densities for the infrared bands in Table 2.1 are quoted in *Astrophysical Quantities* (Allen; 4th edn, 2000) from Cohen *et al.* (1992).

Radio astronomers, following the physicists and electrical engineers who founded the new science, also use a logarithmic scale, the decibel (see Section 3.2), to express the wide range of power encountered in radio engineering, but no decibel scale for fluxes has been adopted.

For historical reasons the calculation of astrophysical radiation processes will be given in cgs units. The necessary conversions between the practical use of mks and the use of cgs units in astrophysics should cause little difficulty.

2.2 Antenna temperature

Figure 2.1 shows a telescope that is enclosed in a black body, an enclosure filled with radiation in thermal equilibrium with the walls of the enclosure, at temperature T. The output terminals of the telescope are connected to a transmission line that in the radio case is usually a coaxial cable or a waveguide. The signal, with or without amplification, passes through a filter that determines the bandwidth B over which the signal is averaged. The

---

[1] The flux equivalents for commonly used visual bands are presented by Gray (1998); centred on the Johnson and Strömgren V bands at 5480 Å, where the flux of an $m_v = 0$ star is given as $F_\lambda = 3.68 \times 10^9 (5480/\lambda)^2 \, \mathrm{erg \, s^{-1} \, cm^{-2} \, Å^{-1}}$; this is equivalent to 3590 Jy at wavelength 0.556 μm.

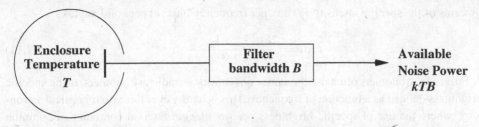

Figure 2.1. An antenna enclosed by a black body at temperature T.

transmission line is terminated by a matched load, a resistor having the same impedance as the line. (There are optical analogues for such passive devices in the infrared and visual parts of the spectrum: optical fibres are rigorously the same as waveguides, and all the properties of waveguide transmission apply; interference filters can determine a receiving band; a resistive sheet that has the impedance of free space, 377 Ω per square, will absorb a plane wave without reflection, acting as a matched load). If the transmission line is terminated by a matched load that is thermally isolated from the outside world, the second law of thermodynamics requires that it will reach an equilibrium temperature T that is identical to the blackbody temperature. Furthermore, the same radio frequency power density P_ν for any narrow band $\delta\nu$ at frequency ν must flow in both directions along the transmission line. This applies whatever the termination of the line, whether it is a resistor, another antenna or an infinite transmission line at temperature T.

If the telescope has a narrow receiving pattern that gathers only the radiation from a small solid angle in the black body, the result is the same, which leads to the conclusion that, if the telescope is viewing a blackbody surface at temperature T that is larger than the receiving pattern of the telescope, a matched load will still reach the temperature T after it has attained thermal equilibrium with the incoming radiation from the blackbody surface. This circumstance is called the *filled-beam* case. It is necessarily an approximation, because all the sidelobes of the gain pattern must also be within the angle subtended by the black body, but the sidelobe patterns fall rapidly enough for most telescopes to make the approximation valid. There is one case that must be recognized: if one is viewing an extended object in a region of low surface brightness, when there is a region of high surface brightness nearby, the approximation might not be valid, and the full sidelobe pattern must be taken into account. For example, this is necessary when studying the faint hydrogen emission in the outer parts of galaxies, where the more intense emission from the inner parts, picked up by the sidelobes, results in the addition of a false signal to the signal being studied.

The radiation density u_ν in the black body is described by the Planck distribution[2]

$$u_\nu \, \mathrm{d}\nu = \frac{8\pi h \nu^3}{c^3} \frac{1}{e^{h\nu/(kT)} - 1} \, \mathrm{d}\nu. \qquad (2.10)$$

[2] Recommended values of constants are Boltzmann $k = 1.3807 \times 10^{-23}$ J K^{-1} and Planck $h = 6.626 \times 10^{-34}$ J s.

In terms of the specific intensity I_ν (flux per frequency interval per solid angle)

$$I_\nu \, d\nu = \frac{2h\nu^3}{c^2} \frac{1}{e^{h\nu/(kT)} - 1} \, d\nu. \tag{2.11}$$

Infrared astronomers often use the same conventions as radio astronomers, using specific brightness defined as a function of frequency. This is hardly ever the case for optical astronomy, where the use of specific brightness per *wavelength* interval continues in common usage. The specific intensity $I_\lambda \, d\lambda$ in terms of wavelength is related to I_ν by

$$I_\lambda \, d\lambda = (c/\lambda^2)I_\nu \, d\nu. \tag{2.12}$$

In the classical derivation of the Planck formula one calculates the density of resonant modes within a blackbody cavity at temperature T with frequencies between ν and $\nu + d\nu$, and associates with each mode an average energy depending on its frequency. When this energy is small compared with kT, i.e. $h\nu/(kT) \ll 1$, the distribution of energy among the modes follows a Rayleigh–Jeans (RJ) law, with flux density increasing as the square of the wavelength, while at frequencies well above kT/h, in the regime of Wien's law, the flux falls exponentially with increasing frequency (a situation commonly met in X-ray astronomy). At millimetre and sub-millimetre wavelengths the full Planck law is often needed, but for frequencies well below kT/h the RJ law suffices:

$$I_\nu \, d\nu = (2kT\nu^2/c^2)d\nu. \tag{2.13}$$

It becomes important to use the full Planck formula at short radio wavelengths and at low temperatures. The maximum of Equation (2.10) occurs at $\nu_{max} = 58.8T$ GHz, while the maximum of the corresponding equation in terms of wavelength occurs at $\lambda_{max} = 0.2898T^{-1}$ cm. For example, if $T = 2.73$ K as in the microwave background, $h\nu/(kT) = 1$ at $\nu = 57$ GHz, that is, at wavelength 5.3 mm (note that the actual peak value of I_ν is at 1.87 mm). The spectrum of the microwave background radiation (Chapter 14) agrees beautifully with the full Planck law; the best fit is given for $T = 2.726 \pm 0.010$ K.

The corresponding derivation for the noise power flowing in a single-mode transmission line connected to a black body at temperature T leads to the one-dimensional analogue of the Planck law:

$$P_\nu \, d\nu = \frac{h\nu}{e^{h\nu/(kT)} - 1} \, d\nu. \tag{2.14}$$

Here P_ν is the *power density*, which is the power per unit bandwidth flowing in each direction along the transmission line. In the RJ approximation this reduces to

$$P_\nu \, d\nu = kT \, d\nu, \tag{2.15}$$

showing again that, within any given narrow band, noise power is proportional to temperature. The noise generated in a resistor R, known as *Johnson noise*, has the same r.m.s. voltage as in a matched transmission line with characteristic impedance R; giving $\langle v_n^2 \rangle = 4RkT$.

We shall use the RJ approximation freely except where explicit recognition of quantum effects is required.

Thermal noise powers are small by everyday standards. For example, consider a radio telescope with a narrow beam pointed at the Moon, and let the beamwidth be much smaller than the angular width of the Moon, so that sidelobe effects can be neglected. The Moon at radio wavelengths is nearly an ideal black body, at a temperature of 216 K, so the antenna temperature will be the same, 216 K, independently of the beamwidth of the radio telescope. (This is the filled-beam case, since the angular size of the black body is much larger than the beamwidth of the radio telescope.) The antenna output then equals that from a resistive load at 216 K. With a typical bandwidth of 1 MHz and at wavelengths for which the RJ approximation (Equation (2.13)) holds, the thermal noise power will be 0.003 picowatts.

For an interesting comparison let a small transmitter, generating power P_{tr}, be placed on the Moon, at a distance D_m from Earth, radiating its power through a small antenna with a gain G_{tr}. According to Equations (2.3) and (2.4), a radio telescope of diameter d_{tel} will receive an amount of power P_{rec} given by

$$P_{rec} = (1/16)G_{tr}(d_{tel}/D_m)^2. \tag{2.16}$$

Here it is assumed that the effective area of the telescope is its geometrical area, a goal approached but never reached in practice. To match the solid angle of the Moon the telescope diameter must be approximately 120 wavelengths, which for wavelength 3 cm would be 36 m. As an example, take a one-watt transmitter, radiating from an antenna that has a gain of 4 over isotropic, and let the distance to the Moon be 300 000 km. These numbers give a received power of 0.037 microwatts, more than 10^4 times the thermal noise from the Moon's surface.

We can now use a more precise argument for Equation (2.7) to relate antenna gain and effective area. In Figure 2.1 let the antenna have an isotropic gain pattern, i.e. it has a constant gain of unity in all directions, and its effective area is also constant and independent of direction. The antenna is bathed in a radiation field of specific intensity I_ν and the power density that it gathers in and delivers to its output will be (dividing by 2 since an antenna can deliver radiation in only one polarization to a given output port)

$$P_\nu \, d\nu = \frac{1}{2} \int_{4\pi} I_\nu A_{eff} \, d\Omega \, d\nu = 2\pi A_{eff}(2kT/\lambda^2)d\nu. \tag{2.17}$$

This, by the second law of thermodynamics, must equal the power flow in one direction in the transmission line from the matched load at temperature T to the antenna port. Using Equation (2.15),

$$kT \, d\nu = \frac{4\pi}{\lambda^2} kT A_{eff} \, d\nu \tag{2.18}$$

and, since the gain of the isotropic antenna is unity, it follows that for any antenna

$$A_{eff} = \frac{\lambda^2}{4\pi} G \tag{2.19}$$

as in Equation (2.7). This relation must be valid for any direction $\hat{n}$.

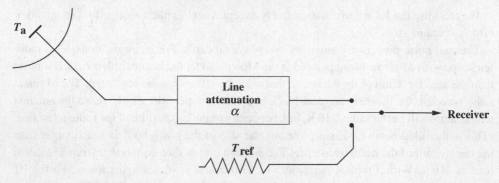

Figure 2.2. A simple radiometer, showing the effect of line attenuation. The receiver can be switched from the antenna to a resistive load with effective noise temperature T_{ref}.

These considerations lead to the convenient concept of *antenna temperature*, T_a, which is the temperature that a resistor would have if it were to generate the same power density at frequency ν as that observed to be coming from the antenna port. Note that the definition is not restricted to blackbody conditions: one considers only the narrow band of frequencies in which the measurement is made. The power received from the radio source need not have a blackbody spectrum either, for the same reason.

This leads to another useful concept, the *brightness temperature*, T_b, which is the temperature at which a black body would have to be in order to duplicate the observed specific intensity of an object at frequency ν. As in the case of antenna temperature, this has an operational convenience, since black bodies and resistive loads at a well-determined temperature are the measurement standards.

The antenna temperature obviously depends on the brightness temperature in the beam of the antenna, and this may vary within the beam. This may have a large effect on system noise, for example, if the Sun were to occupy part of the telescope beam. At short wavelengths the brightness temperature of the Sun is in the range 6000–10 000 K, so a serious degradation of sensitivity may occur even if the Sun is only in the near sidelobes of the antenna pattern.

The process of measurement is schematized in Figure 2.2, which shows a simplified radiometric system that compares the antenna temperature to the temperature T_{ref} of a reference resistor. In such a system, the antenna temperature is often a key parameter that determines the sensitivity of a radio-astronomical measurement, since it is the source of the output noise in any ideal receiver. For a practical system, as shown in Figure 2.2, there will be some loss in the transmission line to the receiver, and extra noise will be generated by the receiver. If the line is at temperature T_1, and the signal is attenuated by a fraction α, then in the ideal case of matched antenna and line the antenna temperature becomes

$$T_a = (1 - \alpha)T_b + \alpha T_1. \tag{2.20}$$

In practice, in addition to the background brightness, T_b, there will be two more contributions to the antenna temperature: the contribution T_{atm} from the Earth's atmosphere (which will depend on antenna elevation) and the contribution T_g from the Earth's thermal emission picked up by the antenna sidelobes (also dependent on antenna orientation). Thus T_b is actually a sum of all three terms.

The total system noise at the input of the receiver in effect includes the noise generated in the receiver itself; by a simple extension, the total may be specified as a *system noise temperature* (Section 3.4). For example, the system noise temperature of a receiver system at 5 GHz, connected to a radio telescope directed towards the Milky Way, might be 30 K, of which 10 K would be the brightness temperature of the sky, 15 K the receiver noise and 5 K the noise due to loss in the transmission line from the antenna to the receiver.

2.3 Electromagnetic waves

Radio-astronomy observations ultimately consist of measuring the energy received from a distant source (as do all astronomical observations), but before the radio energy is measured (the process is usually called *detection*), the electromagnetic wave properties are retained and operated on. An electromagnetic wave has both amplitude and phase and, in the simple detection process, which takes the square of the amplitude, the phase information is lost. In radio astronomy, when the radio telescope gathers the incoming waves, it transforms them into electrical signals on a transmission line, and these can be amplified, shifted to different frequency bands, and cross-correlated with other radio-frequency signals. By these means, even after detection, the relative phase properties of the incoming electromagnetic waves can be inferred.

In free space, the propagation of electromagnetic waves is governed by the wave equation; for the electric field $\mathcal{E}$ this is

$$\nabla^2 \mathcal{E} = \frac{1}{c^2} \frac{\partial^2 \mathcal{E}}{\partial t^2}. \tag{2.21}$$

The magnetic field $\mathcal{B}$, obeying a similar equation, lies perpendicular to the electric field, with the cross-product $\mathcal{E} \times \mathcal{B}$ determining the direction of the propagation, $\hat{\mathbf{n}}$. (If a plasma is present, in which a current j can flow, a term $(4\pi/c)\partial j/\partial t$ must be added to the right-hand side of Equation (2.21); this gives rise to interesting propagation phenomena discussed in Chapter 7.)

The electromagnetic wave equation has a rich set of solutions, and simplifying principles are needed. Near the source, Equation (2.21) requires spherical waves, but astronomical sources are at great distances, such that the curvature of the wavefront is negligible. This means that plane waves are a sufficiently good approximation; a monochromatic transverse wave, with $\mathcal{E}$ in the direction of a unit vector $\hat{\mathbf{a}}$, propagating in the z-direction with inverse wavelength $k = 1/\lambda$, frequency ν and phase ϕ, will take the simple form

$$\mathcal{E}(x, y) = \hat{\mathbf{a}} \mathcal{E}_0 \cos[2\pi(\nu t - kz) + \phi] \tag{2.22}$$

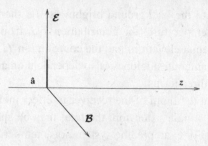

Figure 2.3. Electric and magnetic fields in a linearly polarized wave.

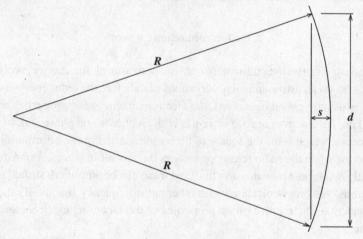

Figure 2.4. The far-field domain. If $R > d^2/\lambda$, the wavefront is plane within $\lambda/2$.

or, in complex notation, with $\mathcal{E}_0$ now containing the phase,

$$\mathcal{E}(x, y) = \hat{\mathbf{a}}\mathcal{E}_0 \exp[\mathrm{i}2\pi(\nu t - kz)]. \tag{2.23}$$

The vector $\hat{\mathbf{a}}$ lies in the x, y-plane, perpendicular to the direction of the propagation, and is parallel to $\mathcal{E}$, as illustrated in Figure 2.3. This defines the polarization direction: if $\hat{\mathbf{a}}$ has a constant orientation as the wave propagates, the wave is *linearly polarized*, with a plane of polarization defined by the electric field. The general polarization case is discussed in Section 2.4.

The requirement that a plane wave should be a sufficiently good approximation to a spherical wave can be put in quantitative terms. Consider a telescope aperture of dimension d, receiving waves with radius of curvature R, as shown in Figure 2.4. If the depth s of the chord d is sufficiently small compared with the wavelength λ, the plane-wave approximation is a good one. In this case, the instrument is making a measurement in the *far-field* domain.

This is always the case for sources at astronomical distances, but the limitation must be borne in mind when a terrestrial source is used to calibrate the performance of a radio telescope.

Conventionally, if the chord has a depth $s < \lambda/2$, this sets the far-field distance R_{ff}, and the sagitta approximation for the depth of a chord gives the condition

$$R_{ff} \geq \frac{d^2}{\lambda}. \qquad (2.24)$$

When space is not empty, and there is an index of refraction n, the plane-wave propagation of Equation (2.17) becomes

$$\mathcal{E} = \hat{\mathbf{a}}\mathcal{E}_0 \exp[i2\pi(vt - knz)]. \qquad (2.25)$$

When the index of refraction varies from place to place, an initially plane wave can develop a curved wavefront, and if the variations in index of refraction are sufficiently rapid, solving the propagation problem can become difficult, unless suitable approximations can be used. When the curvature of the wavefront is sufficiently small, the *eikonal representation* is usually applicable. In this familiar approximation, the electromagnetic waves are described by ray paths perpendicular to the wavefront, a representation well known as geometrical optics. The distinction between the regimes of geometrical optics and wave optics is an important one. For example, at the boundary of the receiving aperture of a radio telescope, or of any astronomical telescope, there is a sharp discontinuity, and the phenomenon of diffraction results. This presents a central problem for all radio telescopes, since the diffraction limitations of finite apertures determine the achievable angular resolution: Chapters 5 and 6 describe the methods of wave optics that are used to solve the angular-resolution problem.

The geometrical optics approximation has, nevertheless, a wide range of applicability. In addition to its usefulness in describing the propagation of radiation through an inhomogeneous medium of gently varying index of refraction, the approximation holds at sharp interfaces, provided that the interface is not sharply curved on the scale of a wavelength. The phenomenon of atmospheric scintillation is a particularly illuminating (and important) case to consider, both for optical and for millimetric radio observations. The Earth's atmosphere is inhomogeneous and time-varying, and the quasi-plane light wave from a star undergoes distortions of two kinds. The wavefront is tilted locally, and this means that the apparent position of the star shifts; the star jitters rapidly with time, degrading the image. In this case, the eikonal approximation holds. It may happen, however, that the wavefront distortion across the aperture may be so complex that a simple tilt cannot be used as a description. In this event, which is more severe the larger the telescope, a full wave description is needed. In radio astronomy, the most dramatic scintillation effects are seen at longer wavelengths, at which the ionospheric plasma, the interplanetary plasma and the interstellar medium all induce measurable effects. These are discussed more fully in Chapter 7.

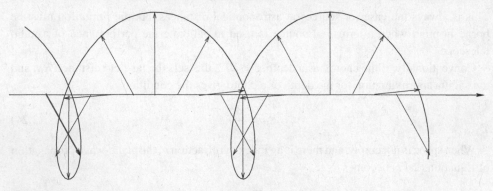

Figure 2.5. A right-hand circularly polarized wave at an instant of time. The tip of the electric vector follows an anticlockwise screw.

2.4 Wave polarization

When the radiation is effectively unpolarized, only the incoming energy flux is measured, and this proves to be sufficient for most of astronomy. Polarization is, however, particularly important in several important classes of observation: the outstanding example is radio emission from pulsars, which may be effectively 100% polarized, while the small degree of linear polarization in the light from stars, and in the radio emission from the galactic background, enables the study of galactic magnetic fields, especially when combined with the Faraday rotation of the pulsar signals.

The simple, linearly polarized wave illustrated in Figure 2.3 is only a special case that must be generalized since cosmic radio sources can exhibit other forms of polarization. They can be partially polarized, or unpolarized, and a systematic description of the polarization state is necessary. Equation (2.23) can be written in a more general form, in which the amplitude is expressed in terms of separate complex amplitudes $\mathcal{E}_x$ and $\mathcal{E}_y$:

$$\mathcal{E}(z,t) = (\hat{\mathbf{x}}\mathcal{E}_x + \hat{\mathbf{y}}\mathcal{E}_y)\exp[\mathrm{i}2\pi(\nu t - kz)]. \tag{2.26}$$

As usual, the real part of $\mathcal{E}$ is taken for the physical quantity.

The *polarization* of the wave is determined by the amplitudes and phases of $\mathcal{E}_x$ and $\mathcal{E}_y$. If the amplitudes are equal and the phases are in quadrature (i.e. differ by 90°), the wave is circularly polarized and the wave vector rotates in the x, y-plane. Figure 2.5 shows how such a circularly polarized wave develops in space; instantaneously the vector follows a spiral with a pitch of one wavelength.

For general values of the amplitudes and phases, the projection of $\mathcal{E}$ on a given x, y-plane erected at a fixed value of z will describe an ellipse in time, and the propagation field amplitude describes an elliptical helical locus in the direction of propagation along the z-azis. The limiting cases are *circular polarization*, with $\mathcal{E}$ rotating while retaining constant amplitude, and *linear polarization*, in which the ellipse degenerates to a straight line. The general elliptical case is, of course, designated *elliptical polarization*. There once were

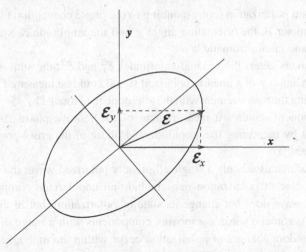

Figure 2.6. A monochromatic elliptically polarized wave. The tip of the electric vector $\mathcal{E}$ traces an ellipse; it may be regarded as the sum of two quadrature-phased components on the major and minor axes, or two components $\mathcal{E}_x$ and $\mathcal{E}_y$ with a different relative phase.

differences in convention between physicists and electrical engineers, but the terminology was finally regularized when the International Astronomical Union (IAU) adopted the IEEE convention that the plane of polarization was the plane of the electric field, and defined right-hand circular polarization as the case when the electric vector, in a fixed plane perpendicular to the ray, rotates in a clockwise direction when viewed in the direction of propagation.[3] A right-hand circularly polarized wave is schematized in Figure 2.5 (note that at any instant the vector traces a left-hand helix in space).

Consider a radio telescope measuring two orthogonal electric-field components on axes x and y; it might, for example, have separate receivers connected to crossed dipoles at the focus. The incident plane wave is monochromatic and elliptically polarized, with the electric vector describing the ellipse shown in Figure 2.6; the ellipse is rotated with respect to the original x- and y-axes by an angle χ. Following Equation (2.26), the x- and y-components of $\mathcal{E}$ have complex amplitudes $\mathcal{E}_x$ and $\mathcal{E}_y$, with the phase of the y-component leading the x-component for right-hand elliptical polarization (propagating into the plane of the diagram). The field can be described in terms of components $\mathcal{E}_a$ and $\mathcal{E}_b$ aligned with the major and minor axes of the ellipse, respectively, with $\mathcal{E}_a$ leading $\mathcal{E}_b$ by $\pm\pi/2$. These components can then be written in terms of a single amplitude $\mathcal{E}_0$ and a parameter β:

$$\mathcal{E}_a = \mathcal{E}_0 \cos\beta; \qquad \mathcal{E}_b = -\mathcal{E}_0 \sin\beta. \tag{2.27}$$

The parameter β specifies the character of the polarization, with $\beta = 0$ describing a plane-polarized wave. Circular polarization results when $\beta = \pm\pi/4$, where the positive sign

---

[3] An early optical convention defined the plane of $\mathcal{B}$ rather than $\mathcal{E}$ as the plane of polarization. Until recently, the definition of handedness of circular polarization also was opposite in optical and radio terminology. The IAU convention is now uniformly agreed upon by astronomers and electrical engineers, but care should be taken in reading the older literature.

defines right-hand polarization (corresponding to the phase convention that $\mathcal{E}_a$ leads $\mathcal{E}_b$ by $\pi/2$). The parameter β, the orientation angle χ and the amplitude $\mathcal{E}_0$ specify completely the state of a plane monochromatic wave.

The two receivers detect the squared amplitudes $\mathcal{E}_x^2$ and $\mathcal{E}_y^2$; the sum of their outputs is $\mathcal{E}_0^2$. The position angle χ of a linearly polarized signal could be measured by the difference $\mathcal{E}_x^2 - \mathcal{E}_y^2$, repeating the measurement with the telescope feed rotated by $45°$, but the circularly polarized component requires a measurement of the relative phase $\Delta\phi$ of $\mathcal{E}_x$ and $\mathcal{E}_y$. This is achieved by measuring the amplitude and phase of the cross-power product $\mathcal{E}_x\mathcal{E}_y$ (cf. Chapters 3 and 5).

We have so far discussed only a single elliptically polarized wave: the full specification of an arbitrary state of polarization must include an unpolarized component. A strictly monochromatic wave does not change its state of polarization, but in the more practical case of a quasi-chromatic wave, comprising components with a range of frequencies and polarizations, random changes of polarization occur within the time taken for a measurement. The polarization is now specified as an average over an observing time that is long compared with the fluctuation time. In an unpolarized or randomly polarized, wave, the random changes result in an average of zero polarization. This component contributes only to the sum $\mathcal{E}_0^2$; it does not contribute to the difference $\mathcal{E}_x^2 - \mathcal{E}_y^2$ or the cross-product $\mathcal{E}_x\mathcal{E}_y$.

2.5 Stokes parameters

Astronomical sources are assemblies of many sources, each of which is emitting radiation in some state of polarization. The vector sum of all these electromagnetic waves at the telescope aperture has a net polarization that is generally stated in terms of the Stokes parameters I, Q, U and V which describe either a single wave or the sum of many waves. For the fully polarized wave

$$I \equiv \mathcal{E}_0^2,$$
$$Q \equiv \mathcal{E}_0^2 \cos(2\beta)\cos(2\chi),$$
$$U \equiv \mathcal{E}_0^2 \cos(2\beta)\sin(2\chi), \tag{2.28}$$
$$V \equiv \mathcal{E}_0^2 \sin(2\beta).$$

It is clear that I must be proportional to the total energy flux. Since β is the parameter in Equation (2.27) that determines the axial ratio of the ellipse, V must be the ellipticity parameter; $V = 0$ describes linear or random polarization, and $V/\mathcal{E}_0^2 = +1$ describes full right-hand polarization. The ratio Q/U specifies the orientation of the ellipse.

The relations (2.28) defining the Stokes parameters can also be manipulated to give

$$\mathcal{E}_0^2 = I,$$
$$\sin(2\beta) = \frac{V}{Q}, \tag{2.29}$$
$$\tan(2\chi) = \frac{U}{Q}.$$

The Stokes parameters can also be expressed in terms of the x- and y-components of $\mathcal{E}$, written as two complex amplitudes $\mathcal{E}_x$ and $\mathcal{E}_y$. These are, with their phases explicitly shown,

$$\mathcal{E} = (\hat{\mathbf{x}}\mathcal{E}_x e^{i\phi_1} + \hat{\mathbf{y}}\mathcal{E}_y e^{i\phi_2})e^{i\omega t} \tag{2.30}$$

and, since one phase is arbitrary, one can with a little manipulation rewrite the Stokes parameters of Equations (2.28) in terms of the real amplitudes $\mathcal{E}_x$ and $\mathcal{E}_y$, and their relative phase $\Delta\phi = \phi_1 - \phi_2$:

$$
\begin{aligned}
I &= \mathcal{E}_x^2 + \mathcal{E}_y^2, \\
Q &= \mathcal{E}_x^2 - \mathcal{E}_y^2, \\
U &= 2\mathcal{E}_x\mathcal{E}_y \cos\Delta\phi, \\
V &= 2\mathcal{E}_x\mathcal{E}_y \sin\Delta\phi.
\end{aligned}
\tag{2.31}
$$

Since the wave is described by two component amplitudes and by their relative phase angle, three parameters in all, the four Stokes parameters of a monochromatic (fully polarized) wave must be related. One sees from Equations (2.30) and (2.31) that the relation is

$$I^2 = Q^2 + U^2 + V^2. \tag{2.32}$$

Cosmic radio sources generally emit broad-band, partially polarized radiation. In a fully polarized monochromatic wave the amplitudes and relative phase of two orthogonal components are constant, but, for a source of radio noise, the amplitudes and phases fluctuate. In an assemblage of incoherent sources the amplitudes and relative phases fluctuate in time; over a band $\delta\nu$ the phase and amplitude will in general change in a characteristic time $1/\delta\nu$. The definitions of the Stokes parameters must therefore be generalized for the radio-astronomy case as time averages.

Cosmic radio sources frequently have complicated polarization distributions and the Stokes parameters are generally used to label the maps that are derived from the data sets observed in various polarizations. The time-averaged Stokes parameters for a broad-band source of radiation are written

$$
\begin{aligned}
I_\nu &\equiv \langle \mathcal{E}_x^2 + \mathcal{E}_y^2 \rangle, \\
Q_\nu &\equiv \langle \mathcal{E}_x^2 - \mathcal{E}_y^2 \rangle, \\
U_\nu &\equiv \langle 2\mathcal{E}_x\mathcal{E}_y \cos\Delta\phi \rangle, \\
V_\nu &\equiv \langle 2\mathcal{E}_x\mathcal{E}_y \sin\Delta\phi \rangle.
\end{aligned}
\tag{2.33}
$$

The four Stokes parameters are now seen to give a complete description of the radiation field. The specific intensity I_ν can be partially polarized, with the partial polarization being linear, circular or elliptical, and four parameters are necessary and sufficient. For completely unpolarized radiation, I_ν is the specific intensity, with Q_ν, U_ν and V_ν all zero. The degree of polarization of partially polarized radiation may be characterized as P/I, where $P = (Q^2 + U^2 + V^2)^{1/2}$. An advantage of the Stokes formulation is that the Stokes parameters of separate sources within a telescope beam add linearly.

Radio telescopes usually take measurements in two orthogonal polarizations, and, when these are perpendicular linear polarizations, Equations (2.33) are of the relevant form; note that both the powers and the cross-power products must be taken (cf. Chapters 3 and 5). When the orthogonal polarizations are the right- and left-hand circularly polarized pair, the power products and the Stokes parameters are related by

$$\begin{bmatrix} \langle RR^* \rangle & \langle RL^* \rangle \\ \langle LR^* \rangle & \langle LL^* \rangle \end{bmatrix} = \begin{bmatrix} I + V & Q + iU \\ Q - iU & I - V \end{bmatrix}, \tag{2.34}$$

where an asterisk denotes a complex conjugate. The formalism carries over to the aperture-synthesis mapping process discussed in Chapter 6. By taking the full set of cross-power products, the cosmic radio sources can be mapped in polarization as well as brightness.

2.6 Radio polarimetry in practice

For radio telescopes, the two orthogonal polarizations have separate outputs, which are amplified before they undergo the transformation to I, Q, U and V. The ready availability of phase-stable amplifiers permits this to be done easily, allowing the signal processing to be done at higher signal levels. The outputs of some radio telescopes use two orthogonal linear polarizations as the fundamental pair, whereas others use right- and left-hand circular polarizations as the orthogonal set. Most large paraboloids use the circular pair, because the symmetry of the telescope beam then avoids some of the effect of changes in parallactic angle as an alt–az-mounted telescope tracks across the sky. In either case, the change with parallactic angle in the recorded polarization of a highly linearly polarized source may be used to calibrate the polarimeter.

Optical telescopes must use a different approach in measuring polarization, since the signals cannot be amplified, for reasons given in Chapter 3. The orthogonal polarizations are separated out by a birefringent filter and then are separately analysed. Except for observations of the Sun, the full set of Stokes parameters is rarely derived, since in most astronomical observations there is only linear polarization to be observed. If all Stokes parameters were to be derived from the available photons received by the optical telescope, and if three of the parameters were not needed, three-quarters of the photons would be lost. Splitting the beam initially into two orthogonal linear polarizations loses no photons, since the sum of the two channels gives the total flux. If the radiation is only partially linearly polarized, note that three parameters must be measured. In the optical domain, this is usually done by observing half the time for a given polarizer orientation, and then rotating the polarizer by $45°$ for the rest of the observation. This circumstance is in marked contrast to the radio case; since optical observations count photons, very little extraneous noise signal is added by the detection apparatus, whereas in radio observations the incoming signal has a dominant quantity of noise added by the receiving apparatus. Infrared astronomy resembles radio astronomy in this instance, since the atmospheric emission dominates the detected signal.

The full range of possible polarization states can be observed in cosmic radio sources, including almost 100% circular in some pulsars and up to 70% linear in synchrotron radiation. The accuracy of measurement depends on calibration of any unwanted cross-coupling between the outputs of the two receivers, which may generate an apparent low level of linear or circular polarization. The effect of this cross-coupling on the four Stokes parameters may be expressed as a 4×4 matrix known as the Mueller matrix (Mueller 1948) (see Appendix 4). The more complicated instrumental effects which are encountered in interferometry and aperture synthesis require an extension of the matrix approach, which we discuss in Chapter 6.

Instrumental cross-coupling effects may occur at any stage of the receiver system, from the antenna and its feeds through filters, amplifiers and detectors. The design and performance of any component may be expressed in terms of another matrix formulation, the *Jones matrix*, familiar in optical polarimetry, which describes the action of any linear system on the field components of a polarized wave (note that the Mueller matrix acts on scalar quantities, whereas the Jones matrix acts on wave components that may be complex). The cross-coupling may vary across the bandwidth of a receiver, or across the antenna beam, and calibration becomes difficult at very low levels of polarization. The characterization of instrumental polarization using matrix methods is described by Tinbergen in *Astronomical Polarimetry* (1996).

The most difficult polarization measurements are those in which an unwanted signal is picked up from outside the main beam of the telescope. The spurious polarization terms in the telescope response inevitably increase off axis, and the apparently polarized signal from an intense source within the sidelobe pattern may be difficult to remove. In the special case of the CMB (Chapter 15), the telescope is required to measure polarization in an almost uniformly bright sky, and the sidelobe problem is minimal. The WMAP satellite has achieved useful measurements of the polarized component of the $3°$ background signal even though the polarized brightness temperature is only a few tens of microkelvins.

Further reading

Astronomical Polarimetry, Tinbergen J., Cambridge, Cambridge University Press, 1996.
Astrophysical Quantities, Allen C. W., ed. Cox A. C., Berlin, Springer-Verlag, 4th edn, 2000.
The Tools of Radio Astronomy, Rohlfs K. and Wilson T. L., Berlin, Springer-Verlag, 4th edn, 2004.

3

Signals, noise, radiometers and spectrometers

The electromagnetic signals that give information about the Universe have the character-istics of random noise. More specifically, in the radio part of the spectrum, the signals are composed of Rayleigh, or Gaussian, noise, the result of an assemblage of many random oscillators with random frequency and phase. As one moves to shorter wavelengths, through the infrared and into the optical, ultraviolet and X-ray bands, the discrete character of photons becomes increasingly dominant, and the random noise obeys Poisson statistics, sometimes called shot noise. Throughout the spectrum, the process of detecting and measuring the signals gathered by a telescope is almost always electronic; for the optical astronomer the eye and the photographic plate are not sensitive enough, and at both the radio and X-ray ends of the spectrum electronic means have always been essential. The device that measures the power of the incoming signal is a *radiometer*; when it measures power as a function of frequency, it is a *spectrometer*.

At wavelengths shorter than about 100 μm, immediate detection of the received power is almost always forced on the observer because the laws of quantum mechanics require any amplifier to add extraneous noise. For the radio astronomer, the incoming signal is amplified before its power is measured in a detector, and the construction of low-noise amplifiers has become an art. The use of amplifiers defines the boundary between radio and infrared astronomy; the cosmic signal can be amplified and manipulated by the radio astronomer, but in infrared astronomy (and for shorter-wavelength bands) the energy of the incoming stream of photons is counted, either in a photon-counting device or by a *bolometer*. No matter which wavelength band the astronomer is working in, there will always be extraneous noise signals that corrupt the observations. The nature of the interfering noise differs, according to the observing band; here, the problems faced by the radio astronomer will be emphasized. Almost always, the cosmic radio signal is far weaker than the extraneous noise, and how the desired signal can be extracted from the noise is the principal subject of this chapter.

3.1 Gaussian random noise

The natural noise sources and the cosmic sources themselves emit Rayleigh, or Gaussian, noise. The amplitude expectation for any time is described by a Gaussian distribution; *Gaussian random noise* is the most common form met in practice and the simplest to deal with quantitatively. The accuracy of radio measurements will depend upon the statistical properties of the voltage amplitude $v(t)$; the noise power $P(t)$ will be proportional to the square of the amplitude, and for much of this discussion the units can be chosen so that $P = v^2$. Although the average power can vary with time, in most cases (man-made interference excepted) the variation is much slower than the coherence time of the radiation. Formally, therefore, $v(t)$ can be treated as a *stationary random variable*; its statistical properties will not depend on the epoch, but they will depend on the particular characteristics of the system. The probability that the value of $v(t)$ will fall between v and $v + dv$ is the *probability density function* $\mathcal{P}(v)$; this must be normalized so that its integral is unity. (Note the distinction from *probability distribution function*, which is the probability that $v(t)$ will take on a value *less* than v.) For a given probability density function, three quantities will have particular interest: the mean value $\langle v \rangle$, the mean square value $\langle v^2 \rangle$ of $v(t)$ over a given interval and the autocorrelation $R(\tau)$ of $v(t)$ (see Appendix 1).

For Gaussian random noise, the probability density function is

$$\mathcal{P}(v) = \frac{1}{\sigma\sqrt{2\pi}} \exp\left(-\frac{v^2}{2\sigma^2}\right). \tag{3.1}$$

The signal amplitude has an equal probability of being positive or negative, and its mean value is zero.[1] The power is the mean value of $v(t)^2$ and, for the given probability density, it is σ^2.

The power spectrum of a random noise signal may be obtained from the autocorrelation of its amplitude by Fourier transform (see Appendix 1: the Wiener–Khinchin theorem). In the real world, the observer can only estimate the signal power, or its autocorrelation, by evaluation over a finite timespan $\mathcal{T}$, so one arrives at an *estimate* of the power spectrum; the estimate becomes more precise as $\mathcal{T}$ increases. The autocorrelation $R(\tau)$ is defined as the product of $v(t)$ with itself, shifted by a lag τ and integrated over all time. For the present, finite case

$$R_{t,\mathcal{T}}(\tau) = \int_{t-T/2}^{t+T/2} v(t')v(t' + \tau)dt', \tag{3.2}$$

where the subscript $t, \mathcal{T}$ indicates that the integration extends over an interval $\mathcal{T}$, centred on time t. The estimated power spectrum, $S_{t,\mathcal{T}}(v)$, for a given epoch and averaging time, is the Fourier transform of the autocorrelation (note that $S(v)$ is identical to P_v as used in Chapter 2; we are following the usual terminology in autocorrelation theory):

$$S_{t,\mathcal{T}}(v) \overset{\text{FT}}{\Longleftrightarrow} R_{t,\mathcal{T}}(\tau). \tag{3.3}$$

In the limit of the averaging time $\mathcal{T}$ going to infinity, the relation converges to

$$S(v) \overset{\text{FT}}{\Longleftrightarrow} R(\tau). \tag{3.4}$$

---

[1] In other contexts the time sequence, $v(t)$, might be regarded as a highly pathological function since it is neither continuous nor differentiable, but the analysis requires only integration, and the integrals can be evaluated by using the Lebesgue integral.

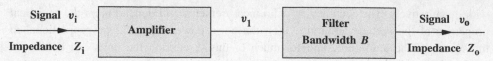

Figure 3.1. A linear receiver.

The spectrum of Gaussian noise is evaluated easily by inspecting Equation (3.3) for the cases of zero and non-zero time lag. When τ is non-zero the integral goes to zero since $v(t)$ is completely uncorrelated from one instant to the next. When $\tau = 0$, however, the autocorrelation, Equation (3.2), becomes the mean square voltage $\langle v^2 \rangle$; i.e. it is the standard deviation of the Gaussian noise distribution. For a sufficiently long integration time, therefore, the autocorrelation for a Gaussian random signal is a Dirac δ-function (technically, a distribution with an infinite spike at $t = 0$, zero elsewhere, but with an integral whose value is 1). Let $t = 0$ (the noise is stationary and the result should converge to the same answer for any epoch):

$$\lim_{T \to \infty} R(t) = v^2 \delta(t) \quad \text{(for } v(t) \text{ Gaussian).} \tag{3.5}$$

The Fourier transform of the δ-function is a constant, from which it follows that Gaussian noise has a flat power spectrum with a power spectral density equal to the square of the standard deviation σ of the probability density $v(t)$. When the signal has gone through a band-limiting filter, Equation (3.5) no longer holds; this case is treated in the next section.

A flat power spectrum is often called a white spectrum, from the colour analogy. Its quality is familiar as the hiss from a radio that is not tuned to a station; the hiss contains all audio frequencies, with no noticeable tonality.

3.2 Band-limited noise

We now examine the response of a radio receiver, with its finite bandwidth, to a noise signal. The receiver amplifies the input signal, and contains filters that define the bandpass. The amplifier will have a bandwidth $\mathcal{B}$ (more generally, it will amplify the input signal by some function of frequency, $G(v)$, but we can still talk of an effective bandwidth $\mathcal{B}$), while the detector will average the power for an *integration time $\mathcal{T}$*. A good receiver should be *linear*, which means that it amplifies signals within the bandpass faithfully, with no distortion. A linear receiver, in its simplest form, can be represented as a combination of an *ideal amplifier* and a *passive filter* that attenuates frequencies outside the bandpass, as illustrated in Figure 3.1. The input signal amplitude, $v_i(t)$, is amplified to become $v_1(t)$ at the input of the filter (the signal amplitude is designated by its associated voltage, but the current $i(t)$ could equally well be used). The filter modifies the spectrum of the signal, yielding an output signal $v_2(t)$ that is amplified further to give an output amplitude $v_o(t)$. In present practice, the separate functions shown in the diagram are sometimes integrated into a single chip, but they can be individual chips on a single board or separate elements connected by waveguide or cable. It is also sometimes advisable to insert a filter in front of the amplifier, as discussed in Section 3.3.

The power gain $G(\nu)$ of a receiver is a function of frequency, but its basic properties can often be summarized by citing its effective gain, G, and its bandwidth, $\mathcal{B}$. The simple receiver in Figure 3.1 is shown with an input load of impedance Z_i and an output load Z_o. The input voltage $v_i(t)$ comes from a source that could be a radio antenna or a signal generator; all the incident power is transferred from the source to the amplifier, with no reflection, if the source impedance is matched to the input impedance of the amplifier. (*Matched impedances* occur when one is the complex conjugate of the other.) Ordinarily, one tries to obtain impedances that are purely resistive, or at least have as small an imaginary component as possible. As bandwidths have grown in modern-day radiometers, achieving an impedance match across the operating band has become an engineering challenge. When one remembers that power is voltage squared divided by the resistance, it follows that the square of the ratio of output and input voltages will give the power gain only when the matched impedances at input and output are equal; in general

$$G = \frac{v_o^2}{v_i^2}\frac{Z_i}{Z_o}. \tag{3.6}$$

The usual unit of power gain is the *decibel*, abbreviated dB and defined as ten times the $\log_{10}$ of the power ratio. A power gain of 10 is therefore 10 dB, a gain of 100 is 20 dB, and so forth. It is worth noting that, for the logarithmic magnitude scale that astronomers have used since the time of Hipparcos (but defined quantitatively much more recently), 1 magnitude is exactly 4 dB.

The complete description of the receiver must recognize, however, not only that a sinusoidal signal will be amplified, but also that its phase will be shifted. The phase shift is usually frequency-dependent, so the increase in signal amplitude will be a complex function of the frequency. The complete description of the receiver is given by the *transfer function* $H(\nu)$:

$$V_o(\nu) = H(\nu)V_i(\nu), \tag{3.7}$$

where $H(\nu)$ is complex, to express the phase shift (note the convention of contrasting the frequency and time domains by using upper-case and lower-case symbols for these Fourier-transform pairs such as $[v_i(t), V_i(\nu)]$).

The transfer function has an easily recognized property: its Fourier transform $h(t)$ is the response of the receiver to a δ-function impulse at the input. This can be seen by noting that, if $V_o(\nu) = V_i(\nu)$, $H(\nu)$ must be unity, and its Fourier transform is a δ-function. Therefore, it follows that

$$v_o(t) = h(t) \qquad \text{if } v_i(t) = \delta(t). \tag{3.8}$$

One sees immediately that, for a pure noise signal, with no correlation from one instant of time to the next, the receiver imposes a finite correlation time, since $h(t)$ will be extended

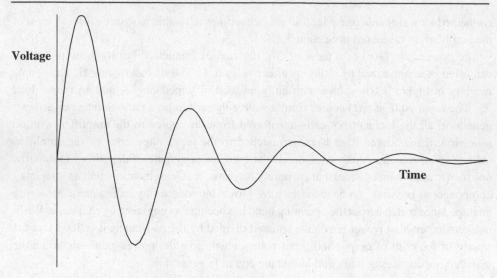

Figure 3.2. The impulse response of an *RLC* filter.

in time if its transfer function $H(\nu)$ has a finite bandwidth. An impulsive signal may be stretched in time, possibly with oscillations; this is generally referred to as *ringing*.

An example of the response of a filter to an impulse $\delta(t)$ is shown in Figure 3.2. This filter is a singly tuned *RLC* filter (an inductance L and a capacitance C, with damping resistance R, with associated resonance frequency ν_0 and sharpness of resonance Q (damping coefficient $1/Q$)). This response shows strong ringing, which dies out in a time of order $1/\mathcal{B} = Q/\nu_0$. This is a property of all receivers; the ringing time is generally of the order of $1/\mathcal{B}$. A pulse shorter than $1/\mathcal{B}$ will be lengthened by the receiver, and might not be properly resolved; this is a severe problem in measuring the width of micropulses from some pulsars (Chapter 12).

3.3 Detection and integration

The response of the receiver to its own random noise, and to the random-noise signal from the radio telescope, is still a random quantity, but with persistence in time of the order of $1/\mathcal{B}$, as the preceding section has shown. Two examples of filtered white noise are shown in Figure 3.3, the first being relatively broad-band with a bandwidth of half the centre frequency, whereas the second example shows the output for a narrow-band filter with the bandwidth only one-tenth of ν_0. The greater correlation in time for the narrow-band case is easily recognized. Nevertheless, the signal is still a random variable, and a measurement of its power, taking an average over some averaging time $\mathcal{T}$, is only an estimate of the true value, with an associated uncertainty. First, the output amplitude must be converted to power, which is proportional to the square of the amplitude. The signal must, therefore, be

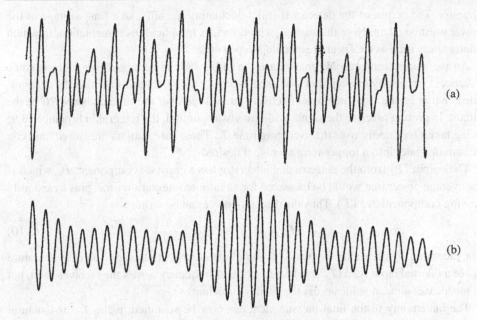

Figure 3.3. Filtered white noise: (a) broad-band and (b) narrow-band.

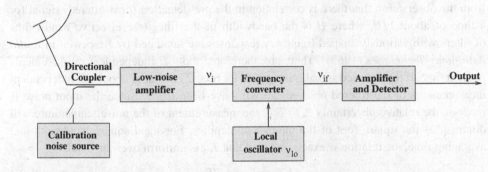

Figure 3.4. The basic components of a radiometer. The directional coupler allows a calibration noise signal to be injected into the receiver input.

multiplied by itself, in a *square-law device*, usually known as a *detector*:

$$\langle P \rangle = \int_{t-T/2}^{t+T/2} p_{\mathrm{d}}(t)\mathrm{d}t = \int_{t-T/2}^{t+T/2} [v_{\mathrm{o}}(t)]^2 \,\mathrm{d}t. \tag{3.9}$$

In this expression the instantaneous power $p_{\mathrm{d}}(t)$ measured by the detector is the square of the output amplitude $v_{\mathrm{o}}(t)$ at the output port of the receiver. Figure 3.4 is a block diagram of the basic components of a radio-astronomy receiver.

The square-law detector is often a simple diode, with associated filters to prevent the radio-frequency signal from proceeding on and possibly interfering with the following circuitry. The output of the detector is still a fluctuating quantity, so a time average of the power must be taken. Note the analogy with detectors in optical instrumentation, in which photometers such as CCDs are responding to power.

An averaging circuit can take many forms. In earlier radiometers (and even today in some instances), an analogue integrator was used, typically an RC circuit in which the accumulated charge on the capacitor was effectively an average over the time constant RC of the circuit. In present practice the signal is almost always digital, the integral in Equation (3.9) being taken repeatedly over the averaging time $\mathcal{T}$. These integrations are stored, and can be summed later into a longer-term average if desired.

The output $\langle P \rangle$ from the detector and integrator has a constant component, P_0, which is the average power that would be measured for an infinite integration time, plus a randomly varying component $\Delta P(\mathcal{T})$. Thus the detector power can be written

$$\langle P \rangle = P_0 + \Delta P(\mathcal{T}). \tag{3.10}$$

The incoming signal is almost always measured with respect to the equivalent temperature T_{eq} of a thermal input load (T_{eq} can be a function of frequency across the receiver band, but in most cases a mean value across the band is sufficient).

The uncertainty in the final measurement can now be specified, using T_{eq} as a natural measure of power density. The detailed analysis of $P(\mathcal{T})$ as a function of a random variable is outlined in textbooks such as that by Rohlfs and Wilson. A heuristic explanation starts from the observation that there is correlation in the pre-detection (post-filtered) signal for a time of about $1/\mathcal{B}$, where $\mathcal{B}$ is the bandwidth of the filter (the effective bandwidths of filters with variously shaped frequency responses are tabulated by Bracewell in *Radio Astronomy Techniques* (1962)). There are, therefore, about $\mathcal{B}$ independent measurements per unit time. If the detected signal is integrated for time $\mathcal{T}$, a total of about $\mathcal{B}\mathcal{T}$ independent measurements of the filtered noise signal will have been made. Since the input noise is random, the relative uncertainty $\Delta T/T$ in the measurement of the noise temperature will diminish as the square root of the number of samples. For ideal square bandwidth and averaging time, the relation is exact: provided that T_{eq} is uniform over the band,

$$\Delta T = T_{eq}/\sqrt{\mathcal{B}\mathcal{T}}. \tag{3.11}$$

For a digitized signal, a correction factor, dependent on the number of bits per sample, is required. For radio-astronomy measurements, the problem is that of how to distinguish between the astronomical noise signal T_a and the fluctuating noise signal T_s, which is the sum of the system noise temperature T_s and the added astronomical signal T_a. The separation of the astronomical signal from system noise is the subject of the following sections.

3.4 Radiometer principles

As seen in Figure 3.4, the noise signal, either from an antenna or a dummy load, passes to the receiver through a network that allows the radiometer to be calibrated, usually in temperature units. The receiver may incorporate a *mixer* or *frequency converter* that shifts the initial frequency band (centred at ν_0) up or down in frequency, but otherwise preserves the signal; a *local oscillator* injects a pure sinusoid at frequency ν_{lo} and the resulting *intermediate-frequency* band is at either the sum or the difference frequency. A receiver of this type, that translates the input signal band to an intermediate frequency, is called a *heterodyne receiver*. The receiver output is detected and integrated, and the results are read out, usually by a computer that records them on tape or disk.

The calibration of a receiver system is achieved by injecting a noise signal, identical in character to the noise from the antenna, into the input, either directly by substituting the source for the antenna, or by injecting a proportion of the calibration signal as in Figure 3.4. A well-matched resistive load in a thermal bath at a well-determined temperature T is a standard source of noise power, allowing a direct calibration of measurements of antenna temperature. Alternative and more easily controlled noise sources are provided by a gas-discharge tube or a back-biased diode; these are *secondary* sources, usually providing a much higher noise power. Another technique that is sometimes used is to cover the antenna or the antenna feed completely with an absorbing enclosure at a known temperature. Usually (but not always) sources of known flux are available in the sky, and these can also be used for calibration (but *their* calibration, at some earlier time, will have been carried out with reference to a thermal load). In the primary calibration process, it is important that both the antenna and the thermal load should be well matched to the receiver, presenting the same impedance.

The secondary calibrator may have a high effective noise temperature, of the order of 10 000 K. This allows the use of a coupler circuit in the receiver input which injects a small proportion of the calibration signal without seriously attenuating the signal from the antenna. This is usually accomplished by using a 20-dB or 30-dB *directional coupler*, thus leaking only 1% or 0.1% of the calibration signal into the receiver. Frequently, there is an adjustable attenuator as well, which allows one to adjust the calibration level to the most convenient value. A small calibration signal injected in this way can be used frequently during an observation to monitor the gain of the receiver system.

With this simple radiometer, the detected output for a given bandwidth and integration time will have a relative uncertainty given by Equation (3.11).

The *system noise temperature*, T_{sys}, the measure of the total extraneous noise generated by the radiometer/antenna system, is the sum of several terms: the radiation from the atmosphere, T_{atm}, the radiation picked up in the sidelobes, T_{sl}, which comes principally from ground spillover in most cases, noise generated by losses in the antenna itself, including the losses in the transmission line to the low-noise preamplifier, T_{loss}, and the extraneous noise of the preamplifier itself, T_n. At the lower frequencies, the galactic synchrotron background can add T_{bg}, and at metre wavelengths this can be the dominant contribution to T_{sys}. Usually, the noise contribution of the source being studied is negligibly small, but, on occasion, a bright celestial source being observed can contribute noise T_c, but this is seldom encountered. We thus define T_{sys}, the total system temperature, as

$$T_{sys} = T_{atm} + T_{sl} + T_{loss} + T_n \ (+T_{bg} + T_c). \tag{3.12}$$

The sum of the first three terms is the total antenna temperature (with the addition of T_{bg} and T_c if necessary), which is the noise generated by the total antenna system presented to the radiometer.

A measurement might consist of moving on and off a celestial source, noting the change in T_{sys}, calibrating the change by turning the secondary calibrator on and off and, at some point, calibrating the secondary calibrator with respect to the primary calibrator. The noise fluctuations contribute to the measurement uncertainty: a good measurement usually requires an integration time long enough to give a signal-to-noise ratio of at least 5σ, where σ is the root mean square (r.m.s.) noise in the output. There are circumstances where this is not sufficient, and the number of degrees of freedom can require a higher ratio. A 10σ result is reliable in nearly all experiments.

Referring again to the receiver in the radio-astronomy system illustrated in Figure 3.4, the signal from the calibration network becomes the input signal to the receiver, $v_i(t)$, passing first through a low-noise preamplifier. This necessarily adds noise in the form of an excess noise temperature, T_n, and the preamplifier design is aimed at adding as little noise as possible. (An absolute lower limit, $T_n > h\nu/k$, is imposed by the laws of quantum mechanics.) After further amplification, the signal goes to a mixer, where it is multiplied by the local oscillator signal. The multiplier itself can be as simple as a diode, which has a non-linear response to the applied signals. It is the second-order term that performs the multiplication; it is the task of the designer to make sure that undesirable effects of higher order are negligible.

The product of two sinusoids at frequencies ν_0 and ν_{lo} creates upper and lower sidebands, at frequencies $\nu_0 \pm \nu_{lo}$, and one of these is chosen as the intermediate-frequency band; a filter rejects the unwanted signals. In an actual receiver, there will be several filters, distributed throughout the system, but, since the system should be linear, the diagram lumps their total effect into one filter with a transfer function $H_r(\nu)$.

Frequently, there is more than one stage of frequency conversion, and more than one intermediate frequency. The final conversion can even be to a band whose lower edge is close to zero frequency, when it is generally known as *video* or *baseband*. This is routinely done when the following stage is an autocorrelation spectrometer, whereas for broad-band radiometry the final intermediate frequency is usually at some tens or hundreds of megahertz. The local oscillator (or oscillators) must be stable. In current practice, the local oscillator is generated by a stable crystal oscillator or by a frequency synthesizer that uses a frequency reference source of high stability, such as an atomic frequency standard.

There is usually an adjustable attenuator somewhere in the system to set the output to an optimum level; this may be before or after the mixer. Whatever the exact configuration may be, the output signal $v_o(t)$ passes to the detector, which, if not a true square-law device, should at least have an output that has a known relation to the power of the input signal. The average at time t_0, lasting for the integration time $\mathcal{T}$, is read out to the storage device.

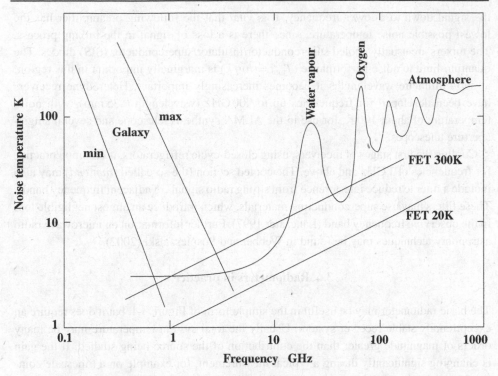

Figure 3.5. Typical antenna and receiver temperatures. Low-noise amplifiers for the first stages of receivers at 10 GHz and above are usually HEMTs cooled to 20 K, providing receiver temperatures comparable to or below the sky background. For later amplifier stages, and for the first stage of receivers at lower frequencies, GaAs MOSFETs are also used; these have the advantage of giving a more linear performance.

3.5 Low-noise amplifiers and mixers

In radio astronomy, the system noise usually exceeds the desired signal by several orders of magnitude. This requires that as small a quantity of noise as possible is added to the system noise. The current state of the art is summarized in Figure 3.5.

The lowest receiver noise at high frequencies is available from indium phosphide high-electron-mobility transistors (HEMTs), cooled to about 20 K. Other types of field-effect transistor (FET) are also used, some of which have better linearity in response to large signals; this is an advantage in situations where large interfering radio signals may occur. At the shortest millimetre wavelengths, where amplifiers are not available, either a bolometer or a mixer must be used. At the lowest frequencies, they need not be cooled, since the Galactic background noise is the dominant noise source: through centimetre wavelengths and into the millimetre domain, they must be cooled for the best performance. At the shortest wavelengths, mixers give the best performance for the first stage, heterodyning the signal down to a lower frequency. It is vital that the following preamplifier has the lowest possible noise temperature, since there is a loss of signal in the mixing process. The mixers are usually cooled superconductor/insulator/superconductor (SIS) diodes. The quantum limit to noise performance ($T_{\min} = h\nu/k$) is marginally important in this region; at sub-millimetre wavelengths it becomes increasingly important. Heterodyne receivers have been developed for frequencies up to 900 GHz (wavelength 0.35 mm), with noise temperature of about 100 K, for use in the ALMA synthesis telescope and several single-aperture telescopes.

Cooling the first stages of receivers, using closed-cycle refrigerators, is common practice for frequencies of 1 GHz and above. The cooled section (the so-called *front end*) may also include a filter to reduce interference from strong radio signals in adjacent frequency bands. These filters may use superconducting materials, which introduce an almost negligible loss in the observing frequency band (Lancaster 1997). Further information on microwave radio astronomy techniques may be found in Webber and Pospieszalski (2002).

3.6 Radiometers in practice

The basic radiometer may be useful in the simple form of Figure 3.4, but it does require an exceptionally stable receiver system. Usually the total system temperature may be many orders of magnitude greater than the contribution of the source being studied. If the gain is changing significantly during a typical measurement, for example on a timescale comparable to the time taken to move the radio telescope on and off the source, a meaningful measurement may be impossible. A practical radiometer system must take account of all sources of noise and also overcome the problem of instabilities in receiver gain.

The power spectral density from the antenna system will be kT_a, and T_a may be a few tens of degrees in typical cases. Figure 3.5 summarizes typical antenna temperatures as a function of frequency. At low frequencies, the intense Galactic background noise predominates; as the frequency increases, this background temperature decreases. At centimetre wavelengths, radiation from the atmosphere and stray radiation from the ground are the dominant noise sources unless one is observing an intense source such as the Sun. At millimetre wavelengths, radiation from water vapour and oxygen in the atmosphere causes the antenna temperature to rise dramatically (note the effects at line frequencies such as the water line at 22.3 GHz and the extensive oxygen complex at 55–60 GHz). At sub-millimetre wavelengths, absorption by water vapour becomes increasingly severe, and high, dry observing sites are required (see Figure 7.9 later). With some interruptions, observations at favourable sites are feasible out to 460 GHz (0.65 mm); there are windows at 460–510, 620–720 and 790–910 GHz, but thereafter, except for a few dim windows, the atmosphere is effectively opaque until the 20-μm infrared window is reached.

In the centimetre wavelength region, at 6 cm (5 GHz) for example, a carefully designed antenna will have a total noise temperature of the order of 15–20 K. The preamplifier will add its excess noise. Figure 3.5 shows typical values of T_n with current practice, and one can expect that a modern 5-GHz cooled preamplifier using an InP HEMT will have an excess noise temperature of less than 5 K. The total system temperature, therefore, might be of the order of 25–30 K, although it can be less if care is taken. A typical bandwidth would be in the range 50–100 MHz (it could be as much as 2 GHz or greater), so for these nominal numbers the total system noise power, kTB, would be about 5×10^{-14} W, or 133 dB below a watt (dBW). In most systems, one would like to have a power level of 1–10 mW presented to the detector, and one would therefore want a total amplifier gain of about 110 dB. Typically, there might be a gain of 80 dB before the first conversion, a loss of 10 dB in the mixer and gain-setting attenuator and another 40 dB of gain in the remaining stages, although the amounts of gain before and after the mixer might easily be interchanged.

The random noise appearing at the output of an ideal radiometer, which we will assume has been calibrated in units of temperature, will necessarily fluctuate, with an r.m.s. uncertainty given by Equation (3.11). The power density received by an antenna of effective area A, observing an unpolarized radio source of flux S, will be $SA/2$ (the factor 2 appears because only one polarization can be observed by a single radiometer). When this power is expressed in units of antenna temperature, the limiting r.m.s. flux sensitivity for a point source, ΔS, is

$$\Delta S = \frac{2kT_{\text{sys}}}{A\sqrt{BT}}. \tag{3.13}$$

One sees, therefore, that the quotient T_n/A is a measure of the sensitivity of a radio-telescope system.

Given a total system temperature of 50 K and a bandwidth of 100 MHz, an averaging time of 1 s would result in an r.m.s. uncertainty of 5 mK for a given measurement by the fundamental relation, Equation (3.11). This is only one part in 10^4 of the system noise, so the fractional gain stability must be much better, despite the total gain of about 110 dB (10^{11}). This is difficult to attain in practice, and it is often necessary to use a switching radiometer, usually known as a *Dicke radiometer* after the inventor, Robert Dicke.

A simplified block diagram of a Dicke radiometer is given in Figure 3.6. There are two key differences from the system shown in Figure 3.4. A comparison switch is interposed between the calibration network and the receiver, switching rapidly between the signal $v_i(t)$ and a reference load generating a noise signal $v_r(t)$, equivalent to a black body at T_{ref}. In addition, the detector is replaced by a *lock-in amplifier*, also known as a *phase-sensitive detector*. The switched input signal to the receiver is illustrated as a function of time in Figure 3.6, together with the associated power. The power envelope will be a square wave with an amplitude that is proportional to the difference between T_a and T_{ref}. If this square wave has a given phase when $T_a > T_{\text{ref}}$, its phase will be reversed when $T_a < T_{\text{ref}}$. The function of the lock-in amplifier is to give a signal that is proportional to the modulation amplitude, but whose sign reflects the two possible phase conditions. When $T_a = T_r$, the output of the lock-in amplifier will be zero. When this condition is approximately met, the measurement of the flux from a weak source will be relatively insensitive to gain

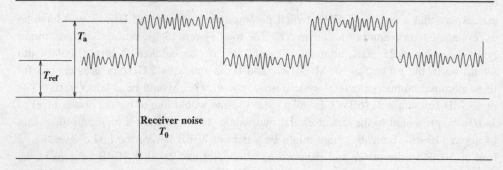

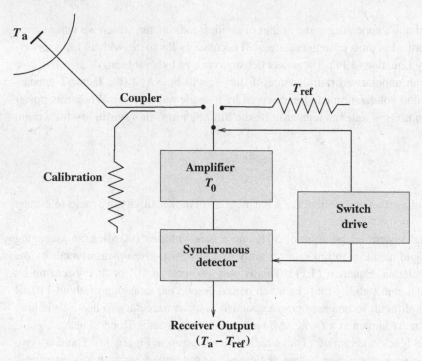

Figure 3.6. The Dicke switched radiometer, showing the switched signal as a function of time.

fluctuations, although thermal noise fluctuations will still be present as an unavoidable limit to the accuracy of measurement.

A small price is paid for this advantage: the antenna temperature is only being measured for half of the time. This means that the relative accuracy of measuring either the antenna temperature or the reference load temperature will be degraded by a factor of $\sqrt{2}$ and, since the difference between two uncorrelated random variables is being taken, there is a further loss in accuracy of $\sqrt{2}$. The net fractional accuracy of the measurement, therefore, is

$$\frac{\Delta T}{\langle T_{sys} \rangle} = \frac{2}{\sqrt{\mathcal{B}T}}. \tag{3.14}$$

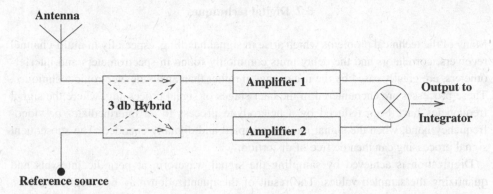

Figure 3.7. A differential radiometer, using a hybrid circuit and multiplier, measuring the difference between the noise power from an antenna and a reference source.

Despite the loss in accuracy by a factor of 2, the greater stability of measurement is often worth the price.

There are several ways of constructing a lock-in amplifier, but functionally one simply takes the product of the detected signal with the reference square wave that drives the comparison switch, as shown in Figure 3.6.

In order to take full advantage of the inherent stability of the Dicke radiometer, the antenna and load temperatures should be nearly equal. When this is not possible, the gain of the receiver can be modulated synchronously to give a null output when one is not observing a source. An alternative to the Dicke switch takes advantage of the coherent nature of radio signals before detection. The key component is shown in Figure 3.7. Here both the input source and the reference are connected to two amplifiers through a hybrid circuit. If the voltages of the noise signals from source and reference are designated v_s and v_r, the hybrid outputs are $v_s + v_r$ and $v_s - v_r$. The amplified signals are multiplied together, giving the required output $v_s^2 - v_r^2$. The required difference has still to be detected against a background of system noise, and is again distinguished by a switching system. The WMAP differential radiometer using this principle is shown in more detail in Figure 15.2 later.

One should note that the measurement of a discrete source is relatively straightforward, since it is a measurement of the difference between the receiver outputs when the telescope is directed on and off the source, but when an extended source such as the Galactic background is being measured, an absolute measurement of antenna temperature must be carried out. In such a case, one starts by replacing the antenna by a dummy load at a known temperature and measuring the difference in temperature between it and the reference load. The antenna is then substituted for the dummy load, and the difference signal is measured. A careful measurement is not carried out easily; the switches may be asymmetric, or the loads might not present the same impedance to the receiver. Measurements at very low temperatures are especially difficult and demanding; the ultimate test is the measurement of the CMB, which will be described in Chapter 14.

3.7 Digital techniques

Many of the technical problems which arise in signal handling, especially in multi-channel receivers, correlators and the delay units commonly found in spectrometers and interferometers, are greatly eased by the use of digital rather than analogue electronic techniques. These processes are encountered in the later stages of signal processing, where the signal frequencies have been reduced by a heterodyne process to an intermediate- or video-frequency band, when the signal may be sampled and digitally encoded. The subsequent signal processing can then be free of distortion.

Digitization is achieved by sampling the signal waveform at periodic intervals and quantizing the sampled values. The result of the quantization may be simply two bits, representing positive or negative values of the signal waveform, or the quantization may be into a number of signal levels. In any case the sampling must be sufficiently rapid that no spectral information is lost.

The theory of digital sampling relies upon two basic concepts: the Nyquist sampling criterion and the van Vleck digital approximation. The Nyquist criterion states that a signal coming from a receiver of bandwidth B can be completely represented by discrete samples taken at the Nyquist frequency, $2B$. When the sampling is digital and the noise signal obeys Gaussian statistics, the relative power spectrum can be deduced even if only the sign of the signal amplitude is measured. This surprising result, in which the noise signal is represented by a string of one-bit samples, is known as the van Vleck quantization approximation. The total noise power in the band can be determined by a separate measurement, thus restoring the spectrum from a relative to an absolute basis. Single-bit quantization of the signal also degrades the signal-to-noise ratio, which is only $2/\pi$ that of an ideal correlator, but this is often a small price to pay for the great simplicity of single-bit signal handling.

Faster, large-scale integrated circuits allow multibit representations, which are increasingly in use; single-bit correlators are being phased out, and two- to five-bit systems are becoming the norm. A two-bit system represents the signal in terms of four signal levels. The two bits represent the sign of the signal and whether its absolute value exceeds a number that is very nearly the r.m.s. level of the noise signal. This means that there are four possible signal levels, and the two-bit system is sometimes called a four-level system. The degradation of the signal-to-noise ratio is less than that of the single-bit system; the value achieved is 0.88 that of an ideal correlator. A larger number of bits can be used; a three-bit representation would give eight levels of quantization, with a value of 0.95 for the signal-to-noise quality factor. The relation between quantization levels and signal-to-noise ratio is set out by Bowers *et al.* (1973). For this quality factor only, the point of diminishing returns is rapidly reached, since the complexity of the digital circuitry increases rapidly, and most modern systems give only an option of one-bit or two-bit quantization.

Digital sampling offers great advantages in signal handling, but the process of quantization is inherently non-linear, which may have important effects. A potential disadvantage is that an interfering radio signal in any part of the observed spectrum may affect the output over a much larger range of the output spectrum; this effect is less with more levels of

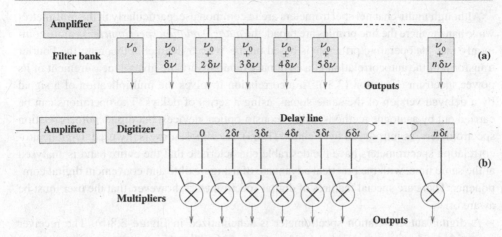

Figure 3.8. Radio spectrometers: (a) using a bank of narrow-band filters; (b) a digital autocorrelation system. The longest delay $n\,\delta t$ is approximately equivalent to the spectral resolution $(\delta\nu)^{-1}$.

quantization. Single-bit sampling, on the other hand, has the advantage that a large impulsive interfering signal may affect one sample only, giving an output limited to a single digit, with negligible effect in the receiver output. A comprehensive discussion of the digital sampling is given by Cooper (1970).

3.8 Spectrometry

The broad continuum spectrum of a radio source is usually determined by making step-by-step flux measurements at as many frequencies as possible, using the entire instantaneous bandwidth of the receiver for each measurement. When the power spectrum includes spectral lines, however, such a procedure would require a narrow bandwidth and would become time-consuming and imprecise. A receiver bandpass is usually wide enough to contain one or more spectral lines, whose detailed profiles are the object of study, and the band can be subdivided into a number of filter channels. The most direct approach to such a multi-channel system, illustrated in Figure 3.8(a), is to have a large number of narrow-band filters in parallel, each with its own detector or phase-sensitive detector, measuring the flux in discrete bins. The individual outputs can be accumulated in integrators that are polled by the data-acquisition computer, which then stores and organizes the data. Usually there is a buffer amplifier in front of each filter to guarantee that there will be no cross-talk between channels. In the diagram, each filter is tuned to a separate frequency, but sometimes the filters are identical, with each channel having its own local oscillator and mixer. In this type of multi-channel receiver, one can reconfigure the filter array, tuning groups of filters to individual lines in the spectrum, or spacing the channels more broadly when searching for a line whose frequency is only approximately known.

Although multi-channel spectrometers are in common use, particularly in the millimetre-wave bands where the line profiles are broad, the *autocorrelation spectrometer* is more commonly met. Its operating principle is based on the Wiener–Khinchin theorem: the Fourier transform of the autocorrelation of a stationary random variable gives a measurement of its power spectrum (Equation (3.3)). Autocorrelation involves the multiplication of a signal by a delayed version of the same signal, using a series of delays. The operations can be carried out by analogue methods, using acousto-optical devices, but digital autocorrelation spectrometers, as introduced by Weinreb (1963), are almost universally used. Digital auto-correlation spectrometers have the desirable characteristic that the entire band is analysed at the same time, with the practical advantage of using flexible and convenient digital components. There are special features of digital spectrometers, however, that the user must be aware of.

A digital autocorrelation spectrometer is schematized in Figure 3.8(b). The receiver output, $v_0(t)$, has been converted to video frequencies (baseband) and goes to a digitizer that samples the input signal at the Nyquist frequency, producing a series of binary packets, each representing the amplitude v_j of the video signal at an instant of time t_j. The signal is then delayed in a series of identical delay units, so that the inputs to a set of multipliers each comprise the direct signal and the same signal delayed by multiples of the unit delay. If there are n separate lags $\delta t, 2\,\delta t, \ldots, n\,\delta t$, the initial signal is multiplied by each of the delayed signals (including a zero-lag multiplication, the signal multiplied by itself), to give a total of $n + 1$ outputs. The output numbers are read out at the clock rate (the Nyquist frequency $2B$) and the results are accumulated in counters. The accumulated totals are read out at a more leisurely rate by a computer that carries out the longer-term accumulation. One therefore assembles a set of $n + 1$ estimates of the discrete correlation function R_n, each a set of M samples of the signal (equivalent to an integration time $M\,\delta t$):

$$R_n = \frac{\sum_{j=1}^{M} v(t_j)v(t_j + n\,\delta t)}{M}. \tag{3.15}$$

Note that division by M is necessary in order to normalize R_n properly. The finite estimate of the autocorrelation must also be corrected for the bit quantization in order to arrive at a true equivalent to $R_{t,T}(\tau)$ in Equation (3.2). The corrected autocorrelation $R_{n,c}$, for one-bit quantization of Gaussian noise, is

$$R_{n,c} = R(0) \sin\left(\frac{\pi}{2} R_n^{(1)}\right). \tag{3.16}$$

Most of the time, the signal under study is small compared with the total system noise, which means that R_n is small compared with 1. Usually, therefore,

$$R_{n,c} \approx R_n^{(1)}. \tag{3.17}$$

The continuous Fourier transform that was used to give Equation (3.3) must now be replaced by a discrete Fourier transform (DFT) operating on the corrected autocorrelation:

$$S_k(\nu_k) \overset{\text{FT}}{\Longleftrightarrow} R_{n,c}(\tau_n). \tag{3.18}$$

The power spectrum S_k is now a discrete set of $n + 1$ values, but if there is a sufficient number of lags, the result will be a representation of the spectrum. There is a caveat, however, arising from the finite number of lags, that is, the finite time over which the autocorrelation is evaluated. This means that the spectrum S_k is logically equivalent to scanning the input spectrum with a filter whose bandpass is the Fourier transform of the lag window. The DFT weights all lags equally in calculating the values of S_k, which means that the true autocorrelation is multiplied by a gating function $\Pi(\tau)$ whose Fourier transform is a sinc-function. This generates relatively high sidelobes (about 22% of the peak) for every line; in effect, the filter is ringing. Although all the information in the spectrum is present, the appearance of such a spectrum can be confusing. The solution is to apply to the autocorrelation a weighting function that tapers the lags and reduces the sidelobes; this is similar to the problem of excessive sidelobe levels in a radio telescope, where the solution is to taper the illumination of the aperture (Chapter 4). Several choices of weighting function are in use. Using integrals instead of sums to make the weighting procedure more easily recognized (although sums are used in digital systems), the calculation amounts to taking the convolution of $w(\tau')$ and $R(\tau')$ to obtain the weighted convolution $R_w(\tau)$:

$$R_w(\tau) = \int w(\tau') R(\tau' + \tau) d\tau'. \tag{3.19}$$

The convolution operation is usually given a more compact form, as in Appendix 1:

$$R_w(\tau) = w(\tau') \otimes R(\tau'), \tag{3.20}$$

where $\otimes$ denotes cross-correlation. The weighting function is frequently the raised cosine, or Hanning, function

$$w_H(\tau') = 1 + \cos(2\pi\tau'). \tag{3.21}$$

This reduces the frequency resolution to 60% of the sinc-function value, but the spurious responses are reduced from 22% to 2.6%.

The concept of determining a power spectrum by autocorrelating the signal amplitude, while it is a powerful and generally used technique, has to be understood more thoroughly than straightforward multi-channel spectrometry. One must beware, in particular, the complications introduced by non-Gaussian signals, such as manmade interference. The radio astronomer must also consider the choices of quantization and lag weighting and their effects on signal-to-noise ratio, referred to above.

3.9 Cross-correlation radiometry: interferometry

In Chapter 5 we introduce the concept of interferometry, in which signals from separate antennas are brought together in a single receiver, called a correlator. The interferometer pair generates interference fringes, and the correlator output gives the fringe amplitude and phase. This output is in effect a complex number that contains information on the angular distribution of the signal from an extended source; this is the basis of interferometry and aperture synthesis, as described in Chapters 5 and 6. We introduce here the formal approach to the characterization of the output of an interferometer.

The interferometer baseline is projected onto the u, v-plane (discussed in Section 5.4; this is simply the plane perpendicular to the source direction, with the coordinates u and v measured in wavelengths). The convolution of the signal amplitudes $x(t)$ and $y(t)$ from the two antennas yields the *cross-correlation function* $R_{x,y}(\tau)$:

$$R_{x,y}(\tau) = x(t) \otimes y(t). \tag{3.22}$$

By analogy with the relation between signal and spectrum in Equation (3.3), the Fourier dual of this measured function is the *cross-spectrum power density* $S_{x,y}(\tau)$. When an interferometer observes a continuum source, the instrumental time delay τ_i is chosen to compensate for the geometrical time delay τ_g. In practice one must be able to update τ_g continually (there are several methods to accomplish this) in order to obtain a proper time-average. If the object being observed is a spectral-line source (the hydrogen distribution in a galaxy, or a complex of molecular masers, for example), the instrumental time delay can be made to differ from τ_g, with an effect similar to the action of the autocorrelation spectrometer in the previous section. The receiver system then produces a set of *cross-correlations*, which contain the spectral information in addition to the spatial information contained in the set of baselines (the set of values in the u, v-plane). The time-delay operation is equivalent to taking data in the w-direction, towards the source and perpendicular to the u, v-plane (see Section 5.4). The cross-correlation data set $R_{x,y}(u, v, \tau)$ is a three-dimensional data set, commonly referred to as a *data cube*. The two-dimensional transform of the u, v-plane data gives the angular distribution of the source brightness, and a transform of the time-delayed data sets gives the variation in frequency. The three-dimensional Fourier transform of $R_{x,y}(u, v, \tau)$ thus gives a data cube $S(\theta, \phi, \nu)$ that contains the angular distribution of brightness as a function of frequency.

The noise limitations of an interferometer can be derived from considerations similar to those outlined earlier. For an integration-time–bandwidth product BT, the r.m.s. uncertainty depends upon the receiver technique. The cross-correlation method gives the best possible signal-to-noise ratio; if the two elements of the interferometer have system noise temperatures T_{n_1} and T_{n_2}, and the antennas have effective areas A_1 and A_2, the r.m.s. fluctuations in power correspond to an uncertainty in the equivalent point-source flux

$$\Delta S_{x,y} = 2k\sqrt{\frac{T_{n_1} T_{n_2}}{2A_1 A_2 BT}}. \tag{3.23}$$

This can be compared with the single-antenna radiometry equivalent, given in Equation (3.12), and it is seen that the system temperature and effective area are replaced by the geometric means of these quantities for the interferometer pair. The extra factor of $\sqrt{2}$ appears because the cross-correlation process is comparing the values of the radiation field at two different places in space. There is an additional total-power term that is, in effect, an autocorrelation term for each antenna, and a total-power radiometer has a peak response that includes this term, avoiding the $\sqrt{2}$ loss. This does not mean that a total-power interferometer would be better, however, for two reasons. Firstly, the total-power interferometer has serious practical disadvantages, as noted in Section 3.6, and, secondly, it is not the instantaneous total power that is desired, but the fringe visibility. No technique gives a better signal-to-noise ratio than cross-correlation in determining the fringe visibility.

In present-day practice, the cross-correlation operation is carried out digitally. Equation (3.15) generalizes to

$$R_{xy,n} = \frac{1}{M} \sum_{j=1}^{M} v_x(t_j) v_y(t_j + n\,\delta t). \tag{3.24}$$

Most commonly, the collection of discrete correlations is usually transformed to a uniform grid, giving a source cube $R(u_i, v_j, \tau_n)$ that can be Fourier-transformed to a data cube $S(x_i, y_j, \nu_k)$. The later operations of data manipulation and self-calibration are carried out on these discrete data sets. Note that this gives spectrum information, in addition to angular structure.

Further reading

Imaging at Sub-millimetre Wavelengths, Dent W. *et al.*, ed. Mangum J. and Radford S., *ASP Conf. Ser.* **217**, 2000.

Microwave Instrumentation for Radio Astronomy, Webber J. C. and Pospieszalski M. W., *IEEE Trans. Microwave Theory and Techniques*, **50**, No. 3, March 2002.

Passive Microwave Device Applications of High-Temperature Superconductors, Lancaster M. J., Cambridge, Cambridge University Press, 1997.

Radio Astronomy Techniques, Bracewell R. N., in *Handbuch der Physik*, **54**, p. 42, ed. Flugge S., Berlin, Springer-Verlag, 1962.

Tools of Radio Astronomy, Rohlfs K. and Wilson T. L., Berlin, Springer-Verlag, 4th edn, 2004.

4

Single-aperture radio telescopes

A radio telescope intercepts the radiation coming from celestial sources, usually separating it into its two polarization components. The telescope sends the energy it receives through transmission lines to a receiving system where the signals are amplified and transferred to the detection system. The radio telescope must meet two basic requirements, sensitivity and angular resolution. The following three chapters are concerned with optimizing these two fundamental parameters. Sensitivity depends on having the largest collecting area possible, while minimizing the contributions of extraneous noise; this depends upon both the telescope design and the quality of the receiving system. The angular resolution is determined by the overall dimensions of the telescope.

In this chapter, we consider single-aperture telescopes, for which large area and high angular resolution go together. It is economically impossible to get the highest angular resolution by extending the size of a single aperture indefinitely, so it is necessary to use widely spaced single-aperture telescopes in an array. Such arrays, and the means by which their data are analysed, are discussed in Chapters 5 and 6. Arrays, which are now usually aperture-synthesis arrays, are made up of individual elements, usually paraboloidal radio telescopes, although they may themselves be arrays of dipoles if intended for long wavelengths. The individual elements instantaneously observe a patch of sky, determined by the beam pattern; this is called the field of view (FOV). When the elements are steerable, as they usually are for paraboloids, their design (or thc horizon) defines the available sky. Paraboloidal radio telescopes can have focal-plane arrays that give multiple beams, and dipole arrays can have multiple outputs that similarly allow many beams simultaneously; this enlarges the field of view. There can be further restrictions imposed by an aperture-synthesis array; these are treated in Chapters 5 and 6.

4.1 Fundamentals: dipoles and horns

The previous chapter noted that, because of reciprocity, a radio telescope can be analysed either as a transmitter or as a receiver. For the transmitting case, the radiated electric field is generally described by the power gain G (proportional to the square of the field strength), whereas for the receiving case it is given by the effective area, A_{eff}. Both are functions of the direction of reception or transmission, and they are related by the fundamental relation

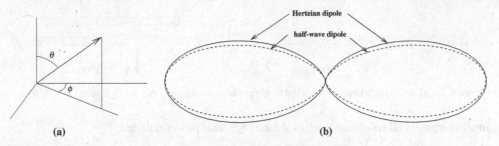

(a)   **(b)**

Figure 4.1. (a) The coordinate system and (b) radiation patterns (polar diagrams) of a Hertzian dipole (full line), and a half-wavelength dipole in free space (broken line).

between gain and effective area (Equation (2.19)):

$$A_{\text{eff}} = [G/(4\pi)]\lambda^2. \tag{4.1}$$

The simplest case to analyse is the Hertzian dipole, an infinitesimal length Δl of current varying sinusoidally at angular frequency ω. The geometry and the radiation pattern are shown in Figure 4.1 for a dipole axis at $\theta = 0°$. Note that the angles θ and ϕ used in this section are convenient for the isolated dipole because of the azimuthal symmetry; consequently, the radiation pattern does not depend on ϕ. In Section 4.2, the convention will change.

The electric and magnetic fields each exhibit two parts to the radiation pattern, with different dependences on both the radial coordinate r and the polar angle θ. The first part is called the *near field* or induction field; it has two terms, diminishing inversely as the second and third powers of r. The second part is the radiation field, or *far field*, which varies inversely with distance; it has an electric-field component only in the θ-direction, and a magnetic-field component in the ϕ-direction. These two field components are related by

$$\mathbf{E} = Z_0 \mathbf{H} \times \hat{\mathbf{n}}, \tag{4.2}$$

where $Z_0 = 388\ \Omega$ is the impedance of free space and $\hat{\mathbf{n}}$ is the unit vector in the r-direction.

The electric field in the far field, for a short sinusoidally varying current element of amplitude I_0 and length Δl, is[1]

$$E_0 = [Z_0/(4\pi)](I_0\,\Delta l)k^2 \sin\theta \exp[\text{j}(\omega t - kr)]/(\text{j}kr), \tag{4.3}$$

where $k = 2\pi/\lambda$. The electric field is an expanding spherical wave, accompanied by an orthogonal magnetic field. Both fields vary as $1/r$, so the average value of the Poynting vector $\langle \mathbf{S} \rangle = \frac{1}{2}\mathbf{E} \times \mathbf{H}$ (including a factor $\frac{1}{2}$ for the time average) is

$$\langle \mathbf{S} \rangle = \frac{1}{8}Z_0(I_0\,\Delta l/\lambda)^2 \sin^2\theta/r^2. \tag{4.4}$$

---

[1] For the imaginary quantity we use both j and i; j for electrical-engineering usage, and i otherwise.

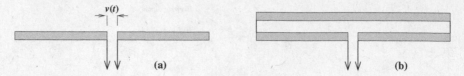

Figure 4.2. (a) A dipole antenna connected to a transmission line. (b) A folded dipole.

Integrating over all solid angles gives the average total power radiated:

$$\langle P \rangle = (\pi/3) Z_0 I_0^2 (\Delta l/\lambda)^2. \tag{4.5}$$

This reminds one of the average power dissipated in a resistor, $\langle P \rangle = \frac{1}{2} I_0^2 R$, and leads to the introduction of another physical quantity, the *radiation resistance* R_r, since the power radiated by the Hertzian dipole in Equation (4.5) is equivalent to the power dissipated in a radiation resistance R_r equal to

$$R_r = (2\pi/3) Z_0 (\Delta l/\lambda)^2 \quad (=789 (\Delta l/\lambda)^2 \, \Omega). \tag{4.6}$$

It is immediately clear that, if the dipole is very short compared with a wavelength, its radiation resistance will be very small.

In Section 2.1 the power gain of an antenna was defined, for a given direction, as the ratio of the power radiated to that radiated by an isotropic radiator. Equation (4.4) leads to the result that the power gain of a Hertzian dipole is

$$G = (3/2) \sin^2\theta \tag{4.7}$$

and so the maximum gain in the direction perpendicular to the dipole is $3/2$. Thus, from Equation (4.1), the maximum effective area of the Hertzian dipole, perpendicular to the dipole at $\theta = 90°$, is

$$A_{eff} = [3/(8\pi)]\lambda^2. \tag{4.8}$$

It may seem paradoxical that a very small wire should have an effective area that is an appreciable fraction of a square wavelength, but it should be remembered that displacement current in electromagnetic theory acts like a real current, and the near-field components, while playing a negligible role in the far-field properties of the antenna, are real, and are important over distances of the order of a wavelength.

Short dipoles are sometimes encountered in radio astronomy, but the *half-wavelength dipole* is more commonly met in practice. When it is excited by a current that is at a frequency whose wavelength is twice the length of the dipole, it is said to be excited at its *resonance*.

The general case of radiation from a cylindrical conductor carrying a current is treated in the standard textbooks. The dipole has to be fed by a transmission line, as illustrated in Figure 4.2, so the dipole has to be broken at the centre, where a voltage appears. The solution for the radiated electromagnetic field depends on matching the boundary conditions. These are that (1) the integral of the electric field across the gap (assumed to be negligibly small) must equal the voltage; (2) the tangential component of the electric field must go to zero at the surface of the cylinder; and (3) the current must go to zero at the end of the cylinder. For the half-wave dipole, the current along the dipole is distributed like a half-sinusoid, zero at the ends and maximum at the centre.

The resulting radiation field is given by an integral equation that can be solved analytically if the radiating cylinder is a sufficiently thin wire. In this case, the radiated power flux at a distance r, as a function of θ, is given by

$$\langle S(\theta) \rangle = [Z_0/(8\pi^2)](I_0/r)^2 \left[\cos\left(\frac{\pi}{2}\cos\theta\right) \Big/ \sin\theta \right]^2. \tag{4.9}$$

The power gain and radiation resistance of the half-wavelength dipole at its resonance wavelength can be calculated by the same methods as were used above for the Hertzian dipole. The gain, $G(\theta)$, is

$$G(\theta) = 1.64 \left[\cos\left(\frac{\pi}{2}\cos\theta\right) \Big/ \sin\theta \right]^2, \tag{4.10}$$

so that the gain normal to the dipole, 1.64, is slightly greater than the maximum gain, 1.5, of the Hertzian dipole. The polar diagram $G(\theta)$ for the half-wave dipole is shown in Figure 4.1(b), where it is seen to be narrower than that for the Hertzian dipole.

At resonance, the radiation resistance of the ideal half-wave dipole is $R_r = 73\,\Omega$. In practice the actual antenna impedance Z_a will be different, partly because of the finite thickness and finite conductivity of the dipole, but mainly because it may be used over a wide bandwidth. The antenna impedance will be frequency-dependent, and will not be purely resistive; it will be a complex quantity $Z_a = R_a + jX_a$. Off resonance, the antenna impedance changes rapidly and the reactive term X_a becomes significant. For thicker dipoles, X_a varies more slowly with frequency, so in most practical cases a fat dipole is used, thus increasing its usable bandwidth without introducing a severe impedance mismatch between receiving system and antenna. The folded dipole (Figure 4.2) also has a comparatively wide bandwidth. At resonance it has a radiation resistance of 292 Ω, four times that of the single dipole (this arises because the transmission line is connected to only half the total current, so for a given voltage there will be twice the current flowing in the dipole, and therefore four times the power will be radiated).

A dipole may be connected to a receiver via a twin-wire (*balanced*) transmission line, or the connection may be via a coaxial-cable (*unbalanced*) transmission line. In the latter case a transformer, known as a *balun*, is required.

Figure 4.3 shows four types of balun. The simplest, shown in Figure 4.3(a), consists of two mutually coupled coils, one balanced and the other unbalanced, with one side earthed. This is impractical for most radio-astronomy purposes, since it can be realized only at low frequencies. A second version, Figure 4.3(b), in use at frequencies up to 10 GHz (and probably realizable up to the highest frequencies at which transistor low-noise amplifiers can be used) uses an amplifier with a balanced input, using a pair of input transistors back to back. Internal circuitry transforms the initially balanced transistor configuration to

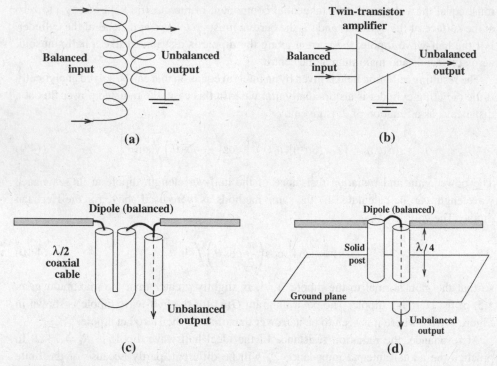

Figure 4.3. Balanced-to-unbalanced transformers, known as baluns: (a) coupled coils, suitable for low frequencies; (b) paired transistors; (c) half-wave coaxial cable; and (d) split coaxial cable above an earth (or ground) plane.

unbalanced circuitry, with an unbalanced output, usually a coaxial cable. This should be realizable up to the highest frequencies at which transistor circuits can be fabricated.

Figure 4.3(c) shows a simple passive balun, in which a half-wavelength coaxial cable provides a phase shift of 180° for one half of the dipole. It also transforms the dipole impedance downwards by a factor of four. Thus, a folded dipole with an impedance of 280 Ω would conveniently present an impedance of 70 Ω to the coaxial cable. The fourth example, Figure 4.3(d), shows a robust structure that has been useful in radio astronomy. Here, the coaxial line is split, with the central conductor attached to one side of the dipole, while the other half of the dipole is attached to the opposite half of the split coaxial line. For a fat dipole, this gives a good impedance match to a 50-Ω cable.

One should note that passive baluns work only over a finite bandwidth. The version in Figure 4.3(a) is limited by the impracticability of making coupled coils at high frequencies, while the two versions in Figures 4.3(c) and 4.3(d) have tuned transmission lines. This is not always a problem, since the dipole itself has a finite bandwidth; however, the need for a very wide bandwidth in modern arrays, such as LOFAR and MWA (Chapter 16), has led to an increasing use of the paired-transistor balun (Figure 4.3(b)).

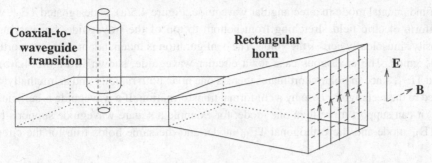

Figure 4.4. A horn antenna, with a transition to a coaxial cable.

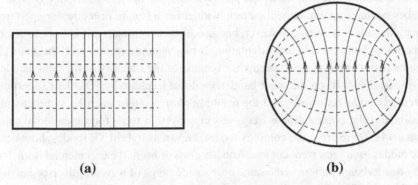

Figure 4.5. The field pattern for transverse electric modes at the aperture of (a) rectangular (TE_{01}) and (b) circular (TE_{11}) horns. Full lines show the **E** field; broken lines show the **B** field.

4.1.1 The horn antenna

We noted the important role of displacement current in radiation from a short dipole; a *horn antenna* is an example of a system where all the current at the aperture is displacement current. Figure 4.4 shows a rectangular, tapered horn connected to a waveguide, into which protrudes the central conductor of a coaxial cable. A received wave propagates down the waveguide, its electric field generates a current in the coaxial probe, and the signal is propagated down the coaxial cable to the receiving device. The horn must be gently tapered, since the wavefront in the horn will be curved, and furthermore unwanted higher modes can be excited.

The mode that is excited in the waveguide approximates the field at the aperture of the horn if the taper is not too abrupt, avoiding a wavefront that is too strongly curved (although some curvature will be present, and this must be accounted for in calculating the actual gain pattern of the horn). The lowest modes in rectangular and circular waveguides, the principal modes, are the two most important simple cases. The field configurations in these two cases are illustrated in Figures 4.5(a) and (b). In both cases, the electric field is transverse, with the magnetic field in the orthogonal direction, in loops along the axis of the waveguide.

The fundamental mode in a rectangular waveguide, Figure 4.5(a), is designated TE_{01}, with a uniform electric field stretching from bottom to top of the waveguide, but varying in intensity sinusoidally across the guide. The configuration is independent of the proportions of the guide. The analogous case for a circular waveguide, shown in Figure 4.5(b), is called TE_{11}, since in polar coordinates there is one node both radially and azimuthally; it is related to the rectangular case by a conformal transformation. If a waveguide is sufficiently large, it can support more than one mode; for example a square waveguide supports both the TE_{01} mode and the orthogonal TE_{10} mode; and the same holds true for the circular case.

The size of the horn aperture, measured in wavelengths, determines the shape of its radiation pattern. In most radio-astronomy applications, an axially symmetrical gain pattern is desired, and this is determined by the proportions of the aperture. A horn fed in the TE_{01} mode does not have a symmetrical pattern, with either a square or a circular aperture. For the square and circular horns, however, because of their symmetry, it is easy to see that a TE_{01} mode can be excited in any orientation. If two separate modes with the same phase, orthogonal to one another, are excited, then because of the linearity of the field their sum is a TE_{01} mode at $45°$. If, on the other hand, they differ in phase by $90°$, their superposition is a circularly polarized mode, and the resulting phase pattern must be symmetrical. For this reason, circular polarization is frequently employed in feeds for paraboloidal antennas (Section 4.14). There are more complex modes, known as hybrid TE modes; horns excited in these modes have seen frequent use. Another class of horn, the exponential horn, has the property that its beam pattern is the same over a wide range of wavelengths (see Section 4.3 below); this type of horn is usually preferred when measuring the CMB, for which a constant beamwidth is desired over a wide spectral range.

4.2 Arrays of radiating elements

4.2.1 Earth planes

Except for the case of dipoles attached to spacecraft, dipoles are usually found in combination with other structures, often an earth plane. A dipole above a conducting earth plane is shown in Figure 4.6. To meet the boundary conditions, the electric field parallel to the conducting surface must be zero. The resulting solution is logically equivalent to a dipole with a reflection, a second virtual dipole, below the earth plane, but reversed in phase. The system no longer has cylindrical symmetry, and the normal to the plane is the natural polar axis. The radiation pattern $G(\theta, \phi)$ will consist of the sum of the radiation patterns of the dipole and its image, multiplied by the radiation pattern of the dipole. The field strength, and therefore the gain, is different in the two planes $\phi = 0$ (parallel to the dipole, commonly called the E-plane pattern) and $\phi = 90°$ (the H-plane pattern, perpendicular to the dipole, shown in Figure 4.6(c)) for the case of a half-wavelength dipole.

The gain is increased, the actual increase depending on the spacing between the dipole and the earth plane. Maximum gain occurs when the dipole is a quarter of a wavelength

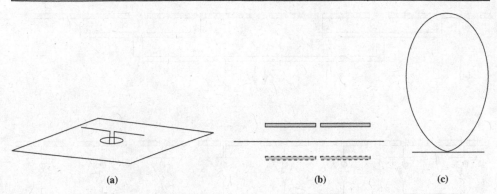

Figure 4.6. (a) A dipole above an earth plane; (b) a schematic representation of the virtual image; and (c) a polar diagram (in the equatorial plane).

above the earth plane; in that case the effective area (and the gain) is increased by a factor of four; one factor of two coming from the system radiating only in the upper half-plane, and the other factor of two coming from the second (virtual) dipole.

4.2.2 Phased arrays

At the longest wavelengths used in radio astronomy, for which it is impractical to use large reflectors, it is possible instead to use flat arrays of dipole elements distributed over the ground, covering a larger area than can be achieved with a steerable reflector. Such an array was used by Anthony Hewish and Jocelyn Bell in the discovery of pulsars (Chapter 12), at a wavelength of 3.7 m. The quarter-square-kilometre array at Kharkov is currently the largest single-aperture array in use; its dipoles are made of mesh, and are 'fat' in order to broaden their usable bandwidth, covering the range 20–30 MHz.

The behaviour of such an array can be visualized easily by considering it as a transmitting array. Consider an input signal, fed into a transmission line that splits symmetrically into transmission lines, each splitting again: eventually each branch is connected to a dipole. This arrangement, illustrated in Figure 4.7, is called a Christmas-tree network. The amplitude radiation pattern, $F_R(\theta)$ (considering the one-dimensional case for simplicity), will be the product of the pattern of the dipole above an earth plane, $F_D(\theta)$, and the vector sum of the radiation field from each element:

$$F_R(\theta) = F_D(\theta) \left[\sum_n i_n \exp(i\phi(n)) \right]. \tag{4.11}$$

For simplicity, consider an even number of dipoles, fed by a Christmas-tree array, with spacing between dipoles $\lambda/2$ (a common case for dense arrays). All the dipoles will have the same current excitation, which we take as unity, since it is only the relative pattern that is being studied. In the far field, the relative phases of a pair of adjacent dipoles will be $\phi = (\pi/2)\sin\theta$ (where $\sin\theta$ can be replaced by θ if the angle is small). The power

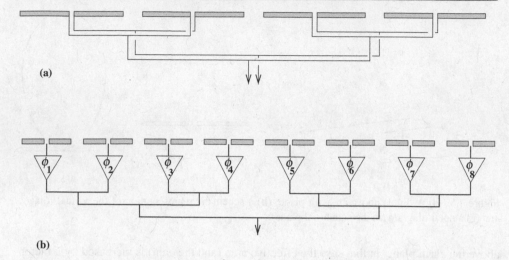

(a)

(b)

Figure 4.7. A phased array: (a) a 'Christmas-tree' network feeds an array of dipoles in phase; (b) phase shifters (and amplifiers) allow a linear gradient of phase, which swings the angle of the radiated beam.

gain will be proportional to the square of the amplitude gain of Equation (4.11) above; a Christmas-tree network has an even number of dipoles, so, with the restriction that N is even, the resulting gain as a function of θ will be

$$G(\theta) = (G_D/N)\left[\sin\left(\frac{N\pi}{2}\sin\theta\right)\middle/\sin\left(\frac{\pi}{2}\sin\theta\right)\right]^2. \qquad (4.12)$$

The gain, as shown in Equation (4.12), at the maximum is N times as great as the gain G_D of a single dipole above an earth plane. It follows, then, that the larger the number of dipoles the narrower the principal beam. Note that an array of N dipoles, spaced by $\lambda/2$, has a dimension of $N\lambda/2$, and the principal beam has an angular width of order $1/N$; this is to be expected from standard diffraction theory, which gives, for an aperture of size D, an angular resolution of order λ/D. This fundamental relation will be considered in greater detail later.

The spacing can be larger than $\lambda/2$, with a consequent sharpening of the beam, but, if the spacing exceeds λ, the periodic nature of Equation (4.12) will mean that other maxima will appear. The one-dimensional case has been used as an illustration, but the same formalism applies for two-dimensional arrays.

In the case of real arrays, phase-stable amplifiers can be inserted to overcome the losses in the transmission lines. Furthermore, the beam can be steered by inserting phase shifters that give a progressive phase delay across the array, as shown in Figure 4.7(b).

Phased arrays are used in two major radio telescopes that are being built to operate in the relatively unexplored low-frequency range below 300 MHz (wavelength 1 m). Among many objectives, these telescopes will attempt observations of the 21-cm hydrogen line,

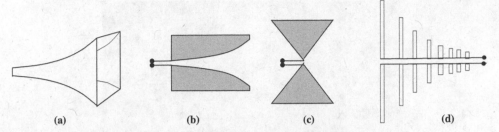

Figure 4.8. Broad-band antennas: (a) exponential horn; (b) Vivaldi exponentially tapered slot; (c) bowtie; and (d) log-periodic.

redshifted by a large but unknown amount from the era after the Big Bang and before reionization (Chapter 15). A high-resolution telescope array with large collecting area and a usable bandwidth of at least 3 : 1 is needed. There are two such arrays that are in the prototype stage and planned to be completed by about 2010 (see Chapter 17). Both are aperture-synthesis arrays (Chapter 6) composed of individual units known as *tiles* that are phased arrays of smaller elements. One of these, the Murchison Wideband Array (MWA), is being built in Western Australia and is designed to cover the bandwidth 80–300 MHz. Each tile of the array is composed of 16 wide-bandwidth bowtie antennas (see Section 4.3 below), each with dual polarization and each with its individual transistor amplifier, having a balanced input and an unbalanced output, as in Figure 4.3(b). No Christmas-tree or other distribution network is used; instead, the individual signals go to an electronic summing network, which has the appropriate phase-shifters to form the beam and swing it about on the sky. The second, the Low Frequency Array (LOFAR), is under construction initially in the Netherlands and will eventually extend through many European countries, using tiles each of which is a phased array of dipoles. Each dipole is an inconspicuous tentlike wire structure, again using the bowtie principle. The frequency range will be covered in two sections, from 30 to 80 MHz and from 115 to 240 MHz (omitting the heavily used FM radio band). Each band is so wide that the impedance match to the amplifier will vary drastically over the band, but the receiver noise will nevertheless be well below the cosmic noise background.

4.3 Frequency-independent antennas

Dipoles and horns have characteristic dimensions, centred on a given wavelength, and this limits their useful bandwidth. The exponential horn, shown in Figure 4.8(a), is an example of an antenna that has no characteristic size and hence a radiation pattern that in principle is independent of wavelength. In practice, of course this is not possible, since the horn has to have finite size: a lower size limit where the radiation is fed into a receiver or a bolometer, and an upper size limit that marks the end of the horn.

The Vivaldi dipole, shown in Figure 4.8(b), is an exponentially tapered slot cut in a conducting plane, and fed by a balanced twin transmission line. As in the exponential horn,

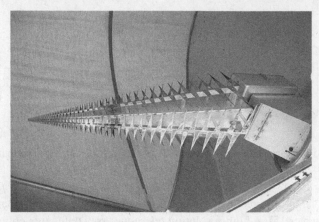

Figure 4.9. The broad-band feed used in the Allen Telescope Array. It consists of extended cross-polarized log-periodic arrays covering 500 MHz to 11 GHz.

a guided wave pattern propagates along the slot until it reaches a slot width of about half a wavelength, when it detaches and is radiated.

These tapered systems are examples of a class of antenna whose properties are described entirely by ratios; they are known as *scalar antennas*. The earliest example, the log-periodic array shown in Figure 4.8(d), was introduced by Rumsey. The array has to be terminated at an upper and lower size, so its wavelength range is still finite, but it turns out that the end effects do not seriously disturb the antenna properties. The pyramidal form of the scalar array is in use as a feed for the elements of the Allen Telescope Array, giving a bandwidth of 20 : 1, the broadest-band feed in existence. It is shown in Figure 4.9.

We include the bowtie antenna among the frequency-independent arrays of various types shown in Figure 4.8; it may be regarded either as a very fat dipole or as a scalar antenna. Its use in the new low-frequency arrays is described in Chapter 17. Figure 4.10 shows the basic pair of bowtie dipoles, mounted over a reflecting sheet, which forms a basic element of the MWA.

4.4 Aperture distributions and beam patterns

We now set out more precisely the relation between the geometry of the telescope aperture (which may be a dipole array, or a horn, or a reflector) and the size and shape of the beam. The appropriate diffraction theory is familiar in optics in terms of light emerging from an aperture; we therefore start with an antenna as transmitter, bearing in mind the precise reciprocity between antenna characteristics in reception and transmission.

In the simple case of Figure 4.11 the aperture is reduced to a line distribution along an axis ξ of excitation currents $i(\xi)$ at a single wavelength λ. An approximate derivation, following the Fraunhofer approximation, illustrates the basic principle. At a large distance from the aperture, in direction θ to the normal, the contribution of each element $i(\xi)\mathrm{d}\xi$ to the

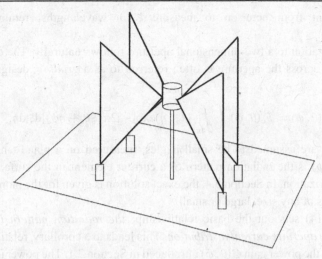

Figure 4.10. An orthogonal pair of bowtie dipoles mounted over a reflecting sheet. In the basic element of the MWA telescope array, each dipole is an open framework, with the long side approximately one metre across, i.e. one half wavelength at the centre of the band 80–300 MHz.

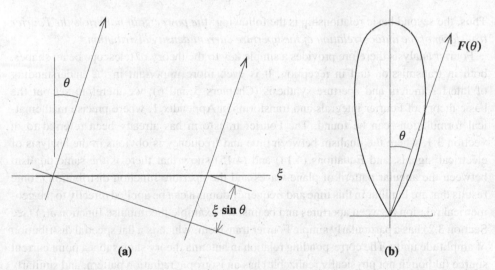

(a)   **(b)**

Figure 4.11. A linear aperture and its radiation pattern. The excitation distribution $i(\xi)$ (a) and the radiation pattern $F(\theta)$ (b) are related by a Fourier transform.

radiation field depends on the phase introduced by the path $\xi \sin \theta$; omitting normalizing factors and making the small-angle approximation $\sin \theta = \theta$, the radiation pattern $F(\theta)$ (referring either to the electric or to the magnetic field) is

$$F(\theta) = \int \exp[-\mathrm{j}(2\pi \xi \theta / \lambda)] i(\xi) \mathrm{d}\xi. \qquad (4.13)$$

It is convenient from here on to measure ξ in wavelengths, removing λ from Equation (4.13).

The generalization to a two-dimensional aperture follows naturally. The distribution of current density across the aperture is often referred to as a *grading*, designated $g(\xi, \eta)$, giving

$$F(\theta, \phi) = F_e(\theta, \phi) \int\!\!\int_{4\pi} g(\xi, \eta) \exp[-j2\pi(\xi\theta + \eta\phi)] d\xi \, d\eta, \qquad (4.14)$$

where θ and ϕ are assumed to be small angles, measured on a plane tangential to the sky, and $F_e(\theta, \phi)$ is the radiation pattern of a current element in the surface, taking due account of polarization. In Section 5.4, the exact solution is given for the more general case involving angles of any size, large or small.

Equation (4.14) sets out the basic relationship: *the radiation pattern is the Fourier transform of the aperture current distribution*. This leads to a corollary, relating the current distribution and the power gain $G(\theta, \phi)$ introduced in Section 2.1. The power is proportional to the square of the field strength, FF^*; the autocorrelation theorem (Appendix 1) then yields the result: for current density $g(\xi, \eta)$

$$G(\theta, \phi) \overset{\text{FT}}{\Longleftrightarrow} g \otimes g. \qquad (4.15)$$

Thus, the second basic relationship is the following: *the power gain pattern is the Fourier transform of the autocorrelation of the aperture current density distribution.*[2]

Fourier analysis therefore provides a simple key to the theory of telescope beam shapes, both in transmission and in reception. It is even more important in the understanding of interferometers and aperture synthesis (Chapters 5 and 6); we therefore set out the basic theory of Fourier integrals and transforms in Appendix 1, where precise mathematical formulations can be found. The Fourier transform has already been referred to in Section 3.1, where the dualism between time and frequency is obvious in the analysis of electrical signals, and Equations (4.14) and (4.15) show that there is the same dualism between the angular pattern of plane waves and the aperture function distribution. Some results that are familiar in this time and frequency domain can be applied directly to the geometrical relation between apertures and beams; for example the impulse function $\delta(t)$ (see Section 3.2) has a particularly simple Fourier transform, which is a flat spectral distribution of amplitude unity. The corresponding relation in antenna theory shows that a point current source (although not physically realizable) has an isotropic radiation pattern, and similarly that an infinitely narrow radiation pattern can be produced only by an infinitely long array or aperture.

We now turn to the practical implications of the Fourier formalism. For simplicity, we use the small-angle approximation, with $\sin\theta \simeq \theta$. First, consider the simple case of a one-dimensional radiator of length D, uniformly excited. For $i = \text{const} = 1$, Equation (4.11)

---

[2] In Chapter 5 we consider two-element interferometer systems, introducing the projected baseline k, equal to $\xi \cos\theta$ in Figure 4.11. In the case treated here, we measure (θ, ϕ) on the sky plane parallel to the array plane.

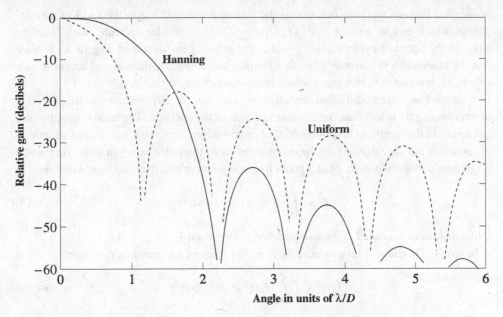

Figure 4.12. The power beam patterns for a circular aperture with uniform and Hanning distributions of aperture current.

gives the radiation pattern

$$F(\theta) = \sin[\pi(D/\lambda)\theta]/[\pi(D/\lambda)\theta]. \tag{4.16}$$

This is recognizable as the sinc function $(\sin x)/x = \operatorname{sinc} x$. In two dimensions, the uniformly illuminated circular aperture of diameter D gives a related result. The two-dimensional integration is a Fourier–Bessel transform, yielding the relative power gain, for an aperture of diameter D,

$$F(\theta) = J_1[\pi(D/\lambda)\sin\theta]/[\pi(D/\lambda)\sin\theta]. \tag{4.17}$$

This, illustrated in Figure 4.12, is recognizable as the *Airy function*. Both expressions show that the main beam has a width of the order of λ/D radians. For the case of the uniform linear feed the first null occurs at $\theta = \lambda/D$, while in the case of the circular aperture the first null occurs at $\theta = 1.22\lambda/D$. As noted in Section 4.2.2, the width of the principal lobe is always of the order of λ/D for apertures of diameter D, with corrections for the individual geometry. In practice, especially for paraboloids, the excitation of the aperture is tapered, maximum near the centre and diminishing towards the edge. This reduces the effective size and area of the aperture, but it has the compensating advantage of reducing the amplitude of the sidelobes. Unwanted sidelobes can pick up unwanted signals, whether astronomical or manmade, so it is normal practice to use some form of tapered aperture distribution. (The corresponding procedure in optics is termed *apodization*. The power gain also has an equivalent term in optical practice; it is called the *point-spread function*.) One convenient form of tapering function is the Hanning taper. This has the form of a raised cosine, which can be written $\cos^2(\pi r/D), r < D/2$. A circular aperture, with Hanning taper in the aperture excitation $i(r)$, yields the power gain shown in Figure 4.12. Note that the beamwidth is noticeably broader than for the uniformly illuminated aperture, as it should be, but also note that the sidelobe level is noticeably reduced.

A tapered aperture distribution obviously implies that the effective area in the direction of maximum gain is less than the geometrical area of the aperture. The ratio is the *aperture efficiency*. In the small-angle (Fraunhofer) approximation, for an effective area as a function of angle on the sky, $A(\theta, \phi)$, the power output from the antenna when observing a sky brightness distribution $B(\theta, \phi)$ at a given frequency, integrated over the bandwidth, is

$$P = \int_{4\pi} B(\theta, \phi)A(\theta, \phi)\mathrm{d}\Omega. \tag{4.18}$$

(Note that, as in Section 2.2, we use matched polarization.)

In terms of brightness temperature $T_b(\theta, \phi)$ the antenna temperature T_a is then

$$T_a = \lambda^{-2} \int_{4\pi} T_b(\theta, \phi)A(\theta, \phi)\mathrm{d}\Omega. \tag{4.19}$$

For a uniform temperature

$$\int_{4\pi} A(\theta, \phi)\mathrm{d}\Omega = \lambda^2 \tag{4.20}$$

so that for a non-lossy antenna with any aperture distribution the all-sky integral of the effective area is one square wavelength.

With the exception of the telescopes with an unblocked aperture, discussed below, all paraboloids must support a feed structure or secondary reflector that partially blocks the aperture. The effect of aperture blockage can be treated by considering the fundamental relation, Equation (4.14), showing that the radiation pattern is the Fourier transform of the current distribution across the aperture. The feed structure causes a gap in the current distribution, so the grading function (the current density) can be written

$$g_{\mathrm{eff}}(\xi, \eta) = g_0 + g_f + g_s, \tag{4.21}$$

where the original grading g_0 is reduced to an effective grading g_{eff} by subtracting the current distribution in the areas blocked by the feed and by the legs of the feed support. If these represent a fraction α of the (weighted) area, the gain will be reduced by a factor 2α, since the gain is proportional to the square of the current distribution. The lost gain appears in unwanted sidelobes, whose pattern is given by the Fourier transform of $g_f + g_s$. The scale of the sidelobes due to a feed structure whose diameter is a fraction β of the dish diameter will be like the radiation pattern of a roughly circular aperture, extended by a factor $1/\beta$ compared with the principal pattern. The sidelobes caused by the supports will be narrow fan beams; these are the familiar diffraction spikes seen in bright star images taken by optical telescopes.

The aperture efficiency of paraboloidal reflectors typically ranges from 50% to 85% depending on the focal ratio and on the polar diagram of the feed system.

For many observations, it is important to minimize the pattern of sidelobes. For example, if a finite disc of brightness temperature T_b is being observed, and if it is approximately the size of the main beam, the observed antenna temperature will be less than T_b, because no radiation is being captured in the sidelobes. The fraction of the temperature observed in the main beam is sometimes called the *beam efficiency*, but it is not a well-defined term. One has to be careful when the term *efficiency* is used; the *aperture efficiency*, defined above, is a well-defined term, but the total efficiency of an antenna depends upon other factors as well, as discussed in Section 4.9.

In practice, in dealing with an extended source such as the Galactic hydrogen radiation, a large field may have to be mapped, including regions of both high and low surface brightness. Observations of a complex region where the surface brightness is low within the primary beam may then be contaminated by the sidelobe pick-up of more intense radiation outside the primary beam. Complete mapping of the surroundings will then be required, allowing a deconvolution of the observations using the techniques described in Chapter 6, and the recovery of the true surface brightness.

4.5 Partially steerable telescopes

It is obviously difficult to construct an enormous collecting area that can be physically steered in all directions. In the early days of radio astronomy, the compromise solution was to have transit instruments that could be steered in elevation only, allowing the sky to drift by as the Earth rotates. A particularly fruitful early approach was to construct cylindrical paraboloids, steerable in elevation, with a linear antenna, or *line feed*, along the focal line. These were useful in their day, but they suffer from complexity of the feed and from relatively narrow bandwidth. Their construction has not been pursued in recent years, although several are still in use.

A most successful radio-telescope design that is steerable while avoiding the necessity of moving a very large reflector surface is used in the 1000-foot Arecibo telescope, shown in Figure 4.17 later. Here the surface is part of a sphere, and the beam is steered by moving the feed system to illuminate different parts of the spherical surface. The surface introduces very large spherical aberration, so that the focus is an axial line rather than a point. The spherical aberration is corrected in different ways for long and short wavelengths. At long wavelengths the feed system is a linear array pointing downwards from a moving gantry; the elements of the array are phased to match the phase distribution along the axial focal line. For shorter wavelengths a reflecting system, analogous to a Gregorian optical telescope system but with an additional tertiary reflector, has been installed. This is shaped to correct for spherical aberration; the feed itself is now a more conventional waveguide feed within the Gregorian. The beam covers angles up to 20° from the zenith, and the effective area corresponds to an aperture about 200 m across. The surface is efficient at wavelengths down to 4 cm. This is the largest single-reflector telescope; it has been used notably for planetary

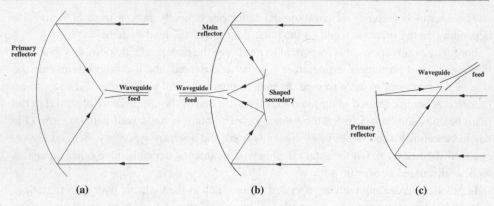

Figure 4.13. Reflector feed systems: (a) prime focus; (b) Cassegrain; and (c) offset prime.

radar, for 21-cm hydrogen-line studies of external galaxies and for deep survey work on pulsars.

4.6 Steerable telescopes

A basic steerable paraboloid with a prime-focus feed is shown in Figure 4.13(a). All large telescopes of this type, including large optical telescopes, are supported by a structure that points the instrument in azimuth and elevation (colloquially, an Az–El mount). The telescope moves in elevation about its elevation axis, and this in turn is supported by the alidade which rotates in azimuth. (Note: most old optical telescopes, and some older radio telescopes, have equatorial mounts.) The Az–El mount has significant structural advantages, at the cost of two complications: firstly, following a source continuously through the zenith would require an instantaneous movement of 180° in azimuth; secondly, at the focus, the image of the sky rotates (i.e. the parallactic angle varies with position). The first complication is met by accepting a dead zone near the zenith, while the second complication requires a rotating focal-plane receiver, or, in the case of a dual-polarization radio receiver, a later correction in the software. The steerable paraboloid is commonly met in present-day radio astronomy, and its radiation pattern is of vital interest. The aperture efficiency must be maximized, and the sidelobe levels minimized, to the maximum feasible extent. The gain pattern of the telescope is given by the prescription in Equation (4.14): the autocorrelation of the field amplitude across the aperture is the Fourier transform of the antenna power gain. The field amplitude across the aperture is determined, in the case of a prime-focus feed, by the radiation pattern of the feed; if the telescope is a Cassegrain or Gregorian, the secondary reflector is illuminated by a primary feed, and it is the gain pattern of that combination that determines the aperture distribution.

The simple paraboloid illustrated in Figure 4.13(a), with a feed at the prime focus, has a drawback that is inherent in the geometry. A typical feed pattern is illustrated in the figure, and it is clear that (i) the field strength will taper towards the edge of the reflector, diminishing the aperture efficiency; and (ii) some of the radiation from the feed will spill over the edge of the reflector; in the receiving mode, the spillover part of the pattern will pick up radiation from the ground. Since the ground is at a physical temperature of 280–290 K, this can contribute significantly to the noise at the input of the low-noise receiver, degrading the system performance. For this reason, the Cassegrain configuration, Figure 4.13(b), is more commonly used. Here there is a secondary hyperboloidal reflector with one focus coincident with the focus of the primary paraboloid, while the feed is located at the conjugate focus (the Gregorian geometry places an elliptical secondary beyond the focal point, with its conjugate focal point near the primary surface, as in the Cassegrain).

The radiation pattern of the feed will have to be narrower than it was for the prime-focus paraboloid, but its spillover will pick up radiation from the sky and the atmosphere, both of which have a far lower brightness temperature than the ground over much of the radio spectrum. (Note, however, that at low frequencies the Galactic background becomes bright, while at short wavelengths, at or near the frequencies of oxygen and water-vapour lines, the atmospheric radiation will dominate.)

A variant of the Cassegrain configuration, the shaped-surface Cassegrain, is in common use. Since all reflectors have to be at least several wavelengths across in order to be useful (five wavelengths is a practical minimum), ray optics can be used. The secondary has a perturbed surface, designed to throw the inner regions of the feed pattern farther out on the primary reflector. This increases the radiation field in the outer parts of the primary, reducing the taper and increasing the antenna efficiency. In order to meet the optical condition of a plane wave across the aperture, the primary has to have a compensating deviation from a paraboloid to equalize the lengths of the ray paths. The gain in aperture efficiency is significant: it is difficult to achieve much better than 60% efficiency with a pure Cassegrain, whereas the shaped Cassegrain can achieve an efficiency of about 85%.

There are two variants of the Cassegrain configuration that are in use: the Nasmyth and coudé systems. In the Nasmyth configuration, a tertiary reflector intercepts the beam and sends it along the elevation axis. In this configuration the receivers are located on the alidade arm, outside the elevation bearing; since they move only in azimuth, they do not tilt as the telescope moves, a design advantage at millimetre wavelengths, for which the receivers may need to be cooled to liquid-helium temperatures. The coudé configuration takes the beam at the Nasmyth focus, and uses a series of mirrors to transport the beam down the azimuth axis to the base of the telescope. This means that the receivers are fixed to the Earth, and can be accessed even when the telescope is moving. Both radio and optical telescopes have used these configurations.

Both in the simple prime-focus and in the Cassegrain systems the feed system and its supporting structure usually block part of the aperture. This brings a double penalty in reduced efficiency: not only is part of the incoming radiation lost, but also the efficiency of the feed system is reduced by uselessly illuminating telescope structure rather than sky. These problems of aperture blockage, resulting in higher sidelobe levels and reduced antenna efficiency, can be eliminated if the primary reflector is an off-axis paraboloid (or shaped paraboloid), with the secondary outside the aperture and the focal plane outside the main

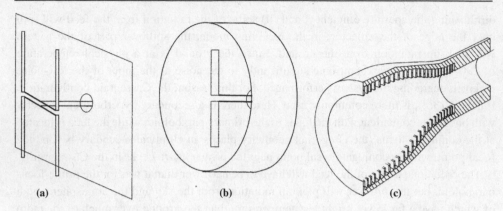

Figure 4.14. Telescope feed systems: (a) a pair of dipoles; (b) hybrid, dipole with waveguide; and (c) the compact horn design used in many Cassegrain radio telescopes (Clarricoats and Olver 1984).

reflector. There are penalties associated with this concept, for the non-circular symmetry of the primary introduces extra complexity and cost, while the supporting structure for the secondary reflector adds an extra structural complication. Nevertheless, there are two examples in use, the GBT 100-m telescope at Green Bank, West Virginia, and the elements of the Allen Telescope Array at Hat Creek in California (see Section 4.3).

4.7 Feed systems

The telescope feed must have a pattern that optimizes the illumination pattern, and also operates over as large a frequency bandwidth as possible. For prime-focus telescopes, dipole systems are usually used. A simple dipole above an earth plane will not do, because it has an asymmetric pattern, narrower in the *E*-plane than in the *H*-plane. Examples of practical systems that have a symmetrical pattern, suited to prime-focus use on paraboloids with an *f*-ratio of approximately 0.4, are shown in Figure 4.14. A pair of dipoles, suitably spaced, can do this (Figure 4.14(a)), although a broader-band system that is a hybrid of a dipole and a waveguide (Figure 4.14(b)) is often used. More frequently, especially on Cassegrain telescopes at higher frequencies, circular horns are in use. One common design, the corrugated circular horn, is illustrated in Figure 4.14(c).

The surface corrugations in the horn feed are designed to control the aperture distribution; at the aperture they are $\lambda/4$ deep, providing a high surface impedance that allows the wave to detach from the wall, while at the junction with the circular waveguide they are $\lambda/2$ deep, providing continuity with the conducting wall of the guide. Clarricoats and Olver (1984) designed compact versions in which the profile of the cone and the spacing of the corrugations are varied; these are physically smaller so that they can be installed more conveniently but nevertheless allow operation over a wide bandwidth. Two orthogonal probes in the circular waveguide feed connect via coaxial lines to the amplifiers. In order to generate a circular polarization, and achieve a symmetrical beam, the probes can connect to the low-noise amplifiers via a hybrid junction that feeds a right-hand circularly polarized signal to one amplifier, and a left-hand circularly polarized signal to a second amplifier, thus giving the ability to measure all Stokes parameters (the design of such a hybrid for wide bandwidth is difficult, and the polarimeters depending on this system need careful calibration).

Most radio-telescope systems in use, such as the various types of horns, use feeds that terminate in an unbalanced transmission line. These systems seldom have a bandwidth of greater than 2 : 1, even though the paraboloid itself may be usable over a bandwidth of more than 100 : 1. The ATA (Figure 4.9) is an exception; its scalar feed gives a usable bandwidth of 20 : 1 (unlike most feeds, this is a balanced system, with a cooled low-noise amplifier, having a balanced input, as the first stage). Some telescopes have separate receivers that can be plugged in when the receiving band is changed. There are feed designs that work at two separate frequencies, and three or even four bands are possible, but seldom seen in use, particularly for prime-focus telescopes. Frequency changing is often achieved by mounting an assemblage of feeds on a carousel at the focal plane, or, as in the case of the VLA, an off-axis secondary reflector can be rotated (nutated), bringing the feed for the desired band onto the optical axis. The Haystack telescope mounts the shortest-wavelength feed horn on-axis, with the longer-wavelength horns displaced from the axis but within the acceptable off-axis range.

4.8 Focal-plane arrays

Multiple feed systems, packed closely side by side in the focal plane, can be used to allow simultaneous observations on a mosaic of adjacent telescope beams. These systems are called *focal-plane arrays* and are coming into general use. They are useful in reducing the time taken in large-scale surveys; for example the 63-m Parkes radio telescope is equipped with an array of 13 feeds on wavelength 21 cm for hydrogen-line surveys and the search for pulsars. At millimetre wavelengths an array of feeds, all contained in a single cryogenic package, is often used, particularly for spectral-line observations where the sources are extended objects with interesting structure.

Developmental work is proceeding in several laboratories, exploring the use of compact arrays to generate multiple beams at the focus. These can be termed *synthetic focal arrays* (in contrast to focal-plane arrays, which are composed of separate feeds, connected to individual receivers), since multiple beams can be synthesized instantaneously by a suitable array. A conceptual version of a synthetic focal array, using four antenna elements and four outputs, is shown in Figure 4.15. Applied to the focal plane, a phased array allows several independent illumination patterns to be connected simultaneously to independent receivers. The corresponding beams may then be steered independently. Within each separate beam, the aperture distribution may be adjusted to compensate for aberrations introduced into the wavefront by distortions of the main reflector.

The antenna elements in a phased array may be Vivaldi dipoles, which can be packed together in a tile, as in Figure 4.16, which shows a two-dimensional dual polarized array.

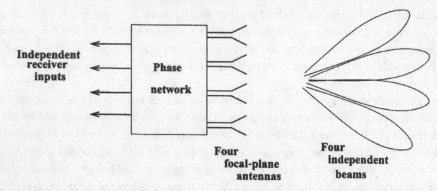

Figure 4.15. A simple synthetic phased array, in which four receiver inputs are constructed from different combinations of four antennas, giving simultaneous operation with four adjacent telescope beams.

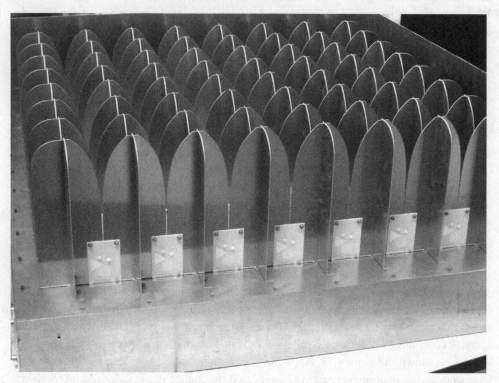

Figure 4.16. An array of Vivaldis forming a tile for use as a synthetic phased array at the focal plane of a telescope (courtesy of Wim van Cappelin, ASTRON).

The Vivaldis generate, in effect, a current sheet across the aperture. Behind the earth plane, there is a distribution network, feeding each Vivaldi through a broad-band balun. Such a tile has electronic phase-shifters built in, to allow beam-steering. Although the Vivaldi array has a bandwidth of approximately 3 : 1, broad-banding the individual beams is an interesting problem, because the excitation has to be variable across the band if the feed is to illuminate the dish uniformly across the band.

The small focal ratios of prime-focus telescopes, usually about $f/0.4$ or $f/0.5$, limit the possibilities for using such arrays of feeds to cover a larger area of sky. The available field of view is only a few beamwidths wide, and the gain rapidly deteriorates off-axis. For Cassegrain (or Gregorian) telescopes, the possibilities are greater. There is more room at the focal plane, and the image deteriorates less rapidly off-axis.

On-axis feeds and their supports inevitably block a small proportion of the aperture, giving both a loss in aperture efficiency and the introduction of noise and interference due to radiation generated or scattered by the structure. This problem is avoided entirely in the offset feed of the GBT 100-m telescope at Green Bank; the cost of this solution lies in the massive structure needed for a rigid support for the feed.

An offset feed system has an important advantage in the very-low-noise systems used for observing the CMB from satellites (Chapter 14). These receivers, operating at millimetre wavelengths, are required to measure a sky brightness temperature of only 3 K, uncontaminated by thermal radiation from any part of the antenna structure. A simple open horn is sufficient for measuring with a wide beamwidth, as in the original COBE measurements, but the systems scanning the CMB with a resolution better than 1° must use a reflector telescope. In systems such as BOOMERANG and MAXIMA, described in Chapter 14, and in the WMAP mission described in Chapter 15, the unwanted thermal noise is avoided by using an offset feed with a narrow polar diagram, which under-fills the primary reflector. The efficiency, as discussed in the previous section, is low, but the extreme tapering of primary illumination gives the very low sidelobe levels which are essential in these very demanding observations. In the CMB measurements with WMAP, the sidelobe level was less than 40–60 dB below peak, giving an integrated contribution of less than 0.5% at W-band energies (95 GHz) (Barnes *et al.* 2002).

4.9 Surface accuracy and efficiency

There is no difficulty in obtaining high reflectivity in the steel or aluminium surface of a reflector radio telescope. (Typically, the modulus of the surface impedance of a metal sheet, or even a wire mesh, is less than one ohm, which is to be compared with 377 Ω, the impedance of free space.) The thickness of the sheet is also unimportant, since the penetration depth is small compared with thicknesses needed for mechanical stability (the skin depth is 1.5 μm at 1 GHz in copper).

A wire mesh has an upper frequency limit, since an electromagnetic wave with a wavelength smaller than, or comparable to, the mesh size can leak through. Even if the leakage is small, the extraneous radiation from the ground, leaking through the mesh, degrades the noise performance of the telescope. In practice, these effects are negligible if the mesh size is finer than one-tenth of a wavelength.

The practical problem is that of how to maintain the correct profile over a large area. The surface of a large steerable telescope is usually built up from separate panels mounted on a deep structural frame. Deformations may be caused by gravitational or wind forces, or by differential thermal expansion. The largest effects are gravitational deformations of the backing structure. One might think, naively, that this problem might be minimized by making the support structure stronger, but this cannot be achieved by increasing the cross-section of the members. This increases the weight, and the increased strength is cancelled out by the increased gravitational torque. The deflection for a given structural geometry is independent of the member cross-sections. The only significant strengthening would be by resorting to exotic composites, such as bonded boron fibres, but the expense is too great for large radio telescopes.

The structural design can minimize the effect of these gravitational deformations by allowing the reflector to deform, but as far as possible constraining the deformations so that the surface remains close to a paraboloid. When the elevation of the telescope changes, the feed system can then be moved to compensate for any change in axis and focal length. This principle of *homologous design*, or *conformal deformation*, first proposed by von Hoerner (1967), was introduced with notable success in the 100-m Effelsberg radio telescope.

In practice, no structural system is perfect, and the remaining corrections can be taken out if the individual surface panels are mounted on motor-driven jacks. For example, the surface of the 100-m GBT radio telescope is continuously adjustable by a system of jacks to compensate for the inevitable distortion due to gravitational stress as the elevation changes. The adjustments can be either 'open-loop', where the deformation is known from prior measurements, or servo-controlled with respect to a reference plane. The latter approach is used in large optical telescopes such as the 10-m Keck telescopes on Mauna Kea.

The effects of any remaining large-scale deformations or inaccuracies in the surface may be estimated by Fourier analysis of the consequent errors in phase across the aperture. Again using the terminology of transmission, phase imperfections take power from the main beam and transfer it to sidelobes. Irregularities on a large linear scale transform into sidelobes on a small angular scale, close to the main beam, while irregularities on a small scale throw power into far sidelobes. Both the near and the far sidelobes can have serious effects on radio-astronomical observations; one practical consideration is that the far sidelobes may be sensitive to terrestrial sources of radio interference that would not otherwise be a nuisance. Interference from satellites can also enter via the far sidelobes, and this will be an increasingly serious problem in the future. This was an important factor in choosing the offset feed design for the 100-m GBT, with its freedom from aperture blockage.

The amount of power taken from the main beam and thrown into the sidelobes by surface irregularities represents a loss of efficiency. This loss is easily estimated for a random Gaussian distribution of phase error across the telescope aperture. A portion of wavefront with a small phase error of ϕ radians makes a reduced contribution to the power in the main beam by a fraction $1 - \phi^2$, or more precisely $e^{-\phi^2}$, and the contributions from the whole surface add randomly. The phase error at a reflector with surface error ϵ is $4\pi\epsilon/\lambda$ for normal reflection. Although normal reflection only applies strictly on the axis of a reflector, the whole error is often quoted as a single r.m.s. error ϵ, which is related to the surface efficiency η_{surf} by the Ruze (1966) formula:

$$\eta_{\text{surf}} = e^{-(4\pi\epsilon/\lambda)^2}. \tag{4.22}$$

An r.m.s. surface error of $\lambda/20$, for example, results in a surface efficiency of 67%. The surface panels of large radio telescopes such as the GBT have r.m.s. surface accuracy of the order of 0.02 mm, but the effects of gravity limit the overall accuracy to 1 mm before any servo control is applied to the surface. Programmed jacks can reduce the errors by a factor of 3 or 4; higher accuracies can be obtained with active servo control. This means that an uncorrected 100-m telescope can operate with reasonably high efficiency at wavelengths greater than 2 cm, that is, at frequencies up to 15 GHz, and at considerably shorter wavelengths given accurate measurement techniques and active panel adjustments.

The mechanical setting of the surface may require a series of repeated measurements of the shape; these are best achieved not by conventional survey methods but, rather, by *radio holography*. The amplitude pattern $F(\theta, \phi)$ is measured by making interferometer observations of a point source, in which a secondary antenna is kept directed at the source while the antenna under survey is scanned across it. This gives full knowledge of phase as well as amplitude in the beam; a Fourier transformation (Equation (4.14)) then yields the surface current distribution, and thus the surface figure.

The discussion above highlights the need for precise terminology in talking about antenna efficiency. Warping of the antenna by gravity disturbs the wavefront phase, surface irregularities disturb the phase on small scales, scattering power into sidelobes, and even though the illumination efficiency, called the aperture efficiency in Section 4.3, is well determined, it has to be considered carefully. A nearly uniform excitation of the aperture can be achieved at the cost of a lot of spillover, which reduces the overall efficiency. A more careful definition, starting at the feed, should be used. For an antenna in free space, we define the *antenna efficiency* as the fraction of power from an on-axis plane wave, impinging on the geometrical aperture, which is actually delivered to the antenna terminals.

Atmospheric losses, or losses due to rain on the structure, are time-variable and frequency-dependent. These have to be applied as separate corrections, to give the total antenna system efficiency.

4.10 Radio telescopes today

The astronomical requirements for single-dish radio observations have dictated a thrust towards ever-higher frequencies. At the same time, the need for greater sensitivity has resulted in the building of telescopes with the largest possible aperture, at the best possible locations. Inevitably, compromises have had to be made. The 1000-ft Arecibo telescope in

Figure 4.17. Four radio telescopes: (a) 1000-ft Arecibo; (b) 100-m Green Bank Telescope; (c) 30-m Pico Valeta; and (d) 15-m James Clerk Maxwell Telescope.

Puerto Rico (Figure 4.17(a)) is an example of such an accommodation. It has the largest collecting area (equivalent to a 200-m conventional paraboloid) and has a short-wavelength limit of 4 cm, enforced by having a mesh surface. The location was set by the need to minimize excavation costs; the karst topography of central Puerto Rico forms natural bowls, with peaks that minimize the height of the feed support towers. It is a humid, near-rain-forest climate, and so is unsuitable for millimetre-wave observations. On the other hand, the large collecting area is ideally suited for metre- and centimetre-wavelength observations of pulsars and of hydrogen emission from distant galaxies. The recent addition of a reflecting secondary gives the telescope great frequency flexibility.

The principle of homologous design has allowed the construction of large single dishes that retain a paraboloidal shape despite the distorting effects of gravity. The Green Bank

Telescope (GBT) of the National Radio Astronomy Observatory (Figure 4.17(b)) is such an example. It can be used effectively at wavelength 2 cm, and its adjustable surface can accommodate to use at wavelengths shorter than 3 mm. It, too, represents a compromise in location, for it makes use of the existing facility at Green Bank WV, greatly reducing support costs. Its location is well shielded by mountains, protecting it from manmade interference despite its situation only 200 km from urban, industrial areas, since it is situated in a radio quiet zone 100-km square. In winter, it is an excellent site for millimetre-wavelength observations because of the low water-vapour content of the atmosphere, but in summer, the humid atmosphere dictates lower-frequency operation. The telescope has an unblocked aperture, with an offset feed support carrying the secondary reflector. The large receiver laboratory visible below the secondary can accommodate a large suite of low-noise receivers.

The Institut de Radio Astronomie Millimétrique (IRAM) is a joint Franco-German centre with headquarters at Grenoble. It operates a 30-m radio telescope, designed for millimetre-wave observations, on the Pico Veleta near Granada (Figure 4.17(c)). This is a homologous design, with an overall r.m.s. surface accuracy of 70 μm, equipped with a suite of receivers that cover the frequency range 80–280 MHz, intended primarily for spectroscopic studies of the interstellar medium. It has a Cassegrain configuration, with the receiver cabin located behind the primary surface at the focal plane. To overcome the significant atmospheric emission from water vapour, it has a nodding secondary that allows comparison of the observing field with a reference field nearby.

The James Clerk Maxwell Telescope (JCMT), operated by a consortium headed by the UK Scientific and Technical Facilities Council, with the Nederlandse Organisatie voor Wetenschappelik Onderzoek and the National Research Council of Canada, is situated near the summit of Mauna Kea in Hawaii. The JCMT (Figure 4.17(d)) is 15 m in diameter, with a surface accuracy of 30 μm r.m.s., with receivers covering the range 150–870 MHz (2 mm to 350 μm). The receiving bands are determined by the 'windows' in this region of the spectrum, between the deep water-vapour atmospheric absorption lines that prevent full spectrum coverage (see Section 7.11). Even the windows have substantial absorption and hence significant thermal emission. This requires the use of a nodding secondary. Because atmospheric absorption is such a limiting factor, the site was chosen because of the low water-vapour content of the atmosphere overhead at this high (4100 m) location.

Interferometry adds an additional requirement, since the site must not only have low water-vapour content overhead, but must also be large enough and flat enough to meet the specifications of the telescope array (summarized in Sections 5.8 and 6.9). These considerations led the large Atacama Large Millimeter-Wavelength Array (ALMA) project to choose a location in Chile, above the Atacama desert, one of the driest places on Earth. Here, accessibility was sacrificed to obtain a high, dry site that had a large enough area to accommodate an array over 10 km in diameter. The ALMA is an aperture-synthesis array, composed of 64 15-m dishes, one of which is illustrated in Figure 4.17. The elements have

high efficiency up to the high-frequency limit of 870 MHz set by the atmosphere. Like
the JCMT, it is not planned to provide receivers below 150 MHz, since emphasis is on
spectroscopy at the highest available frequencies, although continuum observations will be
conducted also. For ease in servicing, the feed system will have a set of mirrors forming
a coudé train that will carry the radio signal out through the elevation axis, back to the
azimuth axis, and thence to the receiver laboratory at the base.

4.11 Smoothing the response to a sky brightness distribution

Scanning the sky with a single-aperture telescope may produce useful general maps such
as those of Chapter 8, but the finite beam size limits the angular resolution of the map. In
theory, one can obtain extraordinarily high resolution because the brightness is an analytic
function, and by analytic continuation small angular details might be derived. This process,
sometimes called superresolution, is of only limited usefulness, because the first and higher
derivatives of the observed brightness must be derived with great accuracy, a process
limited by the inevitable presence of noise and by limited knowledge of the antenna beam
shape. In practice, the degree of detail is limited to the angular dimension, about λ/D,
the diffraction pattern of a telescope of diameter D. If there is a high signal-to-noise ratio,
this can be improved to $\frac{1}{2}\lambda/D$ (Rayleigh's criterion), but seldom better unless there is
reliable a-priori information about the source. Improvement beyond this limit usually leads
to misleading results. We now quantify the effect of scanning any brightness distribution,
using terminology that will be extended into the following chapters on interferometers and
aperture synthesis.

A sky brightness distribution close to a beam direction may often be expressed in terms
of angular coordinates θ and ϕ; more generally direction cosines l and m must be used,
where $l = \sin\theta$ and $m = \sin\phi$. The antenna temperature T_a caused by a sky brightness
distribution $T_b(l, m)$ depends on the effective area $A(l, m)$ of the telescope as a function of
direction:

$$T_a = \lambda^{-2} \int_{4\pi} T_b(l, m) A(l, m) d\Omega. \tag{4.23}$$

If the telescope is pointed in direction l_0, m_0, the effective area is simply $A(l - l_0, m - m_0)$
for directions close to the beam centre. If the angles are small, the projection factor $(1 -
l_0^2 - m_0^2)$ can be neglected in the expression for the solid angle. The antenna temperature
then becomes

$$T_a = \lambda^{-2} \int \int T_b(l, m) A(l - l_0, m - m_0) dl \, dm. \tag{4.24}$$

This is simply the aptly named *smoothing function* of the sky brightness and the antenna
beam. Apart from a change of sign (see Appendix A1), this is the *convolution* of the two
functions (Equation (A1.13)); more exactly, it is the *cross-correlation*.

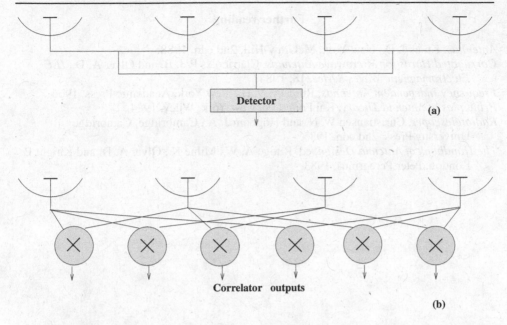

Figure 4.18. A multi-element array connected (a) as a radiometer and (b) to correlation detectors. The single detector connected via a branched feed line receives signals in a single beam, whereas the correlator outputs may be combined to produce multiple beams.

The loss of detail in scanning the sky can now be expressed in terms of Fourier components, since the Fourier transform of T_a is the product of the Fourier transforms of $T_b(l, m)$ and $A(l, m)$. These two are expressed as $\mathbf{t}(u, v)$ and $\mathbf{c}(u, v)$, which are functions of coordinate distances in the plane of the telescope aperture measured in wavelengths. Then

$$T_a(l, m) \rightleftharpoons \lambda^{-2}[\mathbf{t}(u, v) \cdot \mathbf{c}(u, v)]. \tag{4.25}$$

$\mathbf{c}(u, v)$ is the *telescope transfer function*. This applies generally, including to interferometer arrays. Only those Fourier components of sky brightness which correspond to non-zero values of $\mathbf{c}(u, v)$ are recorded in the output of a telescope scan across the sky.

In radio astronomy, the angular resolution of single apertures was extended dramatically, first by adopting a radio version of Michelson's stellar interferometer, and then by using multiple radio telescopes to construct apertures of vast dimensions by synthetic means. In the following chapter, the properties of the two-element Michelson interferometer are examined in detail. In the radio case, the *correlation* between the signal amplitudes received at the two antenna elements is measured, as contrasted with the total-power systems associated with single apertures. The technique of correlating pairs of telescope signals finds wider application in the case of arrays of many radio telescopes (Figure 4.18). These are treated as assemblages of all possible two-element interferometers, the subject of Chapter 6.

Further reading

Antennas, Kraus J. D., New York, McGraw-Hill, 2nd edn, 1988.

Corrugated Horns for Microwave Antennas, Clarricoats P. J. B. and Olver A. D., *IEE Electromagnetic Waves Series*, **18**, 1984.

Frequency Independent Antennas, Rumsey, V. H., New York, Academic Press, 1966.

Principles of Antenna Theory, Kai Fong Lee, New York, Wiley, 1984.

Radiotelescopes, Christiansen W. N. and Högbom J. A., Cambridge, Cambridge University Press, 2nd edn, 1985.

The Handbook of Antenna Design, ed. Rudge A. W., Milne K., Olver A. D. and Knight P., London, Peter Peregrinus, 1983.

5

The two-element interferometer

Radio wavelengths are hundreds to millions of times longer than optical wavelengths. Consequently, all single-aperture radio telescopes are hindered by severe diffraction effects, and their angular resolution is crude by optical standards. The application and development of radio interferometry, building on the rapidly developing arts of electronics and signal processing, overcame this handicap. In their 1947 studies of the Sun, Pawsey, McCready and Payne-Scott recognized that an interferometer's response to an extended source amounted to determining a particular value of the Fourier transform of the source brightness distribution. This insight was broadly recognized in the radio-astronomy community, and informed much of the work in Sydney, Cambridge and Manchester.

As the art of interferometry progressed, interferometers were used with multiple spacings to develop approximate Fourier transforms of extended sources, which could be inverted to give maps of the brightness distribution. Fourier concepts, reviewed in Chapter 3, became the natural language for discussing the brightness distribution across sources. Finally, Ryle formulated Earth-rotation synthesis, in which the rotation of the Earth is used to vary the orientation and length of interferometer baselines, yielding an extensive sampling of the Fourier transform of the sources each day. The *aperture-synthesis arrays*, such as the Westerbork Synthesis Radio Telescope in the Netherlands, the Merlin array in the UK, the Australia Telescope and the Very Large Array in the USA, all use this principle, with each possible pair of array elements forming a separate two-element interferometer.

This chapter presents the principles of the two-element interferometer in this context; the extension of interferometry concepts to multi-element arrays and the full development of aperture synthesis will be given in Chapter 6. Here we include an analysis of the effects of receiver bandwidth and source size, concepts that would have been familiar to Michelson in his classic work on stellar diameters. The discussions will be in a radio-astronomy context, but the formalism would apply equally well to the optical and infrared domain, with relatively few modifications. There is a one-to-one correspondence between the components used for interferometers in the radio and optical spectral domains, even though their physical appearance will differ. Despite the great disparity in wavelengths, the precision attained in the radio domain is considerably greater: we refer briefly to the use of radio interferometers in geodesy, where the relative positions of the elements of an

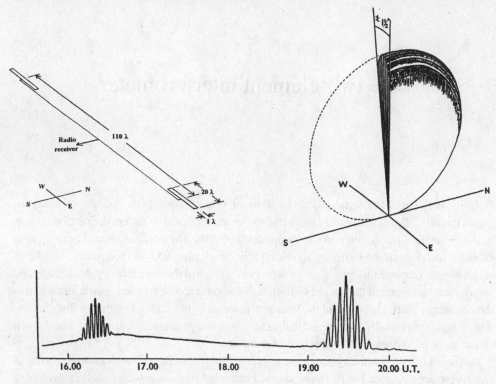

Figure 5.1. The radio sources Cyg A and Cas A recorded with an early interferometer system (Ryle *et al.* 1950). The antenna elements were dipole arrays 20λ by 1λ, giving a fan beam sliced by the interference pattern.

interferometer array thousands of kilometres across may be determined with an accuracy of a few millimetres.

5.1 The basic two-element interferometer

An early example of an interferometric observation of radio sources, shown in Figure 5.1, illustrates the principal features of interferometry. The signals received by two fixed antennas, also shown in the diagram, are added, amplified by a radio receiver, and detected, to give the total power as a function of time. There is a steady background noise, partly generated by the radio receiver itself, and partly coming from the Galaxy. As the Earth rotates, first the extragalactic source Cygnus A and then the supernova remnant Cassiopeia A pass through the antenna patterns. The recording shows the signal from each source, modulated by a quasi-sinusoidal oscillation. This arises from the alternately constructive and destructive interference of the signals from the two antennas, with the average source power and the envelope of the modulation both tracing the antenna gain pattern. The quasi-sinusoidal oscillation is called the fringe[1] pattern, and near the maximum of the antenna response it is sufficiently close to sinusoidal form to be characterized by its frequency, amplitude and phase. Alternatively, it can be represented by a complex number of given amplitude and phase. In present-day usage, the complex representation is most commonly used, and is known as the *complex fringe amplitude* when it is normalized with respect to the total power of the source.

In most practical cases, the fringe amplitude, not the total power, is the desired quantity. Furthermore, a total-power system, such as the one illustrated in Figure 5.1, has undesirable properties, since the stability of its output will be affected by gain fluctuations. There are various ways in which the complex fringe amplitude can be extracted, and these have been summarized by Rogers (1976). The most common practice in present-day installations is to cross-correlate the signal amplitudes, or to use methods that are closely related to cross-correlation. In most cases, the pairs of signals are multiplied after correcting for a geometrical time delay; it will be seen that in VLBI and in spectroscopic aperture synthesis it is necessary to use many time lags, developing the cross-correlation function $R(\tau)$. For an introductory example, the source will be regarded as monochromatic, at frequency ν, and it will also be assumed to be a point source. These restrictions will then be lifted, in Section 5.2, to cover realistic cases.

The basic elements of a two-element Michelson interferometer are shown in Figure 5.2. The baseline vector **b** connects the phase centres of the two antennas; if these are identical, any convenient reference point such as the antenna vertex will do. The radio source under observation is in the direction given by the unit vector **s**. The two antennas track the source as it moves, and one of the antennas is designated the reference antenna. The signal arrives at the second antenna delayed by the geometrical time delay τ_g:

$$\tau_g = \mathbf{b} \cdot \mathbf{s}/c. \tag{5.1}$$

The two signals, at frequency ν, are fed to a voltage multiplier as shown. An additional instrumental time delay τ_i can be inserted to equalize the signal delays; for our initial discussions we set $\tau_i = 0$. The analysis of such a system has been given by Rogers (1976), Thompson, Moran and Swenson (1986) and in the chapters by Clark and by Thompson in the NRAO Summer School Notes, edited by Perley, Schwab and Bridle (1989); in this presentation we largely follow Thompson and Rogers. The cross-correlation $R_{xy}(\tau)$, introduced in Section 3.7, of two amplitudes (voltages) $x(t)$ and $y(t)$ is defined as the time-averaged product of the two amplitudes, with one delayed by time τ:

$$R_{xy}(\tau) \equiv \langle x(t)y(t - \tau) \rangle. \tag{5.2}$$

The cross-correlation of two amplitudes has the dimensions of power, and so it can be called the *cross-power product*. Since we are dealing with a monochromatic source, the amplitudes at the antenna output terminals will be $x(t) = v_1 \cos(2\pi\nu t)$ and $y(t) = v_2 \cos[2\pi\nu(t - \tau_g)]$.

---

[1] The term 'fringe' comes from early optical observations of diffraction at a shadow edge; it was then extended to the optical two-slit interference pattern seen by Young, and is now applied to interference patterns in practically every kind of interferometer.

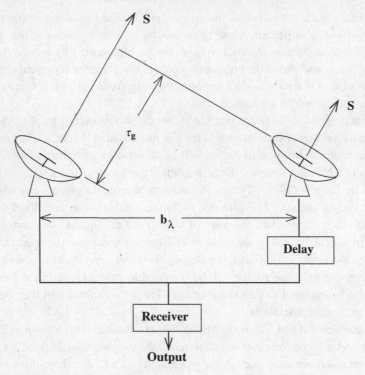

Figure 5.2. The geometry of the two-element Michelson interferometer. The geometrical path delay τ_g is compensated for by the delay circuit in the receiver; an additional delay τ_i can be inserted by this delay circuit.

In forming the cross-correlation, the radio-frequency–voltage product, being the power received from the source, will be proportional to the effective antenna area $A(\mathbf{s})$ and the source flux S; taking the time average, the cross-correlation is

$$R_{xy}(\tau_g) = A(\mathbf{s})S \cos(2\pi \nu \tau_g) = A(\mathbf{s})S \cos(2\pi \nu \mathbf{b} \cdot \mathbf{s}/c). \tag{5.3}$$

In this second form for $R_{xy}(\tau)$, a useful simplification can be made by measuring the baseline vector $\mathbf{b}$ in wavelengths. This dimensionless form of $\mathbf{b}$ will be designated $\mathbf{b}_\lambda = \mathbf{b}/\lambda$, giving a briefer expression for the cross-correlation:

$$R_{xy}(\mathbf{s}) = A(\mathbf{s})S \cos(2\pi \mathbf{b}_\lambda \cdot \mathbf{s}). \tag{5.4}$$

In both representations, the sinusoidal fringe variation is apparent. As the source direction $\mathbf{s}$ changes, the fringe amplitude oscillates, and, for a spacing of many wavelengths, when the source is close to transit, the variation is nearly sinusoidal since $\mathbf{b}_\lambda \cdot \mathbf{s} \approx |\mathbf{b}_\lambda|\theta$. In this small-angle, long-baseline approximation, with an angle θ between source and transit, and also making the simplifying assumption that the baseline is nearly perpendicular to the direction of observation, we have

$$R_{xy}(\theta) \approx A(\mathbf{s})S \cos(2\pi b_\lambda \theta). \tag{5.5}$$

The angle between fringes, in this small-angle limit, is $1/b_\lambda$.

There is an alternative, completely equivalent, way of analysing the two-element interferometer. Instead of looking at cross-correlation in the time domain, between two voltage amplitudes, the process can be described in the radio-frequency domain. For the present case, the alternative analysis in the frequency domain is straightforward, and leads naturally to the complex representation that is commonly used in practice. The Fourier transform of the cross-correlation $R_{xy}(\tau)$, by the convolution theorem, is the product of the transform of $x(t)$ and the complex conjugate of the transform of $y(t)$, and is known as the *cross-spectrum power density*:

$$S_{xy}(\nu) \equiv X(\nu)Y^*(\nu). \tag{5.6}$$

This is used to describe the interferometer output, in contrast to the total power output of a single antenna. The Fourier transform of a monochromatic signal is a delta-function at the frequency ν, and the Fourier transform of the time-delayed signal $y(t - \tau_g)$ has a phase-shift of $2\pi\nu\tau_g$, so it follows that the cross-spectral product described above becomes

$$S_{xy}(\nu) = A(\mathbf{s})S \exp(i2\pi\nu\tau_g) = A(\mathbf{s})S \exp(i2\pi \mathbf{b}_\lambda \cdot \mathbf{s}). \tag{5.7}$$

This is clearly the complex equivalent of Equation (5.4). The interferometric calculations can be carried out in either the frequency or the time domain, depending on which representation is most convenient.

Up to this point, the adjustable time delay τ_i has been set to zero, but now, in preparation for the following section, this condition will be relaxed. We now consider a source whose position is close to a reference position $\mathbf{s}_0$ defined by the condition $\tau_g = \tau_i$. This describes a practical interferometer consisting of a pair of steerable radio telescopes, directed at an arbitrary source position. The reference direction $\mathbf{s}_0$ is called the *phase-tracking centre*. Since τ_i compensates for the geometrical time delay, it can be called the equalizing time delay. The direction to the source, with respect to the phase tracking centre, can be written

$$\hat{\mathbf{s}} = \hat{\mathbf{s}}_0 + \sigma, \tag{5.8}$$

where σ is a small vector, normal to $\mathbf{s}_0$. (It must be normal, since both $\mathbf{s}$ and $\mathbf{s}_0$ are unit vectors.) Since the large geometrical delay associated with $\mathbf{s}_0$ is exactly compensated for by the instrumental time delay, only the small differential associated with σ affects the time delay in most of the analysis that follows.

The analysis of fringe formation by an interferometer has so far assumed that a point source, emitting a single frequency ν, is being observed. Real astronomical radio sources, however, have a finite angular extent, and interferometers have a particularly important role in achieving high angular resolution to map the distribution of brightness across sources. The basic analysis for sources of finite size is presented in Section 5.3, to prepare for the more extensive treatment in Chapter 6. The single-frequency assumption must also

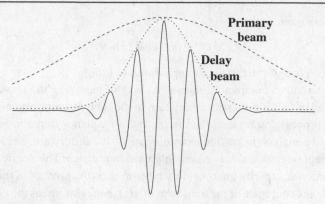

Figure 5.3. The delay-beam effect in an interferometer. The width of the delay beam depends on the bandwidth of the receiver.

be extended to the case of finite bandwidth, which introduces limitations that must be recognized; this formalism is the topic of the following section.

5.2 Interferometers with finite bandwidth

Radio sources emit radio noise over a wide range of frequencies, and all receiving systems have finite bandwidth. In the general case, the formalism can be complicated, but the fundamental effects that are introduced can be understood in simple physical terms. The radio noise is quasi-Gaussian, so the signal at one frequency is uncorrelated with the signals at adjacent frequencies. The radio spectra of continuum sources change only slowly with frequency, and the interferometer usually has a small fractional bandwidth. The radio spectrum across the band is then effectively flat, and this approximation simplifies the analysis.

The effect of finite bandwidth B on the fringe pattern can easily be described: the angular range over which fringes appear is diminished. In Figure 5.2, if the source is exactly normal to the interferometer baseline, constructive interference will occur at all frequencies. If the source is displaced from this direction by a small amount, the phase of the fringe pattern will change, and this phase change will be different at different frequencies across the band; if the displacement is large enough, the signals at one end of the band may be interfering destructively when the signals at the other end of the band are interfering constructively. As a result, the net fringe amplitude will be reduced, and, if the displacement is large enough, the fringe will disappear almost completely. The effect is illustrated in Figure 5.3. Here we see a fringe pattern that would, for a small bandwidth, fill the primary telescope beam pattern; instead the fringes are confined to a narrower beam whose width is determined by a wider bandwidth. The centre of the narrower beam can be moved by changing the delay in one of the interferometer elements; the narrowed interferometric response is therefore known as the *delay beam*.

There is an alternative way of understanding the delay beam by considering the *coherence time* of the incoming signals. The coherence time of the signal will be of the order of the inverse of the bandwidth, so the coherence length will be $c/\mathcal{B}$. If the geometrical delay is greater than the coherence length, the interference between the two signals can no longer occur, and the fringes will disappear.

The calculation of the delay-beam effects, using the flat-spectrum approximation, starts by calculating the interferometer response to a point source when the receiver has a square bandpass $G(\nu)$, unity over a bandwidth $\mathcal{B}$ centred at ν_0 and zero elsewhere (i.e. a gating function $\Pi[(\nu - \nu_0)/\mathcal{B})]$. Since the radio noise is Gaussian, the signal at a given frequency is independent of the signals at other frequencies, and the single-frequency response given in Equation (5.7) can be summed over the bandpass; for simplicity, the source direction is assumed to be nearly normal to the interferometer baseline. The response becomes the *cross-product power* P_{xy}, given by

$$P_{xy} = \int_{-\infty}^{\infty} S_{xy}(\nu, \mathbf{s})G(\nu)d\nu \qquad (5.9)$$

so that P_{xy} is a function of time delay τ_g,

$$P_{xy}(\tau_g) = \int_{\nu_0-B/2}^{\nu_0+B/2} A(\nu, \mathbf{s})S(\nu)\exp(-i2\pi\nu\tau_g)d\nu. \qquad (5.10)$$

Therefore, assuming that the effective area and flux are constant across the bandpass (which is usually a good assumption), the fringe pattern is modulated by the Fourier transform of the band shape:

$$P_{xy}(\nu_0, \tau_g) = A(\nu_0, \mathbf{s})S(\nu_0)\mathcal{B}\exp(-i2\pi\nu_0\tau_g)\frac{\sin(\pi\mathcal{B}\tau_g)}{\pi\mathcal{B}\tau_g}. \qquad (5.11)$$

Since the source flux is approximately constant over the band, but does vary with frequency ν_0, it is convenient to divide by $\mathcal{B}$ to give the power density (power per unit bandwidth). The final term can then be recognized as the sinc function, and Equation (5.11) becomes the *cross-spectrum power density*

$$S_{xy}(\nu_0, \tau_g) = A(\nu_0, \mathbf{s})S(\nu_0)\exp(-2\pi\nu_0\tau_g)\mathrm{sinc}(\mathcal{B}\tau_g). \qquad (5.12)$$

From this it is clear that, when the geometrical time delay τ_g becomes comparable to the inverse of the bandwidth $1/\mathcal{B}$, the interferometer response is severely attenuated. This is the reason for the instrumental time delay τ_i that is shown in Figure 5.2; the extra delay is inserted into the transmission line that carries the signal from antenna 1 to the correlator, in order to compensate for the geometrical time delay in the signal from antenna 2.

In the discussion up to this point, the source has been assumed to be close to transit, but this is seldom the case for most practical interferometers. Usually, the antennas are pointing far from the midline of the interferometer, and an instrumental time delay is an absolute necessity, since the geometrical time delay is far larger than the coherence time. In other words, the source being observed is far away from the delay beam with no compensating time delay. In this more realistic case, Equation (5.12) takes the form

$$S_{xy}(\mathbf{s}) = A(\nu_0, \mathbf{s})S(\nu_0)\text{sinc}[\mathcal{B}(\tau_g - \tau_i)]\exp[-2\pi\nu_0(\tau_g - \tau_i)]. \tag{5.13}$$

In this expression, the fringe oscillations, $\exp(-2\pi\nu_0\tau_g)$, are severely reduced in amplitude by the delay-beam term, $\text{sinc}[\mathcal{B}(\tau_g - \tau_i)]$, when the delay is of order $1/\mathcal{B}$ or larger. The instrumental time delay is chosen to be close to the geometrical time delay for the antenna pointing direction, a condition that receives detailed treatment in the following section. The delay beam still has the form of a sinc function, centred on the compensation direction, if the bandpass is approximately square. Therefore, a source that is well inside the main beams of the antennas, but sufficiently off-axis to be outside the delay beam, will give only weak fringes.

A practical receiving system will not have a perfectly square bandpass, but the delay-beam effects are still present. As noted above, the shape of the delay beam is effectively given by the Fourier transform of the gain function $g(\nu)$. The square-bandpass approximation is usually sufficiently accurate for practical purposes.

5.3 Interferometers and finite source size

Michelson's stellar interferometer was used to measure stellar angular diameters by observing the diminution of the visibility of interference fringes as the two mirrors of his interferometer were moved further apart. He recognized that there is a Fourier-transform relation between the fall of fringe visibility with interferometer spacing and the size and brightness distribution across the stellar source, but he was able to observe only the intensity of the fringes rather than their amplitude. In this section we examine the response of a two-element interferometer to a source of finite size, in preparation for the formal presentation of the Fourier relationship involving both amplitude and phase of the fringe pattern.

The geometry is illustrated in Figure 5.4, showing the radiation received from a small element of an extended radio source of specific brightness $B_\nu(\mathbf{s})$, subtending a solid angle $d\Omega$ in direction $\mathbf{s}$. A direction (usually the direction of maximum antenna gain) is chosen as the phase-tracking centre, $\mathbf{s}_0$, and, following the definition in Equation (5.8), the vector from the phase-tracking centre to the source element is σ. The instrumental delay τ_i will be non-zero, so the cross-spectrum power density produced at frequency ν_0 by this element of the source will have the same form as Equation (5.13), with the source flux $S(\nu_0)$ replaced by the radiating source element $B_{\nu_0}(\mathbf{s}_0 + \sigma)$:

$$S_{xy}(\nu_0, \mathbf{s}_0 + \sigma) = A(\mathbf{s}_0 + \sigma)\text{sinc}(B\tau_g)B_\nu(\mathbf{s}_0 + \sigma)\exp[i2\pi\nu_0(\tau_g - \tau_i)]d\Omega, \tag{5.14}$$

where the delay-beam effects are included as a modification of the effective area $A(\mathbf{s})$. The relative antenna area, $\mathcal{A}$, with value unity in the direction $\mathbf{s}_0$, will therefore be used in expressing Equation (5.14) in simpler form. The total correlator output will be this cross-spectral density integrated over the entire radio source:

$$S_{xy}(\nu_0, \sigma) = \int_{4\pi} \mathcal{A}(\sigma)B_{\nu_0}(\sigma)\exp[i2\pi\nu_0(\tau_g - \tau_i)]d\Omega. \tag{5.15}$$

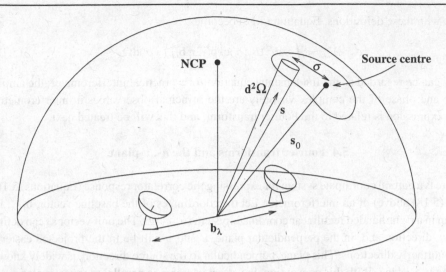

Figure 5.4. The contribution of a small receiving element in the direction **s**, solid angle $d^2\Omega$, to the response of an interferometer; NCP is the North Celestial Pole.

At this point, several approximations are in order: the overall source size will be assumed to be small compared with the response pattern of the delay beam. This means that the effects of the delay beam can be neglected and the centre frequency of the receiver bandpass, ν_o, can be taken as the defining frequency. In practice, when the field of view covers a large angle on the sky, or when the bandwidth is wide in order to gain sensitivity, these simplifications need further examination; see Section 6.9. In many applications, however, the single-frequency approximation is good enough provided that the width θ of the field of view satisfies $\theta \ll (\nu/\mathcal{B})(1/b_\lambda)$, where b_λ is the baseline length measured in wavelengths.

For simplicity, we let ν (without the subscript) be the centre frequency. Usually, the total flux scale is determined by reference to a standard source, so for the purposes of this discussion we will seek only the correlator output in units of relative power density. The geometrical time delay is written explicitly in terms of the direction vectors and the baseline $\mathbf{b}_\lambda$ is expressed in wavelengths, as in Equation (5.4):

$$S_{xy}(\mathbf{s}_0) = \int \mathcal{A}(\sigma)B_\nu(\sigma)\exp\{i2\pi[\mathbf{b}_\lambda \cdot (\mathbf{s}_0 + \sigma) - \nu\tau_i]\}d\Omega. \qquad (5.16)$$

In the following chapter, which considers arrays of many radio telescopes, a given interferometer pair using the ith and jth elements of the array will have a baseline vector $\mathbf{b}_{ij}$. The baseline is commonly measured in wavelengths, and we define $\mathbf{b}_{ij,\lambda} \equiv \mathbf{b}_{ij}/\lambda$. Furthermore, the instrumental delay is usually set so that it cancels out the tracking-centre delay (hence the name 'tracking centre') and in this limit we can define the *complex visibility* V_{ij}:

$$V_{ij} \equiv S_{xy}(\mathbf{b}_{ij} \cdot \mathbf{s}_0) = \nu\tau_i \qquad (5.17)$$

and, with these definitions, Equation (5.16) becomes

$$V_{ij} = \int \mathcal{A}(\sigma)B_v(\sigma)\exp(\mathrm{i}2\pi\mathbf{b}_{ij,\lambda}\cdot\sigma)\mathrm{d}\Omega. \tag{5.18}$$

This can be regarded as the fundamental equation for a practical interferometer; the amplitude and phase of the complex visibility are the principal observables in interferometry. The expression is related to the Fourier transform, and this will be treated next.

5.4 Fourier transforms and the u-, v-plane

There is a natural coordinate system for expressing the correlator response (Equations (5.16) and (5.18) above) of an interferometer. Let the coordinates of the baseline vector, $\mathbf{b}_{ij,\lambda}$, be given in a right-handed rectilinear coordinate system (u, v, w). The unit vector $\mathbf{s}_0$ can define the w-direction and, in the perpendicular plane, u and v will be in the projected easterly and northerly directions. This plane, perpendicular to the source direction, is widely known as the u, v-plane. With this convention, the offset vector σ is parallel to the u, v-plane. All coordinate distances will be expressed in wavelengths (spatial frequency), and Equation (5.18) becomes

$$V_{ij}(\mathbf{s}_0, u, v) = \int \mathcal{A}(l, m)B_v(l, m)\exp[\mathrm{i}2\pi(ul + vm + wn)]\mathrm{d}\Omega. \tag{5.19}$$

The integral is taken over the radio source; and l, m, n are the direction cosines of the unit vector $\mathbf{s}$, with respect to the u, v-plane. Thus, the coordinates of σ are (l, m), and, since $\mathbf{s}_0$ is perpendicular to the u, v-plane, $w = 0$.

The element of solid angle, in terms of the direction cosines of $\mathbf{s}$, is

$$\mathrm{d}\Omega = \frac{\mathrm{d}l\,\mathrm{d}m}{\sqrt{1 - l^2 - m^2}}. \tag{5.20}$$

The denominator of the expression above will never become imaginary, because of the direction-cosine closure requirement; the l and m angles are always close to $90°$ in any event. The fringe visibility is then written

$$V_{ij} = \int_{4\pi} \mathcal{A}(l, m)B_v(l, m)\exp[\mathrm{i}2\pi(ul + vm)]\frac{\mathrm{d}l\,\mathrm{d}m}{\sqrt{1 - l^2 - m^2}} \tag{5.21}$$

and one can see immediately that $V(u, v)$ is the Fourier transform of a modified source brightness:

$$V_{ij} \overset{\mathrm{FT}}{\longleftrightarrow} \frac{\mathcal{A}(l, m)B_v(l, m)}{\sqrt{1 - l^2 - m^2}}. \tag{5.22}$$

This expression is valid for all directions in the sky, but in most cases the offset angle σ is small. Hence, it is convenient to rewrite Equation (5.19) in terms of x and y, the rectilinear coordinates for σ; almost always, x is taken parallel to u. In many instances the source size will be small compared with the antenna beam size, and, if the radio telescopes are pointed well, the variation of gain with angle will also be negligible. On replacing the direction

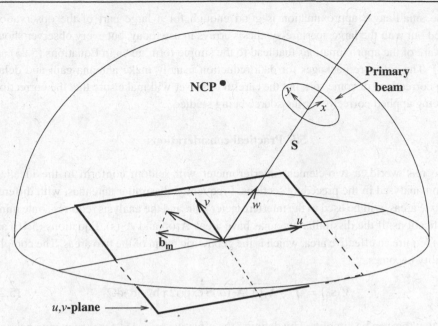

Figure 5.5. The geometrical relationship of an interferometer, a celestial source and the u, v-plane, seen in relation to the celestial sphere: x and y are the components of the offset σ with x parallel to u; NCP is the North Celestial Pole.

cosines by the angular offset coordinates x and y (measured from s_0), Equation (5.22) then takes on the small-angle approximation that is most familiar to the observer, with

$$V(u, v) \approx \mathcal{A} \int B(x, y)\exp[i2\pi(ux + vy)]dx\,dy \tag{5.23}$$

or

$$V(u, v) \xleftrightarrow{\text{FT}} B(x, y). \tag{5.24}$$

Therefore, a single interferometer observation, in which both the amplitude and the phase of the complex visibility are measured, evaluates the Fourier transform of the source brightness distribution for a particular value of the spatial frequency, given by the spatial frequency of the baseline vector, projected onto the u, v-plane. The geometry of the entire construct is illustrated in Figure 5.5, which shows the source plane (the celestial sphere), with coordinates (l, m) that become the celestial angular coordinates (x, y) in the small-angle approximation. The u, v-plane is shown, and the Fourier transform of the source distribution is measured in this plane. A single interferometer observation gives only one value, but a complete assemblage of observations will develop the entire Fourier transform of the source brightness distribution, $b(u, v) = V(u, v)$, which can then be inverted to yield the source brightness distribution $B(l, m)$. The ways in which this can be done are explored further in Chapter 6.

The small-angle approximation is good enough for a large part of the observations carried out with the large aperture-synthesis arrays in use today, but every observer should be aware of the approximations that lead to the simple form given in Equations (5.23) and (5.24). The software packages for data reduction usually make antenna-gain and delay-beam corrections of some sort, but the careful observer will make sure that the corrections are being applied correctly for the source being studied.

5.5 Practical considerations

In the real world, a two-element interferometer will seldom conform to the idealized version analysed in the preceding sections. Frequently, dissimilar antennas, with different effective areas, will be used as the interferometer pair, and the analysis requires some minor modifications. If the dissimilar antennas have areas $A_1(\mathbf{s})$ and $A_2(\mathbf{s})$, Equations (5.16) and (5.18) require an effective area, which is the geometric mean of the two areas. The complex visibility becomes

$$V(\mathbf{s}_0) = \int \sqrt{\mathbf{A}_1(\sigma)\mathbf{A}_2(\sigma)}\mathbf{S}\exp(\mathrm{i}2\pi\mathbf{b}_\lambda \cdot \sigma)\mathrm{d}\Omega. \tag{5.25}$$

Note that this can be simplified by defining the effective area A_{12}^{eff} for the antenna pair:

$$A_{12}^{\text{eff}}(\sigma) = \mathcal{D}(o)\sqrt{A_1(\sigma)A_2(\sigma)}, \tag{5.26}$$

where $\mathcal{D}$ is a reminder that delay-beam effects may have to be considered. An important precaution that must be heeded is to define the baseline vector correctly; this must connect the phase centres of the dissimilar antennas. In practice, a reference point, not necessarily the phase centre, is used to define a reference baseline. The offset of the phase centre from the reference point is calculated as a function of the pointing direction, and the physical baseline is calculated in the software. Although it will be assumed that the antennas are identical in all the following analysis, one need only replace the antenna area by the geometric mean of the two antennas, and recognize that the baseline is changing with time, for the formulae to apply to the dissimilar case.

A practical interferometer may use antennas separated by large distances, and there may be several antennas forming an array of interferometer pairs. It is usually impractical to transmit the signal directly to the correlator. Almost always, a heterodyne receiver is used (Figure 5.6). Except for the shortest millimetre wavelengths, the signal is usually amplified at its original frequency by a low-noise amplifier, but it is then translated to another frequency (the intermediate frequency, ν_{if}) by combining it with a local oscillator ν_{lo} in a mixer, which multiplies the two signals. This produces signals at both sum and difference frequencies:

$$\nu_{\text{if}} = \nu_0 \pm \nu_{\text{lo}}. \tag{5.27}$$

The sum frequency is sometimes used, but it is usually the difference frequency that is subsequently amplified. The local oscillator can be either below or above the initial

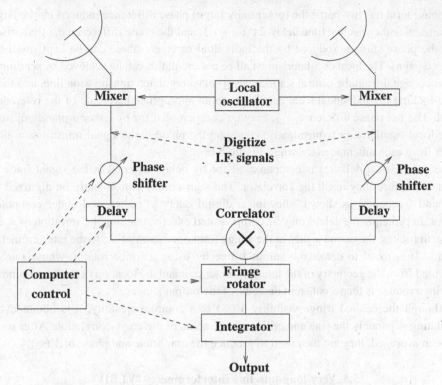

Figure 5.6. The next-stage approximation to a practical interferometer system.

frequency; these are known as upper-sideband and lower-sideband conversion, respectively. In many cases, the frequency conversion will be made more than once, and sometimes the final conversion results in the frequency band extending from (nearly) zero frequency to the upper limit of the original bandpass. This final option is known as conversion to baseband, or conversion to video. If there are many conversions, there is effectively only a single local oscillator frequency, since the heterodyne operation translates only the initial frequency band; attention must be paid, however, to the spectrum inversion that occurs when there is lower-sideband conversion.

The amplified and converted signals are sent to the central station, where the appropriate time delay is inserted to compensate for the geometrical time delay. The signals are then cross-correlated. If the baselines are sufficiently short, the signals can be sent via coaxial cable or waveguide, but in most installations today they are sent via optical fibre, as a modulation of laser light. The signals may be sent in analogue form or they may be digitized before transmission.

The response of the frequency-conversion system is a modification of Equation (5.16):

$$V(s_0) = \int \mathcal{A}_{\text{eff}}(\sigma) B_\nu(\sigma) \exp\{i[2\pi(ul + vm) + 2\pi\nu(\tau_g - \tau_i) + \theta_{12}]\} dl\, dm. \quad (5.28)$$

The phase term has two parts: the (potentially large) phase difference induced by the large geometrical and corrective time delay $2\pi(\nu_g - \nu_i)$, and the phase difference θ_{12} that arises from the phase shifts introduced by the individual receivers (these can be kept small by proper design). The local oscillators must all be coherent; this can be achieved by sending a reference signal from the central station over a servo-regulated transmission line, and then phase-locking the individual local oscillators to an appropriate harmonic of the reference signal. The net phase difference, θ_{12}, may be compensated for by either a phase adjuster in the local oscillator, or (equivalently) changing the phase in the signal-transmission line length from each antenna, as shown in Figure 5.6.

The geometrical delay τ_g is compensated for by delays inserted in the signal lines at baseband, before they reach the correlator. The signals may conveniently be digitized at baseband frequency, as shown, allowing a digital delay to be inserted under computer control. In principle the delays may be compensated exactly, tracking the variation of τ_g as the Earth rotates; the output is then at the fringe frequency appropriate to the interferometer spacing. It is usual to detect the output fringe by using a fringe rotator, whose rate is computed from the geometry of the interferometer pair and the location of the beam centre; the fringe rotator is then a coherent detector of the output fringe.

Although the desired fringe visibility $V(\mathbf{s}_0)$ is a complex quantity, it is obtained by calculating separately the sine and cosine components of the cross-correlation. After each has been averaged, they are then used to produce the amplitude and phase of $V(\mathbf{s}_0)$.

5.6 Very-long-baseline interferometry (VLBI)

The interferometer shown in Figure 5.6 is a 'hard-wired' system, in which all signals travel along transmission lines, optical fibres, or radio links. This need not be the case; the stability of atomic frequency standards and the wide bandwidth capabilities of modern digital recorders permit the interferometer telescopes to be separated by arbitrarily large distances, including the case of spacecraft carrying radio telescopes. The technique of very-long-baseline interferometry (VLBI) is illustrated in Figure 5.7: the local oscillators at each station are synthesized from a master signal generated by a stable atomic frequency standard (generally a hydrogen-maser oscillator), which is also used to generate a set of accurate timing pulses, which control the digitization of the celestial radio signal and its recording on a wide-bandwidth tape recorder. Their cross-correlation S_{xy} is evaluated subsequently at a central facility, giving a set of fringe visibilities V_{ij} that are analogues of the expressions given in Equations (5.16) and (5.28). Because the VLBI stations have independent timing, each S_{xy} must be evaluated for multiple values of the system parameters, deriving the maximum value in each case. Many aspects of the VLBI technique are discussed at length by Zensus, Diamond and Napier (1995), and by Thomson, Moran and Swenson (1986): see also the review by Moran in Perley, Schwab and Bridle (1989). For space VLBI see Schilizzi *et al.* (1984) and Hirabashi and Hirosawa (2000).

Curiously, three sets of astronomical motivation provided the impetus to the development of VLBI: the rapid time variations of quasar fluxes implied that the regions in quasars that

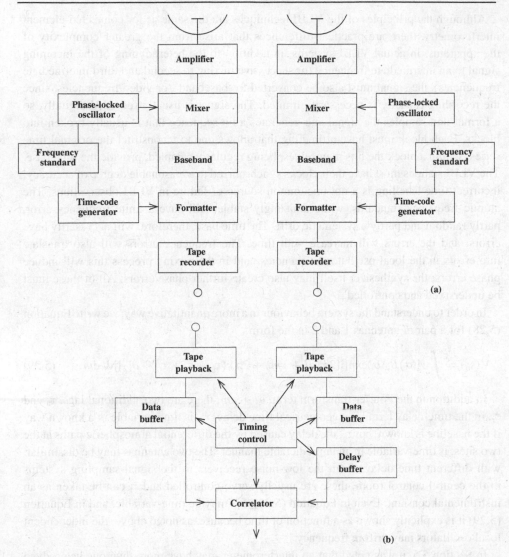

Figure 5.7. An outline diagram of a VLBI system, showing (a) two independent receiver stations and (b) the correlator system.

emitted the radiation were of the order of light-days in size; the scintillations of quasar radio signals passing through the irregularities of the solar corona implied the same thing; and the OH-maser sources, while more local, were unresolved by hard-wired interferometers. Much longer baseline length, and hence much higher angular resolution, was needed for the study of both quasars and OH masers. Both the stable frequency standards and the recorder technology were available.

Although the principles of the VLBI techniques are the same as for connected-element interferometry, there are practical differences that arise from the greater complexity of the apparatus. In actual VLBI systems, in addition to the heterodyning of the incoming signal to an intermediate frequency (or successive mixing to second and third intermediate frequencies), the signal must also be converted to 'baseband', or video frequencies, since the recording capacity is necessarily limited. The signal is usually recorded digitally, so a format unit samples the signal and generates a bit sequence that is usually broken into blocks. Each block must have a time tag that allows one to reconstruct the original time series; within a block the bits themselves, being regularly sampled, provide the time base. The VLBI stations must have their clocks synchronized to a reasonable degree of accuracy; incorrect time labelling is a not uncommon source of failure in VLBI observations. The atomic frequency standards, while amazingly stable, still have a finite frequency error, partly random and partly a systematic drift. The time base, therefore, will necessarily have errors, and the errors will increase with time. The frequency errors will also translate into errors in the local oscillator frequencies, and in the mixing process this will induce phase errors; the synthesizer itself may also create further phase errors. All of these must be understood and controlled.

In order to understand the system behaviour in a more quantitative way, we write Equation (5.28) for a pair of antennas 1 and 2 in the form

$$V(s_0) = \int \mathcal{A}(\sigma) B_\nu(\sigma) \exp\{i[2\pi(ul + vm) + 2\pi\nu(\tau_g + \tau_a + \tau_i) + \theta_{12}]\} dl \, dm. \qquad (5.29)$$

In addition to the Fourier-transform term, $ul + vm$, there are two additional lags, τ_a and τ_i, in the time-delay term. The geometrical time delay, τ_g, is time-variable in a known way if the baseline is known, but τ_a, the delay caused by the differential atmospheric paths at the two sites, is time-variable in an unpredictable manner. The two antennas may be dissimilar, with different time delays from the low-noise receivers to the signal-sampling systems in the central control room; these are usually servo-controlled and τ_i can be taken as an instrumental constant. Even in Equation (5.28), θ_{12} may be time-variable, and in Equation (5.29) it is explicitly shown as a function of time because, as noted above, the independent local oscillators may drift in frequency.

In Section 5.5, it was noted that an interferometer may have more than one heterodyne conversion, but even though one can effectively treat the frequency conversion as if there were only a single local oscillator frequency, close attention must be paid to the details of the conversions, since a lower-sideband conversion inverts the frequency spectrum. There are further differences between VLBI and connected-element interferometry, however, that should be noted. In VLBI, there is always a conversion to baseband. Nevertheless, the sensitivity to phase at the front end (atmospheric, instrumental, and local oscillator phase variations) is retained at baseband. The VLBI baselines can be so large that the geometrical time delay may be changing rapidly, so adjustments in instrumental time delay must be swift and precise, usually through computer and tape manipulations. Furthermore, each atomic frequency standard, while remarkably precise, is always in error by some small frequency offset $\Delta \nu_j$ from the correct frequency ν_{lo}^0, leading to a systematic error in the time base, usually expressed as a rate $R_j = \Delta \nu_j / \nu_{lo}^0$. There are multiple opportunities for timing errors as well: clock-setting errors, unexpected errors in the instrumental time delay and atmospheric time delays all contribute to the net timing error $\Delta \tau_j$. An interferometer output will be affected only by the difference between the timing and frequency errors at the two antennas, Δ_{12} and R_{12}.

Given these complications, the correlator output will contain a number of perturbations in the phase term ϕ_{12} of Equation (5.29). Some are negligibly small, but, if one retains the dominant terms, and expresses the frequency term in terms of the baseband frequency $\nu_{bb} = \nu - \nu_{lo}$, the correction terms can be included as part of the phase term θ_{12},

$$\theta_{12} = 2\pi (\nu_{bb} R_{12} t - \nu_{lo} R_{12} t - \nu_{bb} \Delta \tau_{12}) + \theta_{12}^r, \qquad (5.30)$$

where θ_{12}^r represents all the phase errors not included in local oscillators and time shifts.

The effect of the difference in the independent frequency standards shows up in the first two terms, but the video frequencies, ν_{bb}, are far lower than the initial radio frequency ν, so the first term is always small compared with the second. Thus the phase variation due to frequency drift in the local time standards shows up as a term varying linearly with time, while a timing error shows up as a term varying linearly with frequency.

The correlator takes the time-average of the cross-power product $\langle R_{xy} \rangle$, but these effects reduce its measured value from the correct value. The frequency offset term, $\nu_{lo}^0 R_{12} t$, increases the phase linearly with time, and, unless $R_{12} t \ll \nu_{lo}^0$ over the integration time, the visibility will be reduced. The time-delay term, $\nu_{lo} \Delta \tau_{12}$, has a phase shift that changes linearly across the receiving band, and, unless $\mathcal{B} \Delta \tau_{12} \ll 1$, there will also be a reduction in the visibility. These two perturbations are called the differential fringe-rotation error and the timing error; both R_{12} and $\Delta \tau_{12}$ must be sufficiently stable over the integration time. The integration time cannot be longer than the stability of the reference oscillators allows: hydrogen masers typically have a stability of the order of a part in 10^{14}, so at a frequency of 30 GHz (wavelength 1 cm) an integration time of the order of 100 s might be permissible if a $10°$ phase shift is acceptable. In correcting for the time delay term $\Delta \tau_{12}$ there are many contributors to consider. The baseline errors can be greatly reduced as one gains experience with the interferometer, and the error in determining the epoch at each station is reduced if the interferometer is in constant use (but surprises can occur without warning). The atmospheric and ionospheric contributions to $\Delta \tau_{12}$ are random, but estimates can be made from knowledge of local conditions. Finally, the random oscillator error, θ_{12}, must be accounted for. This can be a catastrophic source of fringe-visibility reduction and, if it is large, it usually means that there is an instrumentation problem in either the frequency standard or the local oscillator-multiplier chain. A number of further technical details must be coped with, such as fractional-sample delay and digital phase rotation; reference should be made to the exposition by Romney in Zensus, Diamond and Napier (1995).

The fundamental problems of correcting for frequency and timing offsets can be resolved by a two-dimensional search of $\langle R_x y \rangle$, maximizing its value both for frequency offset R_{12}

and for time delay $\Delta \tau_{12}$. This gives an estimate of the fringe amplitude, but the phase is more elusive, unless there is a reference source within the common antenna beams. Although two-telescope, single-baseline interferometers are still occasionally used, arrays of several telescopes are used to carry out nearly all VLBI observations. Having a multiplicity of baselines simultaneously permits much of the phase information to be recovered, using the closure relations described in Chapter 6: the use of three telescopes, giving three baselines, retrieves the relative phase; and using four telescopes, with six baselines, improves the measurement of fringe amplitude.

5.7 Beam switching

The calibration of fringe visibility and phase in a long-baseline interferometer is best achieved by observing a source with well-determined flux density. There may be such a source within the primary beam of the interferometer telescopes, whose signal can be distinguished by its periodicity and phase; this is particularly applicable to aperture-synthesis observations (Chapter 6). In the VLBI case, the delay beam can be small compared with the antenna beam, and, if the reference source meets this condition, it will provide an accurate calibrator. More usually it is necessary to calibrate by observing a known source outside the main beam, but close enough to avoid introducing gain differences due, for example, to different paths through the atmosphere. In the large arrays used in aperture synthesis, such a calibration involves swinging all the telescopes of the array to the calibration source and resetting the delays and fringe rotators, using the computer control. The loss of observing time on the required source is well compensated for by the improvement in calibration and stability. Beam switching every few minutes is often needed to keep track of atmospheric delays and individual instrumental phase shifts. Calibration sources must be chosen with care, especially if the interferometer has a high angular resolution. It is frequently necessary to allow for the angular structure of each calibration source, if accurate synthesis maps are to be made.

5.8 The interferometer in geodesy and astrometry

In the treatment of interferometry in the preceding sections, the geometrical time delay has appeared as a barrier to observations, to be compensated for by the instrumental time delay τ_i. The observed time delay for a given source is, however, a function of the source coordinates and the interferometer baseline orientation. This means that, for a point source of fixed location on the celestial sphere, the baseline vector can be obtained from the measurement of the geometrical time delay. Thus the distance between two points on the Earth's surface can be measured with great accuracy. Conversely, if the distance between the antennas is fixed, the measurement of τ_g is equivalent to a source-position measurement. For VLBI systems, these interferometric properties have been applied directly to the problems of geodesy and astrometry. The source position and interferometer baseline orientation are both unknown initially, and must therefore be solved for simultaneously if absolute solutions are required. Differential measurements can avoid some of the complications of solving the complete problem; measurements of continental drift and of relative motions caused by earthquakes are practical examples.

In VLBI, discussed in more detail in Chapter 6, the instrumental time delay is provided by shifting the time bases of the data streams within the computer, or by shifting the recorder tapes by a known amount. The data are recorded digitally in a 'self-clocking' manner, in which the station clock determines the sampling intervals; thus the data are in precise time order. The accuracy of determining the relative time delays on the two tapes is limited by a combination of the signal-to-noise ratio and the signal bandwidth. One can see that the wider the bandwidth, the more precise the measurement will be. In practice, the VLBI recorder bandwidth is fixed, and this would seem to limit the achievable accuracy. A large improvement can be made by using the method of *synthetic bandwidth*, which consists of rapidly moving the reception band from one frequency to another by switching the local oscillator frequency. The local oscillator frequencies at each station must be switched in perfect synchronism, or more exactly in perfect synchronism with respect to the station clocks. Large synthetic bandwidths, of the order of 1 GHz or more, can be achieved by this method; the overall relative position accuracy that can be determined in a single measurement is at present about 3 cm, but this can be reduced by averaging over many measurements. Already, there are examples in which a year of data has reached an accuracy better than 1 cm.

The complete scheme of measuring geodetic positions by VLBI must recognize the presence of many small perturbations. The most serious of these is the extra delay fluctuation imposed by the Earth's atmosphere. The total delay imposed by the refractive index of the atmosphere at the zenith is about 2 m with respect to free space, but the atmosphere is not uniform, and differences of up to 10% can be imposed on a VLB interferometer having a baseline of thousands of kilometres. Measurements of atmospheric pressure and water-vapour content must be made at each station, in order to calculate the corrections. Earth tides, which may amount to tens of centimetres, also cause systematic changes in the baseline direction, and must be corrected for. The total art of geodesy is far too developed to summarize here; reference should be made to Rogers (1993) for discussions of the methods and results. The rates of continental drift have been measured; North America and Europe are separating at a rate of $17 \, \text{mm} \, \text{yr}^{-1}$ (McMillan and Ma 1994), and many local measurements related to seismic activity have been made.

Interferometric measurements of radio source positions have also been carried out to high accuracy by VLBI methods, and by the use of conventional interferometers as well. The position of the origin of right ascension, which is defined by the solar system (Appendix 2), has to be determined by the best possible measurements of celestial sources at extragalactic distances. Declination is defined by the pole of the Earth's rotation; it is determined by the interferometer measurements themselves. Up to now, the core of the radio source 3C 279 has been chosen as the origin of right ascension for a radio-based coordinate system. The precision of the resulting radio coordinates of the brighter radio sources exceeds that of the best optical determinations of stellar coordinates by an order of magnitude; on the other hand, the relative positions of over 100 000 visible stars are available from the Hipparcos catalogue to an accuracy of about 1.5 milliarcseconds. A small number of radio determinations of quasars that occur also in the Hipparcos catalogue therefore allows the construction of a reference frame covering the whole sky that is based on extragalactic sources (see Appendix 2 and a discussion by Johnston and de Vegt (1999)). The successor to the J2000 system will be based on this work.

5.9 Interferometry at millimetre wavelengths

The problems of interferometry are compounded at wavelengths shorter than a centimetre or so. The effective areas of radio telescopes are smaller, as noted in Section 4.7, while the fluxes from synchrotron sources are diminished as well. The noise contribution of the atmosphere is higher, due to the effects of water vapour and oxygen, so there is a lower limit to the achievable system temperature. This means that the fringes from sources will be more difficult to detect. Furthermore, atmospheric attenuation is not only significant, but can be highly variable. The choice of telescope site is therefore highly constrained; not only must there be a high fraction of days with clear skies, but also the total water-vapour content of the atmosphere overhead should be as low as possible. The existing interferometric arrays have only moderately good performance in this respect, and there is competition for observing time during the seasons with low overhead concentration of water vapour. The site considerations have been summarized further in Section 4.6.

There are two general cases to be considered: the study of continuum sources and molecular-line emitters. The structure of molecular-line objects is generally so complex that arrays with many elements are needed to obtain sufficient coverage of the u, v-plane. Closure relations and self-calibration can be used, as described in Chapter 6, to improve the fringe visibilities. The necessity of carrying out the operations in the full u, v, w data cube, a requirement imposed by the spectral-line nature of the radiation, is a complication, but one that does not differ in principle from the methods outlined earlier. There is a practical limitation, however, that limits the detectability of molecular-line emitters, which are generally extended objects. This minimum-surface-brightness condition can be derived easily, as follows. For an interferometer of spacing D, with telescope elements of diameter d, the minimum surface brightness temperature, T_{min}, that can be detected with a system temperature T_{sys} and a given bandwidth–averaging-time product is of the order of $T_{sys}(\tau \mathcal{B})^{-1/2}(D/d)^2$. The bandwidth is fixed by the Doppler width of the spectral line, the averaging time is fixed by the coherence time of the atmosphere and the system temperature is limited by the atmospheric emission. The square of the ratio D/d, generally referred to as the *beam dilution factor*, is the critical quantity; there is, as a consequence, a natural limit to the interferometer spacing that can be used in the millimetre regime for any given system. The use of N antennas improves the value of T_{min} by a factor of N, so arrays of many elements will be a desirable future development.

The study of continuum sources becomes difficult at millimetre wavelengths, because the source fluxes are weak and the effective areas are small. The VLBI observations, in particular, present special problems, although the results can be of great scientific interest. Millimetre wavelengths are needed to overcome the complications introduced by interstellar scattering at the Galactic centre (Chapter 10), while the optically-thick synchrotron emitters, encountered for example in Chapter 13, can be more easily penetrated at millimetre wavelengths.

The practical difficulties are twofold: calibration of the fringe visibilities is complicated, because of a lack of reference sources, and the integration time is severely limited by atmospheric fluctuations, particularly in the water-vapour content. One is usually reduced, therefore, to employing short coherent integration times, and averaging the short segments incoherently. The signal-to-noise ratio (SNR) improves as the square root of the integration time for coherent averaging, but an incoherent average of n samples only improves the SNR as the fourth root $n^{1/4}$. There are times when more sophisticated methods can be used. Clark (1968) pointed out that between completely incoherent and completely coherent averaging there are intermediate cases. These have been discussed further by Thompson, Moran and Swenson, and a modern summary of all techniques is given by Rogers, Doelman and Moran (1995).

The Atacama Large Millimetre Array (ALMA), which is under construction at present, will be at the limits of technology. The main array will comprise 66 antennas, of diameter 12 m and surface accuracy 25 μm, spaced up to 15 km apart. The phase stability of local oscillators and signal lines, using fibre optics, will reach 1 radian even at the highest frequency of 950 GHz. At 100 GHz the angular resolution will reach 0.01 arcsec, with a spectral resolution of 0.01 km s^{-1}, sufficient to resolve thermal line widths, and the flux limit for point sources will be below one millijansky. The construction and operation of the ALMA, on a remote desert plateau at an elevation of 5100 m, is a major task that is being undertaken jointly by collaborators in Europe, Japan and North America.

5.10 Optical interferometry

Michelson's stellar interferometer was invented for use at optical wavelengths, although the technique has been adopted far more extensively in radio astronomy. The schematic plan of a radio version of the Michelson interferometer, shown in Figure 5.2, is equally valid for an optical interferometer. In the radio case, the radio telescope usually incorporates a stable preamplifier, which amplifies the collected radio flux and sends it into a transmission line. In the optical case, amplifiers are rarely feasible, but the same general scheme applies. Modern optical interferometers have receiving elements (telescopes or coelostats) that convert the incoming light into a quasi-plane wave for transmission to a central station. Optical delay lines are used to equalize the path lengths, and the signals are then cross-correlated as in the radio case; the optical and radio techniques are equivalent but quite different in their realization. Ordinarily the fringes are not observed directly; the output, the fringe visibility, is a complex number, as it is for a radio interferometer, and it is important to record both amplitude and phase.

All the important components of an optical interferometer have radio analogues, but there are two important differences concerned with phase stability and with SNR. Atmospheric scintillation can change the optical path length through the atmosphere by many wavelengths within a few milliseconds, requiring observations to be made with very short integration times. The system noise has an extra component, which is the *shot noise* of the individual photons; this places a limit on measurement of correlation for faint sources, which becomes catastrophically poor when less than one photon is received during the short integration time. It is necessary to use large collecting areas and a long series of short integration times, each yielding a measurement of visibility $V_{i,j}$ (see Section 5.4), and averaging of the results.

The rapid fluctuations due to scintillation mean that the phases of the successive measurements of fringe visibility lack coherence. This cannot be overcome in a simple two-element interferometer, but, if there are three or more elements, giving three or more interferometer pairs, the method of *phase closure* described in Section 6.5 can be used. The routine achievement of stable fringes at optical wavelengths, with phase closure, has been demonstrated by Baldwin *et al.* (1996). The discussion of optical interferometers in Shao *et al.* (1987) has been extended greatly by the SPIE volume edited by Traub (2004); see also the review by Quirrenbach (2001).

When an optical interferometer uses a pair of large telescopes, there will be varying phase shifts across the wavefronts within the individual apertures, which destroy the coherence. The coherence is restored by the use of adaptive optics to correct the wavefront. In each telescope the entrance aperture is then re-imaged, so that the light travels as a plane wave with a much reduced cross-section to the base of the telescope and thence to the beam-combining system. Two such examples of large-telescope interferometers are the 10-m Keck telescopes on Mauna Kea and the four 8-m telescopes of the European Southern Observatory in Chile. These are now in routine use at infrared wavelengths.

The effects of atmospheric scintillation are smaller at longer wavelengths; interferometers working at infrared wavelengths can use larger individual apertures and longer integration times. Infrared interferometers have been used to determine stellar diameters and limb-darkening; the cyclic variation in the diameter of a Cepheid variable star has been observed directly (Lane *et al.* 2000) at wavelength 1.6 μm using a two-element interferometer with a baseline of 110 m.

A dedicated astrometric interferometer array, the Naval Prototype Optical Interferometer (NPOI) at Flagstaff, Arizona, routinely obtains astrometric positions of stars with a precision of better than 10 milliarcseconds. The positions determined by the astrometric Hipparcos space mission were initially more precise, but proper motions have accumulated over the years, and the NPOI positions are now the best available.

Relative astrometric positions can be much more accurate for small angular separations, such as in binary and multiple star systems. When the two stars in a binary system are within the 'antenna beam' of the optical interferometer, the precision can be far greater. The Palomar Testbed Interferometer (PTI) has a dual-beam system, in which one star can serve as a phase reference; by locking one time delay onto the reference star, and then sweeping the time delay in the second beam, the interferometer fringes of both stars can be observed, and the relative phase difference gives the angular separation on the sky with an accuracy of 10–20 microarcseconds (Lane and Colavita 2003). This has enabled the parallax, and therefore the distance, of (several) binary systems to be measured to 0.5%; these are the most accurate parallaxes ever measured (Pan *et al.* 2004).

Further reading

ARAA mm and Sub-mm Interferometry, Sargent A. I. and Welch W. J., *Ann. Rev. Astron. Astrophys.*, **31**, 297, 1993.

Interferometry and Synthesis in Radio Astronomy, Thompson A. R., Moran J. M. and Swenson G. W. Jr., New York, Wiley, 1986.

New Frontiers in Stellar Interferometry, ed. Traub W. A., *Proc. SPIE* **5491**, 2004.

Subarcsecond Radio Astronomy, ed. Davis R. J. and Booth R. S., Cambridge, Cambridge University Press, 1993.

Synthesis Imaging in Radio Astronomy, ed. Perley R. A., Schwab F. R. and Bridle A. H., Washington, Astronomical Society of the Pacific Series No. 6, 1989.

The Intensity Interferometer, Hanbury Brown R., London, Taylor and Francis, 1974.

Very Long Baseline Interferometry and the Very Long Baseline Array, ed. Zensus J. A., Diamond P. J. and Napier P. J., New York, ASP Conference Series No. 82, 1995.

6

Aperture synthesis

The Michelson interferometer, whose basic properties were reviewed in the preceding chapter, was originally designed to measure the angular diameters of stars. It was first used in the radio domain by Ryle and Vonberg (1946) to find the angular diameter of sunspot radiation, and by McCready *et al.* (1947), who showed that interferometry could be used to make a map of the radio emission from the whole Sun. At radio wavelengths of the order of 1 m, no single aperture could map the Sun with enough angular resolution to be interesting, because of diffraction. The use of Michelson interferometry turned out to be an effective tool to obtain the necessary angular resolution. The technique was soon applied to study discrete radio sources, and, from these modest beginnings, large and complex interferometer systems have been built to map the distribution of brightness across small-diameter radio sources, overcoming the limitations of diffraction that are inherent in single-aperture telescopes. The resulting angular resolution now exceeds the resolving power of the largest optical telescopes.

The essential link between interferometer observations and the brightness distribution of a source is the Fourier transform, as the analysis in Section 5.4 has demonstrated: the amplitude and phase of the fringe visibility, defined by Equation (5.18), give one complex Fourier component of the brightness distribution. An array of radio telescopes, their outputs separately amplified and combined pairwise to form all possible interferometric combinations, is called an *aperture-synthesis array*. In simple terms, the source is being observed with a collection of two-element interferometers, each with a sine-wave gain pattern rippling across the source, with frequency and orientation determined by the particular baselines.

The art of calibrating and combining this array of interferometer products to yield a brightness map of a radio source with high angular resolution is the subject of this chapter.

6.1 Interferometer arrays

The concept of aperture synthesis grew gradually in a series of transit interferometer experiments at the Cavendish Laboratory in Cambridge, England, at the Radiophysics laboratory of Australia's CSIRO in Sydney, and at the University of Sydney. At the

Cavendish Laboratory, Ryle and Neville (1962) showed that an individual baseline could provide an extended coverage of Fourier components by continuing observations while the baseline rotated as the Earth rotated on its axis, changing the length and orientation of the baseline. This technique immediately became known as Earth-rotation synthesis, and their work led to the building of the three-element Cambridge One-mile Telescope, the first Earth-rotation-synthesis radio telescope (Ryle 1962). As the technique became widely adopted, the name was shortened, since the rotation of the Earth was understood; aperture-synthesis array or aperture-synthesis telescope is now the common description.

Both the Cambridge One-mile Telescope, and the Westerbork Synthesis Radio Telescope (WSRT) which followed (Baars *et al.* 1973) were disposed along an east–west track. Earth-rotation synthesis was essential in their operation: as the individual telescopes were tracking, the spacings followed an arc in the u, v-plane. The Cambridge One-mile Telescope had a single movable dish, and the u, v-plane was filled in further by moving this single element on separate days. The WSRT had ten fixed and two movable dishes, thus speeding up the data gathering by an order of magnitude; and two more were added later.

As the number of elements increases, and with the addition of north–south baselines, the construction of a sufficiently well-defined Fourier transform becomes easier. If the number of antennas is sufficiently large and the source is sufficiently simple, a brief averaging time will already provide a usable map; this is known as the 'snapshot' mode. Otherwise, the data set must be extended by using Earth rotation and often by using more than one array configuration.

The principle of aperture synthesis was adopted on a larger scale in the Very Large Array (VLA) of the National Radio Observatory, situated in New Mexico. The array, in its most compact configuration, is shown in Figure 6.1. It consists of 28 identical paraboloids, 27 of which are in operation at any given time, with the 28th off-line for refurbishment. The individual telescopes, connected by a hard-wired data network (currently optical fibres) are arranged in a Y configuration (Figure 6.2). All telescopes are movable on railroad tracks to fixed foundations, allowing the spacings between the antennas to be changed (in Figure 6.2 the array is in its most compact configuration, with maximum spacing 1.5 km). At each station, there is a connection to a transmission line that conducts the signals to the central station. The arms of the Y are approximately 20 km long, giving a maximum baseline length of 35 km; there are four basic configurations, although hybrid configurations are often used, and the array can be subdivided into smaller subarrays for simultaneous observations in separate programmes that do not need the full array.

For an array with N elements there are $N(N - 1)/2$ ways of making interferometer pairs; for the 27 elements of the VLA this amounts to 351 separate baselines. The entire array can alternatively be co-phased to serve as a single telescope having the collecting area of a 134-m aperture; in this mode, it has been used to gather telemetry data from spacecraft during planetary encounters.

The MERLIN array (Multi-Element Radio-Linked Interferometer Network) is an example of a more extended array, using six fixed telescopes distributed across a large part of

Figure 6.1. The VLA interferometer array, New Mexico, in its most compact configuration (courtesy of the National Radio Astronomy Observatory).

central England with a maximum baseline of over 200 km (Figure 6.3). The array was originally connected by radio links, but is now linked by optical fibres to the central station at Jodrell Bank near Manchester. The VLA and MERLIN are complementary, with MERLIN having the longest baselines and therefore the highest angular resolution at a given frequency, and the VLA having greater collecting area and denser u, v-coverage but with baselines an order of magnitude shorter. The greater sensitivity of the VLA, with its 351 individual baselines, often allows its use in a single short observation (the snapshot mode), while the 15 instantaneous baselines of MERLIN have to be supplemented by Earth-rotation synthesis in most observations.

The GMRT array in India (Swarup *et al.* 1991), consisting of 30 telescopes, each 45 m in diameter with baselines up to 25 km, is designed for longer wavelengths, for which a large collecting area can be constructed by using telescopes with a light-weight reflector surface.

In the southern hemisphere, the Australia Telescope is also an aperture-synthesis array. Its core array consists of six movable telescopes, distributed along an east–west baseline. All the arrays described so far can be augmented by adding more distant elements using VLBI techniques (see Section 6.11).

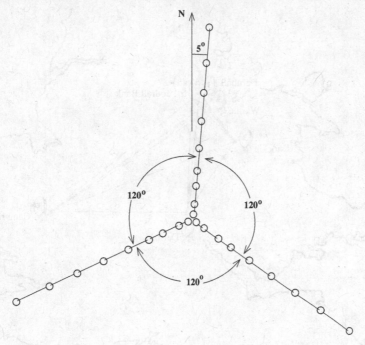

Figure 6.2. The 27 antennas of the VLA. Each arm of the Y is 20 km long.

These arrays were designed to operate primarily at decimetric and centimetric wave-lengths. The available range has since been extended to both longer and shorter wave-lengths: two arrays, LOFAR (based in the Netherlands) and MWA (in Australia), have been built for metre wavelengths, and ALMA (in Chile) for millimetre wavelengths (see Section 5.9 and Chapter 17).

6.2 Cross-power products in an array

We now generalize the analysis of Chapter 5, for a single two-element Michelson inter-ferometer, to the practical case of N radio telescopes forming an aperture-synthesis array. During an observation, the correlator evaluates the cross-product of the voltages from each pair of elements i, j in the array, and takes the average over the pre-set integration time, to form the cross-power product $\langle R_{ij} \rangle$ as in Equation (5.2). This gives an estimate of the fringe visibility, $V_{ij}(u_k, v_k)$ for the kth integration period. In the case of VLBI, treated in Section 5.6, a range of time delays will be evaluated, but for most arrays the time delay is primarily determined by the geometrical time delay τ_g, which can be calculated from the known baseline coordinates u, v and the direction cosines of the known phase center l, m.

The complete set of fringe visibilities $V_{ij}(u_k, v_k)$ for the observation will be a discrete matrix of complex quantities, and the calculation of a brightness map from this set of visibilities will be carried out by taking a discrete Fourier transform. Nevertheless, it is

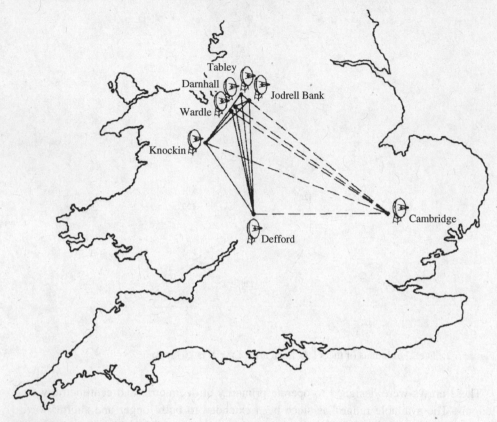

Figure 6.3. The Multi-Element Radio-Linked Interferometer network (MERLIN).

conceptually convenient to view the process in its integral form, since a complete set of visibilities over the u, v-plane is the Fourier transform of the source brightness distribution, introduced in Section 5.4. For a source of sufficiently small angular size, for which the curvature of the celestial sphere can be neglected (or entered as a small correction), the relationship was described by Equations (5.23) and (5.24), which we write as

$$V(u, v) = \int\int B(l, m)\exp[-2\pi i(ul + vm)]\mathrm{d}l\,\mathrm{d}m, \tag{6.1}$$

$$B(x, y) = \int\int V(u, v)\exp[2\pi i(ul + vm)]\mathrm{d}u\,\mathrm{d}v. \tag{6.2}$$

The approximation in this transformation is valid provided that there is little significant emission at large angles from the centre of the map: this is equivalent to ignoring an aberration analogous to field curvature in an optical telescope.

For wide-field arrays such as the MWA and LOFAR, the field may extend far from the reference direction, necessitating the use of direction cosines, but for most synthesis applications the angular distances (x, y), where $l = \sin x$ and $m = \sin y$, are sufficient, because all angular distances from the reference direction are small.

Constructing the correct visibility function $V(u, v)$, and hence the sky map $B(l, m)$ for the observed samples V_{ij}, requires two important and computationally demanding tasks: careful calibration and image reconstruction. The accuracy of calibration is vital to the production of maps with high dynamic range; furthermore, the signal path through each antenna is subject to changes in the atmosphere and in the amplification network, so calibrations must be repeated at frequent intervals, as described in Section 6.3. Both instrumental and atmospheric corrections to τ_g are generally small, and can be included as a phase shift associated with each element, as part of the calibration process. The fully calibrated image requires reconstruction, since it may suffer from severe distortion due to the missing Fourier components; however, remarkable improvements are possible through processes generally known as 'mapping', which we describe in Section 6.5

Each pair of elements i, j is combined as an interferometer, and the correlator will evaluate the cross-power product R_{ij} for the two voltage amplitudes as in Equation (5.2), and perform the time averaging. The geometrical time delay τ_g, which depends on the orientation of the baseline relative to the direction of the source, must be computed relative to a fiducial point in the array, and compensated for by an instrumental time delay; all possible pairs can be compensated for if there is an instrumental time delay τ_i inserted in the signal line from each element, compensating for a plane wave arriving at the array from the source being observed. If each instrumental time delay is set to match the geometrical time delay for that plane wave (having specified a reference point), the time average of the cross-power product, $\langle R_{ij} \rangle$, becomes an estimate of the fringe visibility V_{ij} for all the interferometer pairs in the array (Equations (5.18), (5.21) and (5.23)). When Earth-rotation synthesis is used to obtain a more complete set of visibilities in the u, v-plane, each set of visibilities has a time tag, specifying a particular value of (u_k, v_k) for the kth time interval of integration. This gives a set of values, necessarily incomplete but often sufficient, for the Fourier transform of the source brightness distribution $B_\nu(l, m)$.

6.2.1 *The spectral sensitivity function*

After calibration, the set of $N(N - 1)/2$ visibilities $V_{ij}(u_k, v_k)$ (which will continue to be treated as a set of integral functions for clarity) forms an approximation to the Fourier transform of the source brightness, which may, or might not, fill the u, v-plane adequately. The array elements, if they are steerable paraboloids, will be tracking the source continuously, so that with respect to the array there is a separate tracking direction s_0 associated with each of the k integration samples (although s_0 has a constant orientation with respect to the celestial sphere); recall that the u, v-plane, with coordinates measured in wavelengths, is perpendicular to s_0. With the small-angle approximation, a complete sampling of the u, v-plane, $V(u, v)$, would be the Fourier transform of the brightness distribution, $B(x, y)$. The sampling of the Fourier plane by an array is necessarily incomplete, with missing spacings that result in unknown components in the Fourier-transform plane, and it is necessary to understand the limitation this imposes on reconstructing the source brightness distribution $B(x, y)$. This limited sampling is expressed as a *spectral sensitivity function* $W(u, v)$. (This plays the same role as the transfer function introduced in the discussion of filters in Section 3.2. The transfer function $H(v)$ operates in the frequency/time domain, and its Fourier transform gives the response of the filter to a voltage impulse $\delta(t)$.) If a two-dimensional point source, $\delta(x, y)$, is observed by the interferometer array, the source appears as a spread pattern, which is the two-dimensional Fourier transform of $W(u, v)$. For this reason, the spectral sensitivity function is sometimes called the *spatial transfer function*.

In Section 4.4, we distinguished between the voltage-reception pattern of an antenna, $F(x, y)$ (with complex conjugate $F^*(x, y)$), and its power-reception pattern, $G(x, y)$:

$$G(x, y) = F(x, y)F^*(x, y). \tag{6.3}$$

In an interferometer we are concerned with the voltage-reception pattern, which is related to the excitation $E(\xi, \eta)$ (the current distribution in Section 4.4) across the antenna aperture by the Fourier transform of $C(\xi, \eta)$, the autocorrelation of $E(\xi, \eta)$:

$$G(x, y) \xleftrightarrow{\text{FT}} C(\xi, \eta), \tag{6.4}$$

where ξ and η are measured in wavelengths. The convolution theorem shows that the autocorrelation function of $E(\xi, \eta)$ is the Fourier transform of the power pattern $G(x, y)$.

The same relations hold for an array, where the spectral sensitivity function $W(u, v)$ is similarly given by the autocorrelation function of the excitation:

$$W(u, v) = \int E(\xi, \eta)E^*(\xi - u, \eta - v)\mathrm{d}\xi \, \mathrm{d}\eta \tag{6.5}$$

or

$$W(u, v) = E(\xi, \eta) \otimes E^*(\xi, \eta). \tag{6.6}$$

The sample of the Fourier transform from each interferometer pair is defined by the projected baseline (the baseline length can be defined in either sense, although in geodesic applications the home station must be identified). This means that the Fourier transform of the brightness distribution is defined over half of the u, v-plane, but the pattern of coverage has reflection symmetry through the origin. The response $V(u, v)$ in Equation (5.18), the *visibility* or *fringe visibility*, at each baseline u, v, is the fundamental measured quantity in interferometry.

It is useful to map the spectral visibility function $W(u, v)$ on the u, v-plane to obtain a pictorial representation of the adequacy of the array in mapping the brightness distribution across sources of various angular sizes. It is particularly important to determine the domain of the u, v-plane over which $W(u, v)$ is non-zero; this is the *support domain* outside which no Fourier components are available. (The concept was introduced by Bracewell (1961), who referred to $W(u, v)$ as the *spatial sensitivity diagram*.)

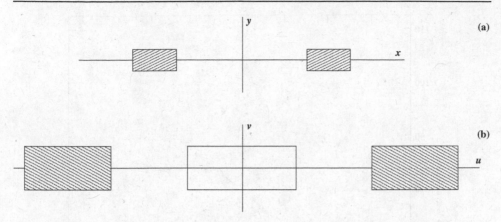

Figure 6.4. The coverage, or 'support', in the u, v-plane for a two-element interferometer: (a) the interferometer antennas in the x, y-plane and (b) the support in the u, v-plane.

In the example of Figure 6.4, a simple interferometer has two equal and uniformly filled rectangular elements. The *support* in the u, v-plane, defined by the cross-correlation $W(u, v)$ of the excitation $E(\xi, \eta)$ of the two antennas, is shown on the same scale, where both ξ, η and u, v are measured in wavelengths. The open area in the centre is the additional support domain for a receiver that is sensitive to total power as well as the interference signal between the spaced telescopes; this area is the range of Fourier components available to each single telescope aperture, detecting the total received power. An example of early observations with such a total-power two-element interferometer was shown in Figure 5.1, where the recording shows the sinusoidal fringes of each radio source lifted by the total power; the extent of the fringes is limited by the reception pattern of the antenna, defined by the Fourier components in the central part of $W(u, v)$. Receiver systems for aperture-synthesis arrays, on the other hand, generally eliminate the central part, since the total power is not recorded.

6.2.2 Coverage of the u, v-plane by arrays

Arrays such as the VLA (in its extended configurations) and MERLIN can often be approximated as arrays of point radiators, and their support will then be a discrete array of points. In the example of a two-element interferometer, as in Figure 6.4, if the elements were represented by a pair of delta-functions, the convolution would be three delta-functions, two at plus and minus the spacing, and a central spike having twice the height (this is the total-power spike, which is ordinarily not present). If the VLA is represented as an array of point radiators, its spectral support function, when observing at the zenith, is as shown in Figure 6.5. Note that it is symmetrical with respect to reflection about the centre; this is necessary since each baseline has length but not direction. The effect of the finite spacings shows as gaps, and the overall support is limited in two ways. Its hexagonal pattern falls

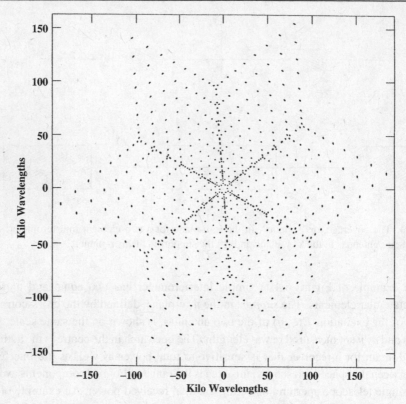

Figure 6.5. The instantaneous coverage, or 'support', in the u, v-plane, of the VLA, for an observation at the zenith.

within a circle, which limits the angular resolution, since there are no spacings larger than the tips of the hexagonal pattern. Another limitation of the support can also be seen at the centre, where there is a hole. This is a necessary consequence of the array having a minimum spacing, and its consequence is that, for any extended source, structures larger than the angular size corresponding to the minimum spacing will go undetected and the map will be incomplete. Thus, both the maximum and the minimum available spacing of an array must be heeded when mapping sources with a synthesis array.

The amplitude of any part of the spatial spectral sensitivity function within the support domain may be varied during the process of synthesizing the final map, by applying gain factors in the addition of the individually recorded components. Including any such modification, and with the addition of the total power recorded at individual antennas (to give the 'zero-spacing' component), the function $W(u, v)$ determines the power-reception pattern $G(l, m)$.

Multi-element interferometers are, however, often used at such large spacings that it is in practice impossible to provide more than a sample coverage of the u, v-plane. Missing spacings in the u, v-plane imply a loss of Fourier components in the synthesized image;

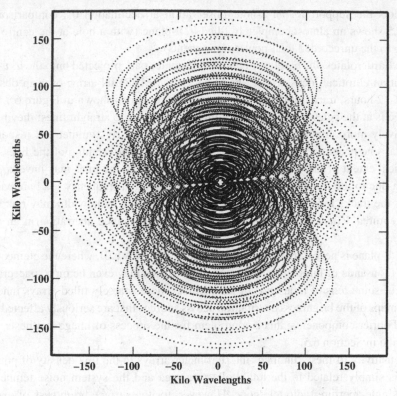

Figure 6.6. The u, v-plane coverage for an 8-hour tracking observation with the VLA, for observations at declination $\delta = 30°$.

nevertheless, remarkably good images are obtained in these 'open', or 'unfilled', arrays, by using the image-reconstruction processes described in Section 6.5.

Figure 6.2 showed the configuration of the 27 antennas of the VLA; as noted, the antennas can be moved to various spacings along the three arms. The array, with its maximum-size configuration, yields a resolution of 0.3 arcsec at wavelength 6 cm, while the other three standard array configurations are successively smaller in scale by a factor of approximately 3. The Y shape of the array was chosen to cover the u, v-plane as uniformly as possible for sources between the north pole and declination $-30°$.

Figure 6.5 shows the instantaneous coverage, or 'support' of the VLA in the u, v-plane for a source at the zenith, i.e. at declination $\delta = 30°$; this coverage is sufficient for useful maps to be made with even a short 'snapshot' observation, taking no advantage of Earth rotation.

The MERLIN array has only six elements, which are spread over distances up to 200 km. Such a sparsely filled network can produce maps only by Earth-rotation synthesis over an extended period, during which the rotation of the Earth changes the orientation of the interferometer baselines; the support then becomes a series of arcs in the u, v-plane. The advantage Earth-rotation gives in covering the u, v-plane is seen for the VLA in Figure 6.6, which plots the support for an 8-hour observation at declination $0°$; comparison with Figure 6.5 shows an almost completely filled u, v-plane (with a hole at the centre corresponding to the unrecorded total-power components).

As the Earth rotates, the baseline of an interferometer pair, projected onto the u, v-plane, traces out an elliptical locus. The 15 baselines of the MERLIN array, over an observing period of 12 hours, trace out the spectral sensitivity functions shown in Figure 6.7. When the source is at the celestial equator, the ellipses degenerate to straight lines; they become progressively more circular (thus generating a more circularly symmetrical response on the sky) as one progresses towards high declinations. The broadening of the lines due to the physical extent of the elements is usually negligible, but the lines may have appreciable width due to the frequency bandwidth of the receiver; at frequency ν a bandwidth $\delta\nu$ broadens the tracks radially by a fraction $\delta\nu/\nu$. A very wide bandwidth may risk smearing the required $V(u, v)$ pattern; this is analogous to chromatic aberration in an optical telescope.

The u, v-plane is necessarily even more sparsely filled in VLBI, where the elements may be some thousands of kilometres apart, and one or more may even be on a spacecraft at a distance of some tens of thousands of kilometres. These sparsely filled arrays inherently produce maps of the brightness distribution across sources that are seriously affected by the missing Fourier components; this is a problem for the process of image synthesis, which we describe in Section 6.5.

The sensitivity of these 'unfilled' interferometer arrays to the flux density of point-like sources is simply related to the total collecting area and the system noise temperature, as for a single-aperture radio telescope. However, for the surface brightness of extended sources there is a dilution factor, noted in Section 5.9, and the maps will have a sensitivity limit in brightness temperature T_b. A typical observation with MERLIN, for example, at 5 GHz may be limited to $T_b > 100$ K in a single resolution area; this is of course sufficient to detect many extended sources such as a typical H II region (Chapter 9). In a very dilute array, as is often the case in VLBI, the sensitivity limit in a single pixel may be 10^6 K or more. When the array is sparser, the coverage is naturally less complete. The seven telescopes of the MERLIN array give 21 baselines, and Figure 6.7 shows the resulting transfer function for $\delta = 60°$ (near the zenith) and $\delta = 0°$, the celestial equator. When using data with this sparse sampling, it is essential that the reduction procedures and limitations be understood.

The maps of the double quasar B0957+561, shown in Figure 6.8, show the effects of incomplete u, v-plane coverage in dramatic fashion. MERLIN obtains higher resolution at the expense of losing small spacings in the u, v-plane, showing the structure in greater detail but losing the extended emission seen in the VLA image.

In Section 6.2.1, it was pointed out that the Fourier transform of a given transfer function, $W(u, v)$, gives the response to a point source. This pattern, which we designate $b_o(x, y)$, is generally known as the synthesis beam or the dirty beam; it is the map that would be generated by the computer that performs the Fourier transform of the visibilities from observing a point source such as an unresolved quasar. Note that this is not the same as

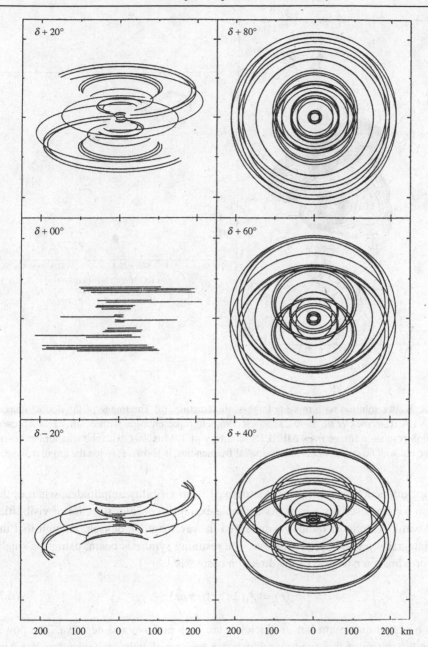

Figure 6.7. The MERLIN interferometer array using six telescopes; the *u, v*-plane coverage for an 8-hour tracking observation at six declinations.

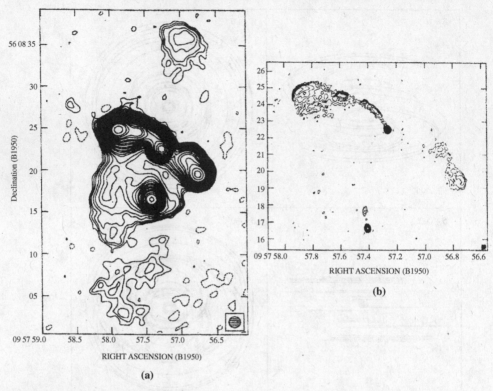

(a)

Figure 6.8. High resolution with missing large-scale structure. (a) The image of the double quasar, from the VLA (Harvanek *et al.* 1997), shows a complete representation with beam size 1.4 arcsec. (b) The high-resolution image from MERLIN (courtesy of T. Muxlow), with the smaller beam size 0.25 arcsec but with no components at low spatial frequencies, is shown at twice the angular scale.

a telescope diffraction pattern, since it is derived from visibility amplitudes, whereas the diffraction pattern describes power. As a simple example, consider the case of visibilities sampled within a circle of radius ρ (measured in wavelengths), densely enough that the Fourier integral is a good approximation. The resulting synthesis beam, using the small-angle approximation r instead of the direction cosine, is

$$b_0(r) = J_1(2\pi\rho r)/(\pi\rho r). \tag{6.7}$$

This has a central maximum but its sidelobes are both positive and negative. The power pattern is the square of this function, which can be immediately recognized as the Airy function, the diffraction pattern for a uniform circular aperture. The synthesis beam plays the role of a diffraction pattern, but with several distinctions; it is a computed function residing in the computer, and it must have both positive and negative values (since the total-power terms are neglected, its integral over the sky must be zero). The term 'dirty beam' derives from the unpleasant-looking character of its sidelobe pattern.

6.3 Calibration

The true visibilities V_{ij} are related to the observed visibilities $V_{obs}(i, j)$ by

$$V_{obs}(i, j) = F_i F_j^* F_{ij} V_{ij}, \tag{6.8}$$

where the complex quantities F_i and F_j are the complex voltage-gain factors for each antenna, including the atmospheric transmission path, and G_{ij} is a baseline-dependent gain, which may be characteristic of the correlator. These factors are determined by observations of a calibration source sufficiently frequently to keep track of any variations. A suitable calibration source should be effectively point-like, and close to the area or object to be mapped.

Two classes of calibration source are needed: phase calibrators and flux calibrators. The phase calibrators are chosen to be close to the target region, so that frequent reference can be made to them. The flux calibrators are a smaller selection of sources with accurately known fluxes that at worst vary only slowly with time. Ideally, the flux calibrators should be reasonably compact, so that they are not resolved by the interferometer; unfortunately, this is frequently not the case. The radio source 3C 286, for example, is a commonly used flux standard, but it is partially resolved both by the VLA and by MERLIN. Ultimately, the flux standards rely upon absolute measurements of strong but less compact sources such as M87, Cygnus A or the Orion Nebula. The most intense source at low frequencies, Cassiopeia A, decreases in flux by about 1% per year. The careful and well-documented measurements of Baars *et al.* (1977) are currently used as the ultimate flux standards. Millimetre-wave arrays can use planets as flux standards, although they, too, may be partly resolved at large interferometer spacings.

The resulting observations consist of two classes of fringe visibilities V_{ij} for each baseline, the sources to be studied and the calibration sources. Note that, to simplify this discussion, we are treating the problem as if only one frequency and one polarization were being observed. Real observations are usually taken for all four Stokes parameters, and for at least two independent frequency bands, so the set of visibilities should read $\{V_{ij}^s\}_v$ where the superscript s indicates the Stokes parameter and v describes the particular band. The primary integration time, the instrumental integration time, is a short length of time that can usually be set by the observer. This time interval, t_o, should be short enough that the errors in V_{ij} do not accumulate significantly, but it should also be as long as feasible in order to limit the size of the data set. The total observation set, for an observation lasting time T, is the set of all fringe visibilities sampled T/t_o times. The aim is to arrive at a set of calibrated fringe visibilities, whose Fourier transform is an approximation to the brightness distribution on the sky.

6.4 Reducing the data

The practical steps by which a data set is reduced and prepared for the mapping process are referred to as flagging, gridding and calibration. The observed raw fringe visibilities are first examined to identify any baselines that are obviously bad, and 'flagged' so that the reduction program can ignore them. The data must then be transferred from the format in which the array records the visibilities to a uniform rectangular grid format that is easier to treat mathematically, since a fast Fourier transform (FFT) can then be implemented easily, with great saving in computing time. Calibration involves the analysis of a large data set obtained from the calibration source; this may indicate that further flagging of contaminated data is necessary, followed by a further cycle of analysis.

The data set from the calibrator is now used to provide, baseline by baseline, the correction factors in Equation (6.8) to the visibilities. There is redundant information in the total data set, and the solutions for amplitudes and gains of the individual signal paths can be optimized by using *closure relations*, which we now describe.

An array of N radio telescopes can form $N(N-1)/2$ baselines, and any group of three baselines provides information about the phases introduced at each telescope. This is possible because of the *phase-closure* property. Consider the set of fringes from a group of three antennas, designated i, j and k, observing a reference point source. Given the fringe phase of any pair, ϕ_{ij}, one can form the sum of the fringe phases for the three baselines:

$$\phi_{ijk} = \phi_{ij} + \phi_{jk} + \phi_{ki}. \tag{6.9}$$

This quantity, the *closure phase* ϕ_{ijk}, is independent of phase shifts caused by the instruments and the atmosphere, provided that they are constant during the time of integration. Values of ϕ_{ijk} from several groups of antennas can then be combined algebraically to extract phase corrections for individual antennas, although only relative to the phase correction of a single chosen antenna.

There is also a closure relation for the fringe amplitudes for any group of four radio telescopes,

$$A_{ijkl} = \frac{|V_{ij}||V_{kl}|}{|V_{ik}||V_{jl}|}, \tag{6.10}$$

and the *closure amplitude* A_{ijkl} is also independent of the atmospheric and instrumental effects. This may be used to find the relative gains of the individual antennas, again providing further information in the construction of the model and its transform.

As the number of radio telescopes in the array increases, the fraction of the recovered phase and amplitude information increases. If these fractions are f_ϕ and f_A their N-dependences are

$$f_\phi = \frac{N-2}{N}, \tag{6.11}$$

$$f_A = \frac{N-3}{N-1}. \tag{6.12}$$

The current reduction software packages make use of these closure relations, and maps of high quality can be obtained when the u, v-plane coverage is sufficiently complete. An extensive discussion of the methods can be found in Zensus, Diamond and Napier (1995).

The cyclical process in finding the calibration factors F_i and F_j involves intensive computation, and furthermore must be repeated at frequent intervals. The process is, however, fully automated in a suite of programs known as the Astronomical Image Processing System (AIPS) introduced and maintained by the NRAO. The system basics are described in the AIPS Cookbook, available on the Web at http://www.aoc.nrao.edu/aips/cook.html. There are other data-reduction systems, including those for MERLIN and the Australian Telescope, also available on the Web but they generally follow the same sequence of steps. If the source being studied is simple, or if only a crude map is required, it is often possible to use a pipeline system in which the steps are carried out automatically; intervention by an experienced observer may nevertheless produce the best results.

6.5 Producing a map

The results of the procedure above, which may have been iterated a number of times to obtain an optimum calibration of the data, is a solution table in which the gain solutions reside. A raw Fourier transform, starting from the recorded visibilities, now produces a 'dirty map', which is a convolution of the true source brightness distribution with the 'dirty beam'.

Without further correction, the point-spread function of a map synthesized by a single transformation of the gridded data, even with weighting, still contains large sidelobes, often in the form of rings and radial lines. This so-called *dirty beam* is the transform of the irregular and patchy distribution of available components in the u, v-plane. The precise form of the dirty beam is, however, known from that distribution, and a correction can be made for it. If a map contained a single prominent point-like source, the first stage would be to subtract from the map a 'dirty beam', or point-spread function, with suitable amplitude and centred on that point source. The prominent sidelobes then disappear, revealing any smaller sources that may have been obscured. If another dominant source now appears, the same process can be repeated, and so on until no further sources can be distinguished. (In practice, this simple procedure is not possible because the process is non-linear, and the ever-present noise components will prevent convergence. Instead, only a small percentage of the peak is subtracted, and the process is iterated thousands of times, until the residuals look like noise.) The map can now be reconstructed by adding all the individual components indentified in the map, using an idealized point-spread function, the so-called *clean beam*.

This process, known as CLEAN and invented by Högbom (1974), is useful over a much wider range of mapping than might be expected from the above description. Even if there are no unresolved sources in the beam, it still proves to be useful to identify successively the peaks in the dirty map and to remove a scaled-down dirty beam sequentially at these peaks. An example of a complex map, showing both the dirty map and the CLEANed version, is shown in Figure 6.9.

Under some circumstances there may be no reliable phase information in the correlator outputs, and it is useful to consider what can be done without any at all. A transform using

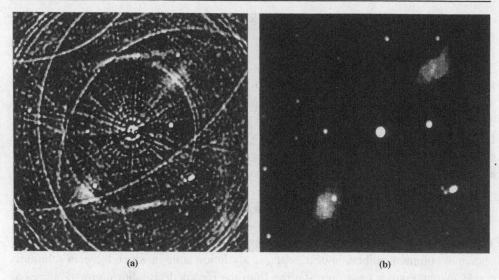

(a)   (b)

Figure 6.9. The CLEAN process: an image of a radio source made by an interferometric array, before (a) and after (b) the process. In the untreated image all sources are surrounded by the sidelobes of a point-spread function. (Courtesy of Jan E. Noordam, NFRA.)

the modulus only of each component may produce a recognizable map, especially if the map is dominated by a small number of point sources; in this case the process gives a central bright spot with the real sources appearing twice, both in their correct places and as mirror images. If there is any prior information about the correct map, such as a rough position of one or two of the brightest sources, this may be used to construct a model map. This model map is then transformed back to give a revised set of visibilities, including a set of phases that provides a first approximation to the missing phase information for each interferometer pair. These phases are used, with the original amplitudes, in a second stage of processing to give an improved map. This procedure, introduced by Baldwin and Warner (1978) and by Readhead and Wilkinson (1978), is known as *hybrid mapping*, the hybrid being the combination of the original amplitude data with phase data obtained from a model.

Hybrid mapping is often used as part of the image-reconstruction process even when most of the phase information is available. The two processes, CLEAN and hybrid mapping, are combined in a cycle shown in Figure 6.10. The combined process may be repeated many times, and inspection of the final result may suggest a repeat of the whole cycle after re-examination of the calibration of input data in some frequency channels or from some particular antennas. Efficient and fast computer programs are therefore essential. The subtraction involved at each stage of CLEAN may be carried out either on the output map itself, or on the grid of Fourier components (the u, v data). The second process, known as the Cotton–Schwab algorithm, proves to be more efficient and is incorporated in most imaging programs.

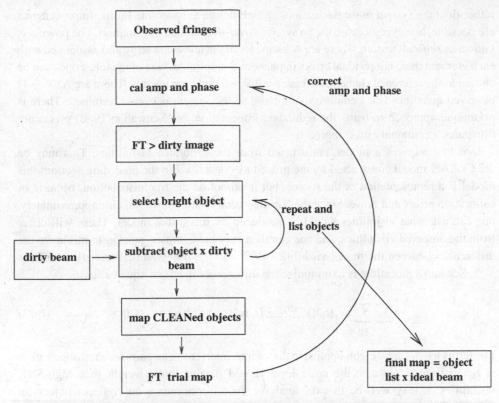

Figure 6.10. The cyclical mapping process, involving CLEAN and hybrid mapping.

Another process for improving a dirty map is known as the maximum-entropy method (MEM), in which one searches for the most probable solution to the determination of the true source structure. Its effect is to force all parts of a map to give a positive brightness temperature and to compress the range of pixel values. It may give a more reliable representation of smooth brightness distributions, while CLEAN is naturally more appropriate for a distribution of point sources. The two processes can, of course, be used successively for one image or for a specific part of a large image; the MX routine, for example, takes specified boxes within the field of view, and performs a routine similar to CLEAN. Neither CLEAN nor MEM has so far been justified by rigorous theory, but there have been many tests of their reliability using artificial data, from which they generally emerge with credit. CLEAN is the more generally used, although MEM is preferred for very large data sets because it is more economical in computer power.

6.6 Self-calibration

After the procedures outlined above, acceptable maps may be obtained but, for the highest contrast ratios, the residual phase and amplitude errors will still be the limiting factor, rather than the system noise fluctuations, which define the absolute limits. Improvements are possible, however, because the array gives over-determined information. The process is known as *self-calibration*. There are N elements in the telescope array and associated with each element there are residual errors in phase and amplitude. One of the telescopes can be chosen as the reference element, so there are $2N - 2$ error parameters. There are $N(N - 1)$ observed quantities (the complex visibilities), so the system is over-determined. There is no unique approach to using the redundant information, but Schwab's (1980) procedure illustrates a commonly used approach.

One first requires a model, constructed from the calibrated visibilities. This may be the CLEAN model constructed by the procedures described in the preceding section; this model is a representation of the source, but it cannot be the true distribution, because of calibration errors and noise. Since the baseline vectors are known, at least approximately, one can ask what visibilities would be produced by this initial model. These will differ from the observed visibilities, and one can then vary the complex gains g_i to minimize the differences between the model visibilities V_{ij}^{mod} and the observed (calibrated) visibilities V_{ij}^{cal}. Schwab's procedure is to minimize the differences in a least-squares sense:

$$\sum_k \sum_{i,j} w_{ij}(t_k) |V_{ij}^{\text{cal}} - g_i(t_k)g_j^*(t_k)V_{ij}^{\text{mod}}(t_k)|^2 \longrightarrow 0. \qquad (6.13)$$

The terms $w_{ij}(t_k)$ are weights applied to the various interferometer pairs for each observation at t_k. Visibilities close to the noise level should receive lower weight than high-SNR visibilities; it may well be that the analysis should concentrate on compact objects in the map that dictate the exclusive use of high-spatial-frequency interferometer pairs. The corrections to the complex gains can then be applied to the initial data set, to give a better approximation to the corrected visibilities, and the mapping procedure is then applied to the new data set.

For a more extended discussion, see the chapter by Cornwell and Fomalont in the Perley, Schwab and Bridle compendium (1989). They point out that no convergence proofs have been given for self-calibration (it is a non-linear procedure, carried out on a data set that has already been subjected to a non-linear procedure), but successive iterations seem to converge in practice, at least asymptotically. Empirically, the procedure works best when only the phases of the complex gains are varied in the initial iterations; when there seems to be strong convergence, the amplitudes can then be varied.

6.7 Frequency diversity

The limited coverage of the u, v-plane in unfilled interferometer arrays, especially for VLBI, can be improved by observing on adjacent frequencies and adding the maps after independent analyses. This procedure, which is of course applicable only to sources with a continuum spectrum, spreads each point on the u, v-plane radially, and can make a useful

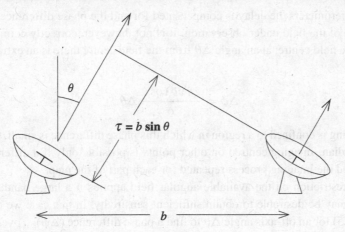

Figure 6.11. The delay τ in a simple interferometer.

improvement if a frequency range of, say, 10% is available in the receiving equipment. For some arrays this extension of coverage is sufficient to fill most or all of the gaps in coverage of the u, v-plane. However, extending the frequency bands outside the allocations for radio astronomy carries the risk of suffering from interfering radio transmissions; in practice this may limit the use of frequency diversity for central frequencies below 1 GHz, but the technique has been found useful for higher-frequency bands. Note that frequency diversity involves independent observation and analysis on adjacent frequency channels; it differs from the use of a wide bandwidth for a single channel, whose effects we now analyse.

6.8 Wide fields and wide bandwidths

Although aperture synthesis, and especially VLBI synthesis, provides maps of the radio emission from individual sources with an angular resolution that easily matches that of the best optical images, the complexity of the imaging process and the large volume of data handling involved in the analysis have usually confined the radio images to individual objects. Improvements in sensitivity and in computer speed now allow images to be made of wider fields of view; for example detailed maps have been made of a large part of the star-burst galaxy M82, and of the radio counterparts of the visible galaxies in the Hubble Deep Field (Chapter 13). These maps are, of course, restricted to the area of sky within the beamwidth of the individual antennas of the array. The use of wide bandwidths in wide-field imaging has a fundamental constraint, which we now examine.

In a simple interferometer (Figure 6.11) with baseline b observing at angle θ the delay in the longer path is τ and the phase difference ϕ is

$$\phi = 2\pi\nu\tau = \frac{2\pi\nu b \sin\theta}{c}. \tag{6.14}$$

In radio interferometers the delay is compensated for and the phase difference is removed for the centre of the field under observation. It is not, however, correctly compensated for away from the field centre; at an angle $\Delta\theta$ from the field centre there is an extra phase $\Delta\phi$, where

$$\Delta\phi = \frac{2\pi \nu b \cos\theta}{c} \Delta\theta. \qquad (6.15)$$

Correct imaging is confined to a region in which this phase difference is small, for example less than a radian. Imaging centred on other points is possible only if a different delay is introduced and the imaging process repeated for each part of the map.

A similar restriction on the available angular field applies if a large bandwidth $\Delta\nu$ is used, which may be desirable to obtain sufficient sensitivity. In this case we differentiate Equation (6.15) for an off-axis angle $\Delta\theta$ to find a phase difference $(\Delta\phi)_{\Delta\nu}$ over bandwidth $\Delta\nu$:

$$(\Delta\phi)_{\Delta\nu} = \frac{2\pi b \cos\theta}{c} \Delta\theta \, \Delta\nu. \qquad (6.16)$$

The angular resolution, i.e. the width of the synthesized beam θ_{beam}, is $c/(\nu b \cos\theta)$, so for the phase difference over the field $\Delta\theta$ to be small we require

$$\frac{\Delta\nu}{\nu} \ll \frac{\theta_{\text{beam}}}{\Delta\theta}. \qquad (6.17)$$

This can be a severe limitation; for example, if a field of view one arcsecond across is to be observed with a beamwidth of three milliarcseconds, the relative bandwidth $\Delta\nu/\nu$ must be less than 0.003. At the widely used frequency of 1660 MHz this restricts the bandwidth to 5 MHz, whereas observing techniques commonly allow the use of bandwidths of 200 MHz or more. Since the sensitivity is proportional to $(\Delta\nu)^{1/2}$, this entails a severe loss of sensitivity. The consequence of this limitation is clear: the receiver must be divided into many independent adjacent bands, each with its own independent correlator system, and each analysed independently.

At low frequencies the antenna beamwidth may be very large, as in LOFAR (see Chapter 17); in this case the synthesis process is considerably complicated by the need to extend the geometry of the Fourier transform to spherical coordinates rather than the simple use of a planar approximation. A particular problem of wide-angle observation is that intense radio sources (such as the Sun and Cas A) may produce large signals even if they are far from the area under analysis. However, individual restricted areas can still be analysed by the simple process, and it may be necessary as an initial stage to analyse selected areas in which there are intense sources. The removal of such sources (a process known as *peeling*) from the observed interferometer outputs should then reduce the problem.

In Chapter 15, which treats the CMB, which covers the whole sky, the ultimate challenge in wide-field imaging must be met. This is accomplished by expressing the brightness as an expansion in spherical harmonics, the spherical analogue of the planar Fourier representation. The program SEALfit accomplishes this task for the CMB, but it can also be applied to a large but limited area of sky when the small-angle approximation is inappropriate.

6.9 Wide fields: mosaicing

Making an image of a wide field often requires separate observation of a number of adjacent fields, whose images must then be merged to give a large image in which the geometrical coordinates are continuous and the calibration and other characteristics of the image components are nearly identical across the boundaries. The problems of making such a mosaic are of course familiar in many astronomical observations, but they are particularly important in radio-synthesis observations because of several effects of wide-field imaging.

Apart from the geometrical effects of wide-field imaging, in which the analysis usually assumes a two-dimensional rectangular coordinate system as an approximation for small angular distances from the beam centre, the field covered in a single pointing may be restricted to considerably less than the beamwidth by the bandwidth effect described in Section 6.8 above. There may also be a considerable dependence of polarization calibrations on position within the beamwidth.

It may therefore be necessary to analyse the observed data from a single pointing in several areas, each smaller than the beamwidth, using different phase centres and different calibrations. These must then be merged together as a mosaic, with careful tests of consistency at the borders. Planning a survey of a considerable area of sky may involve a choice between multiple pointings and a more complex analysis procedure in which a single wide beam is analysed in a mosaic of phase centres. There are many examples of observations involving both types of mosaicing: we mention here an observation of a selected area of sky, 4 square degrees, for compiling a catalogue of sources at the comparatively low frequency of 610 MHz (Garn *et al.* 2007). This survey used the GMRT synthesis telescope, with 27 telescopes forming 351 baselines. Each telescope pointing was analysed with 31 different phase centres, each giving a separate facet within the main beam, and an array of 19 separate pointings was used to cover the required area of sky. It is instructive to read the considerations which led to the observing strategy adopted, and the techniques involved in constructing the final mosaic image.

6.10 Signal-to-noise limitations and dynamic range

The previous sections have dealt with the mechanics of how data from an interferometric array can be used to construct radio source maps. The maps must necessarily be corrupted by the presence of uncorrelated noise, and the set of fringe visibilities V_{ij} will be corrupted to some extent by measurement errors, even when the data are calibrated carefully and self-calibration has been used to minimize the instrumental errors. Furthermore, most arrays are incompletely filled, and the complete set of observed visibilities will yield an incomplete sampling of the Fourier transform of the brightness distribution across the field of view.

The effects of noise on source detectability will be considered first. Each observed visibility will be a sum of the source visibility plus a noise signal, usually exhibiting Gaussian statistics. This thermal noise component, having r.m.s. value $\sigma = \delta S$ (measured in flux per beamwidth) determines the minimal detectable flux in a map having a given field of view. If one takes as an example a map of 1000×1000 resolution elements, it is obvious that an apparent source whose flux is only 3σ above the noise has a good chance of being spurious. In a Gaussian sample, there is about one chance in 1000 that a 3σ fluctuation will occur, and this would indeed be the chance of error if one were looking for a source at a given position. Searching for random sources, however, is an entirely different matter; in the given field of a million resolution elements there will be a thousand 3σ random fluctuations, and there will even be a few 5σ fluctuations. Searches for sources, therefore, must choose a significance level that is appropriate to the a-priori probability of having a chance fluctuation. As a matter of practice, unless the statistical properties of the data set and the a-priori hypotheses have been carefully defined, 10σ detections are seldom false, 5σ detections are not necessarily convincing, and 3σ detections are often spurious, because of the large number of a-priori possibilities that may be possible and unaccounted-for. One should always remember that once one has looked at a given data set, it is no longer a random sample.

We next consider the effects of errors and difficulties that are inherent in the aperture-synthesis process. A set of cross-correlation observations gives only an approximation to the actual values of the Fourier transform for three reasons: firstly, as noted in Section 6.4, there will be gaps in the coverage of the u, v-plane; secondly, the measured visibilities will have residual errors in both amplitude and phase, despite the most careful calibration; thirdly, there is always a noise signal present that corrupts the observations. The concept of an incompletely filled array was introduced in Section 6.2, where the support domain, determined by the spectral sensitivity function $W(u, v)$, determines how well the brightness distribution of a given source will be reconstructed by observations with a given array. The self-calibration procedure tries to minimize the errors, using the redundancy in the observations, but there is necessarily a remaining noise signal in the derived map (see Kulkarni 1989). We can thus regard each observed visibility as a sum of two terms, the true source visibility multiplied by an unknown gain factor, which we represent by the complex number g_{ij}, and a noise term N_{ij}:

$$V_{ij}^{\text{obs}} = g_{ij} V_{ij}^{\text{true}} + N_{ij}. \tag{6.18}$$

The map that results from taking the Fourier transform of these visibilities, therefore, can be expressed in the following form:

$$B(x, y) = B^{\text{true}} + B^{\text{sp}} + N(x, y), \tag{6.19}$$

where B^{true} is the actual brightness distribution, B^{sp} is the sum of the spurious responses that result from incomplete sampling of the u, v-plane, and $N(x, y)$ represents the noise background generated by the system noise. The precautions that must be taken to avoid mistaking a noise maximum for a signal were discussed above; the implications of the spurious-response term will now be considered.

A completely filled array of diameter D illustrates the spurious-response problem in a particularly simple way. All of the values of the projected baseline vector **b** with components (u, v) and magnitude $|k|$ are sampled up to a maximum value $|k| = D/\lambda$. The source may have structure with small angular size, giving components at larger k, but these components are undetected because they are not within the support domain (in Bracewell's nomenclature, these are *invisible solutions*). In principle, these might be recovered by analytic continuation, because $\mathbf{b}(u, v)$ is an analytic function, but the various procedures to do this, under the name of *superresolution*, are of limited usefulness because of the unknown noise signal that is always present. If there are good reasons to suppose that a structure is simple (e.g. two close point sources), the resolution limit may be better than Rayleigh's criterion (half the half-power beamwidth), but if the structure hypothesis is wrong, the map constructed by superresolution can be wildly wrong.

The effect of the incomplete support of an aperture-synthesis array can now be understood. If the array has a total extent D, but only a fraction α is occupied by the elements of the array, the effective area of the array is αD^2 and an effective area $(1 - \alpha)D^2$ has a response spread over the sky in the form of sidelobes. The form of the sidelobes will depend upon the geometry of the unfilled area (in fact, they will have a configuration proportional to the Fourier transform of the unfilled area). Regularities in the unfilled area can lead to concentrated sidelobes; the WSRT, which is based on a regularly spaced array of ten telescopes (plus four outriggers) has grating sidelobes that show up prominently in the 'dirty maps' (Figure 6.9). If there is a point source in the field, these grating sidelobes are reduced when the CLEAN routine (or its equivalent) is used to subtract the point source. The subtraction is never perfect, because the gain correction factors g_{ij} are never exactly unity, so there is always a residual set of sidelobes present. These 'error sidelobes', being unpredictable, increase the noise background in the resulting map.

When there are many small holes in the support (as in the case for the 'snapshot' VLA pattern shown in Figure 6.5) the sidelobe pattern is effectively that of a large set of uncorrelated oscillators, and this leads to a sidelobe background of fluctuations of angular scale λ/D, spread randomly over the sky, with approximately Gaussian statistics. As in the case of the WSRT, these will be reduced by CLEAN, but the errors g_{ij} will always lead to a residual background that can never be removed. When there is a strong source in the field, the limitations of the sidelobes induced by array errors, as opposed to the spurious signals of the random noise background $N(ij)$, determine the effective SNR. This is known as the *dynamic range* of the map, and is a critical quantity when high-quality maps are desired.

Knowing the properties of the Fourier transform can be useful in diagnosing problems in aperture-synthesis maps. The 'dirty map' shows the normal sidelobe structures, but, if there are gain problems with individual elements of the array (g_{ij} much greater or much less than unity), the offending element can be identified, and 'flagged out' of the array. For

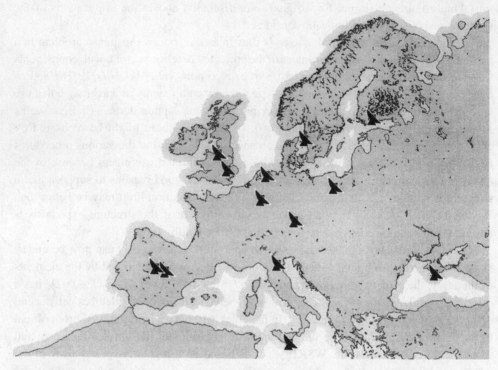

Figure 6.12. The European Network of radio telescopes available for VLBI observations.

radio maps of the highest quality, especially when dynamic range is a crucial consideration, scrutiny of the data from each element is essential. For the VLA, with its large number of elements, one can be quite ruthless in removing dubious array elements from the data, and still retain a large number of visibilities, quite sufficient for the task.

6.11 VLBI arrays

The principles and complexities of long-baseline interferometry were described in Section 5.6. There are now several VLBI systems in the world that regularly produce synthesized images of radio sources. Their impact has been significant in a number of fields, as demonstrated by the descriptions of the maser phenomena in Chapter 11, and of quasar and AGN structure in Chapter 13. There are currently two principal VLBI arrays in the world. The only dedicated array, the Very Long Baseline Array (VLBA) of the USA consists of ten identical radio telescopes, extending at northern latitudes from New Hampshire to the state of Washington, and at southern latitudes from St Croix in the Virgin Islands to the island of Hawaii. The European VLB Network (EVN, Figure 6.12) consists of a cooperative arrangement of radio telescopes in the UK, Netherlands, Germany, Italy,

Poland, Russia, Ukraine, China and Japan. The EVN and the VLBA cooperate to form a world-wide network; in addition, the Australian Telescope is partly a VLBI array and cooperates with the other VLBI arrays of the world. Radio telescopes that have been used for VLBI, either in ad-hoc combinations or as part of an established array, are numerous and geographically widespread.

The smaller synthesis arrays, such as MERLIN, are connected by fibre-optic cables, and some of the baselines of the EVN have been connected to operate in this way. However, with such a wide dispersion of array elements, the signals from the separate telescopes often cannot be connected directly, and are instead recorded on broad-band digital tapes or discs for subsequent correlation. The local oscillators at the separate receivers are not linked in phase, and the relative phases of the signals are therefore unknown. Also, at the large spacings the relative phases introduced by the transmission paths through the atmosphere and ionosphere will differ and will vary in an unknown way. The relative phases must therefore be discovered as part of the synthesis and imaging process. Although the imaging techniques which we described in Section 6.5 do very largely overcome these difficulties, it is important to bear in mind the fundamental limitations on the accuracy and even the reliability of synthesis maps which are imposed by these parameters. In particular, a single interferometric pair observing over a period of time might not produce definitive information on the brightness distribution even when, with Earth rotation, it traces out a significant part of the u, v-plane. It may, however, be useful for modelling purposes even without phase calibration; for example, the visibility amplitude alone can set a limit on the size of an isolated source.

6.12 Space VLBI

In Section 5.6 it was pointed out that a VLBI station need not be on Earth; a radio telescope in space, with the VLBI electronics of Figure 5.7, can be an interferometer element. The stable frequency standard can be in space, or it can be at a ground station, with the reference frequency being relayed to the satellite, where a secondary oscillator is locked to the reference. Its signal is then returned to the ground, where a comparison with the original standard allows corrections to be made. The astronomical signal is relayed to the ground by a separate radio link and recorded as in a standard VLBI system. The baseline is changing rapidly with time, which complicates the reduction process but also brings a major advantage; the spacecraft sweeps out a wide range of baselines in the u, v-plane so that with only one or a small set of ground radio telescopes a reasonably complete synthesis is possible.

The first space VLBI experiment used a NASA TDRSS satellite as the space radio telescope; its results are described in Section 13.9. The system description was published by Levy *et al.* (1986). A dedicated spacecraft, VSOP/HALCA, was launched by Japan's Institute for Space and Astronautical Sciences in 1997. The system and the first results are collected in Hirabayashi *et al.* (2000).

6.13 Aperture synthesis at millimetre wavelengths

The mapping of radio sources at millimetre and sub-millimetre wavelengths requires special techniques. Radio telescopes for such short wavelengths are necessarily small, with an effective area considerably smaller than that of the largest radio telescopes in use today. Lower sensitivity need not necessarily be a severe disadvantage, since there are many spectral-line sources, including some very bright masers, to be observed at millimetre wavelengths. Despite the severe technical problems, an international consortium is constructing a very large synthesis array (ALMA) for millimetre wavelengths (see Chapter 17).

There are several existing millimetre-wavelength arrays with small numbers of elements. At present, the BIMA array at Hat Creek in California is composed of nine 6-m telescopes, the OVRO array in the Owens Valley of California has six 10-m telescopes, the IRAM array on the Plateau de Bure in France has four 15-m telescopes, and the Nobeyama array in Japan has six 10-m telescopes. These arrays have regularly achieved angular resolutions of about 1 arcsec at 100 GHz (wavelength 3 mm). The 15-m James Clerk Maxwell and 10-m CalTech telescopes on Mauna Kea, Hawaii, have been used as a two-element interferometer with greater angular resolution, but a single baseline has limited usefulness.

Propagation effects, especially the effect of atmospheric water vapour, present severe problems hindering achievement of the necessary phase stability in large millimetre-wavelength synthesis arrays. At a wavelength of 1 mm the fluctuations in path length must be followed to an accuracy of 0.05 mm, whereas the actual path for millimetre-wavelength radio waves through an atmosphere containing 1 mm of precipitable water vapour is 6.5 mm. Even at very dry mountain sites it is usual to find a rapidly varying water-vapour content of several millimetres of precipitable water vapour, so large phase corrections are needed at frequent intervals. Calibration sources within the primary beam appear to be essential: the alternative would be to nod the whole array to and from a calibration source at intervals of only a few seconds.

Another approach is to measure the water-vapour content over each element of the array by observing the strength of the atmospheric emission line at 22 GHz. Provided that this line radiation is optically thin, the intensity is directly proportional to the total water-vapour content, and can be used to correct the phase of each element. It is not yet known how accurately this can be done.

The rewards for a successful millimetre array, with effective beamwidth of about 0.1 arcsec, are substantial. The distribution of molecular gas in our own and other galaxies is complex, and, even in our own Galaxy, the study of molecular clouds with a single telescope gives only a general idea of the large-scale distribution. Meaningful studies of external galaxies require the angular resolution that interferometric arrays provide; single telescopes can detect the presence of molecular gas, but make only the crudest of maps, and those only for the nearest galaxies. Resolution of the order of an arcsecond is regularly achieved at present; the methods are generally those summarized in the preceding section, relying heavily upon fringe fitting and closure rules. The state of the art is advancing rapidly;

adequate frequency allocations have been made by international agreement (Chapter 17), and receiver techniques are continually being improved.

Further reading

Radiotelescopes. Christiansen W. N. and Högbom J. A., Cambridge, Cambridge University Press, 2nd edn, 1985.

Synthesis Imaging in Radio Astronomy II, ed. Taylor G. B., Carilli C. L. and Perley R. A., Washington, Astronomical Society of the Pacific Conference Series, No. 180, 1989.

Very Long Baseline Interferometry and the Very Long Baseline Array, ed. Zensus J. A., Diamond P. J. and Napier P. J., Washington, Astronomical Society of the Pacific Conference Series, No. 82, 1995.

7

Radiation, propagation and absorption of radio waves

In astrophysical contexts, the propagation of radio waves is governed, as for other parts of the electromagnetic spectrum, by the laws of radiative transfer and refraction. In radio astronomy, however, there is an emphasis on classical (non-quantized) radiative and refractive processes. Synchrotron radiation is the dominant radiation process at the longer wavelengths; spectral-line emission is observed mainly at shorter wavelengths. Maser action, the microwave equivalent of lasers, is encountered in several astrophysical contexts: this is due to the low energy of radio photons which can be significantly amplified by small population inversions in rotational and vibrational energy levels. Refraction is important in astrophysical plasmas; even though these are usually electrically neutral, protons have a negligible effect and the electron gas can have a significant effect on the velocity of radio waves. In the presence of a magnetic field, birefringence can lead to Faraday rotation of the plane of polarization.

In this chapter we set out the basic theories of radiative transfer, and outline the processes of radiation that are of particular importance in radio astronomy: free–free emission, line emission (and particularly maser emission) in dilute gas and synchrotron radiation. Free–free emission, or bremsstrahlung, is the main source in ionized hydrogen clouds, whereas synchrotron radiation is responsible for the background radiation in our Galaxy (Chapter 8) and is also practically universal in discrete radio sources from supernova remnants to quasars. The importance of masering action in astrophysical sources lies both in the very high brightness that allows their observation at large distances and in the narrowness of the amplified spectral lines, which provides unique information on the dynamics of such objects as outflowing stellar atmospheres (Chapter 11) and the gas clouds surrounding the central black hole of an active galactic nucleus (Chapter 13). Refraction and Faraday rotation are particularly important in propagation through the interstellar medium; here the polarized short pulses from pulsars act as a probe that can provide unique measurements of electron density and magnetic field strength (Chapter 8).

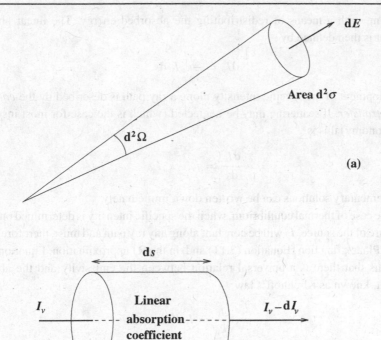

Figure 7.1. (a) Specific intensity. (b) Emissivity and absorption.

7.1 Radiative transfer

The *specific intensity* $I_\nu(\hat{\mathbf{n}})$ of radiation at a point in space, where energy dE is passing through an area $d\sigma$ into a solid angle $d\Omega$ (Figure 7.1(a)) is defined as

$$I_\nu(\hat{\mathbf{n}}) \equiv \frac{dE}{d\Omega\,\hat{\mathbf{n}} \cdot d\sigma\,d\nu\,dt} \tag{7.1}$$

with units of power per steradian, per unit area, per frequency interval. Note that I_ν refers to the value of the specific intensity anywhere along the ray path, whereas the brightness B_ν, introduced in Chapter 2, is reserved for the specific intensity at the observing telescope.

When there is matter along the ray path, energy can be absorbed or emitted, and the specific intensity will not be conserved. The *specific emissivity* j_ν (the symbol ϵ_ν is sometimes used) is the power emitted per unit volume, per frequency interval, per steradian, while the diminution of specific intensity along a ray path depends upon the *linear absorption coefficient* κ_ν. Figure 7.1(b) shows a ray path traversing a volume element with cross-section $d\sigma$ and length ds. The specific intensity entering the volume element, I_ν, may be reduced by absorption or scattering processes, and emerges with a value reduced by dI_ν. The change in specific intensity will be proportional to I_ν provided that the absorption process is a small perturbation of the absorbing system. This implies that the system must be in steady-state equilibrium, with a means of redistributing the absorbed energy. The linear absorption coefficient is then defined by

$$dI_\nu = -\kappa_\nu I_\nu \, ds. \tag{7.2}$$

The development of the specific intensity along a ray path is described by the *equation of radiative transfer*. If scattering may be neglected (which is the case for most instances in radio astronomy) this is

$$\frac{dI_\nu}{ds} = j_\nu - \kappa_\nu I_\nu. \tag{7.3}$$

Two elementary solutions can be written down immediately.

(i) In the case of thermal equilibrium, when the specific intensity is determined only by the temperature of the source, I_ν will be constant along any ray path and must, therefore, take the form of a Planck function (Equation (2.11) and, in the RJ approximation, Equation (2.13)). This means that there is a universal relation between the emissivity and the absorption coefficient, known as Kirchhoff's law:

$$\frac{j_\nu}{\kappa_\nu} = I_\nu(T). \tag{7.4}$$

(ii) A ray with specific intensity I_ν^0 at the origin has, at a position s along the ray path,

$$I_\nu(s) = I_\nu^0 + \int_0^s j_\nu(s') ds' \quad \text{(emission only)}, \tag{7.5}$$

$$I_\nu(s) = I_\nu^0 \exp\left(-\int_0^s \kappa_\nu(s') ds'\right) \quad \text{(absorption only)}. \tag{7.6}$$

The exponential term in Equation (7.6) occurs frequently in radiative transfer problems, and is called the optical depth τ_ν:

$$\tau_\nu(s) \equiv \int_0^s \kappa_\nu(s') ds'. \tag{7.7}$$

When τ_ν is large with strong attenuation, the medium is termed *optically thick*, while for small τ_ν, it is designated *optically thin*.

When a distant source is observed through an isothermal cloud with temperature T and optical depth τ, the observed brightness $I_\nu(s)$ is the sum of the attenuated source brightness I_ν^0 and the emission from the cloud:

$$I_\nu(s) = I_\nu^0 \exp(-\tau_\nu) + \int_0^{\tau(s)} \frac{j_\nu}{\kappa_\nu} d\tau. \tag{7.8}$$

where j_ν/κ_ν is the brightness of an optically thick cloud at temperature T. The source function inside the integral, when there is local thermodynamic equilibrium, is the Planck function $B_\nu(T)$.

7.2 Synchrotron radiation

The process by which electrons with large relativistic energy radiate when they are accelerated in a magnetic field first received attention when the first electron synchrotrons were built, where it was found to be an important mechanism of energy loss. The radiation has most commonly been called synchrotron radiation ever since, although the term magnetobremsstrahlung is sometimes used. In the rest frame of the electron, elementary gyroradiation is produced, and we present this treatment first. The remarkable (and intense) synchrotron radiation is then derived by transforming to the observer's frame by a Lorentz transformation.

The frequency of gyration of a non-relativistic electron circulating in a magnetic field B, known as the *cyclotron frequency*, ν_{cyc}, is given by

$$\nu_{cyc} = \frac{eB}{2\pi mc}. \tag{7.9}$$

Note that this is in Gaussian units (commonly used in astrophysics; 1 gauss = 10^{-4} Tesla), so

$$\nu_{cyc} = 2.80B \, \text{MHz}. \tag{7.10}$$

The cyclotron frequency is independent of pitch angle, i.e. the electron may have a velocity component along the magnetic-field direction.

When the energy $\gamma m_0 c^2$ of the electron is relativistic ($\gamma \gg 1$), the frequency of gyration is reduced, and becomes the gyrofrequency ν_g which is no longer independent of energy,

$$\nu_g = \nu_{cyc}/\gamma. \tag{7.11}$$

For electrons with a velocity component along the magnetic field, moving with pitch angle α, the radius R of the helical trajectory is, for relativistic electrons,

$$R = \frac{\gamma mc^2 \sin\alpha}{eB} \quad (\gamma \gg 1). \tag{7.12}$$

For a relativistic electron, with its energy E_{GeV} expressed in GeV,

$$R = \frac{1}{3} \times 10^5 \frac{E_{GeV}}{B} \, \text{m}. \tag{7.13}$$

The scale of the phenomenon can be appreciated by noting that, for an electron with an energy of 10 GeV and an interstellar magnetic field of 3 microgauss, its helical radius will be 10^{12} m, or about 7 a.u., which is large compared with terrestrial phenomena, but considerably less than a parsec. The gyrofrequency will be 4.2×10^{-4} Hz, but, as we will see, the bulk of the radiation will be produced at much higher frequencies.

A highly relativistic electron radiates in a narrow beam, with width of order $1/\gamma$, in the direction of motion. Figure 7.2 shows the geometry of this headlight radiation as the electron proceeds along its helical path. The observer receives a short pulse each time the beam crosses the line of sight. The time transformation from the accelerated system to

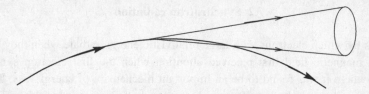

Figure 7.2. The radiation pattern of an electron moving with relativistic velocity. The radiation is concentrated in a beamwidth of approximately γ^{-1} radians.

the observer's frame contracts the sweeping time by $1/\gamma^2$, so the total time that elapses for the main radiation pattern to flit by the observer is of the order of

$$\delta t \approx \frac{1}{\gamma^3 \nu_g}. \tag{7.14}$$

This is a pulse that is very much shorter than the period ν_g^{-1}. If the magnetic field were absolutely uniform and the electron energy constant, the radiation would consist of periodic impulses, and a Fourier series would describe the spectrum as a sum of harmonics of the fundamental gyrofrequency. In real radio sources, the pulses are asynchronous, and the spectrum is the same as that of a single pulse. The spectrum is then concentrated at a characteristic frequency ν_0, which is the inverse of the pulse duration,

$$\nu_0 \approx \gamma^3 \nu_g \tag{7.15}$$

or, in terms of the (non-relativistic) cyclotron frequency ν_{cyc},

$$\nu_0 \approx \gamma^2 \nu_{cyc}. \tag{7.16}$$

The full calculation of the spectrum is well documented (see particularly the reviews by Ginzburg and Syrovatsky (1965 and 1969, the latter correcting several errors in the earlier literature) and the textbook treatment by Rybicki and Lightman). The spectrum is expressed in terms of a natural parameter, x, defined by

$$x \equiv \nu/\nu_{crit} \tag{7.17}$$

where

$$\nu_{crit} = \frac{3}{2} \gamma^3 \nu_g \sin \alpha = \frac{3}{2} \gamma^2 \nu_{cyc} \sin \alpha. \tag{7.18}$$

The angle α is the pitch angle of the electron trajectory; without the trigonometric term, ν_{crit} is known as the *critical frequency*.

It is often convenient to consider the synchrotron radiation from an electron with energy with relativistic factor γ, or energy E_{GeV}, to be concentrated at ν_{crit}. For a magnetic field $B \sin \alpha = B_\perp$ gauss, $\nu_{crit} = 4.2 B_\perp \gamma^2 \, \text{MHz} = 16 \times 10^6 B_\perp E_{GeV}^2$. For example, the electrons responsible for synchrotron radiation at $10 \, \text{MHz}$ from the Galaxy, where the field is a few microgauss (Chapter 8), have relativistic factors of order $\gamma \approx 10^3$, i.e. energies of order

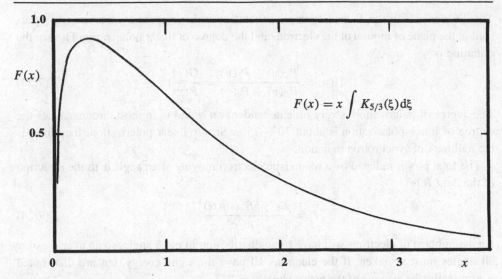

Figure 7.3. The function $F(x)$, which appears in the spectrum of synchrotron radiation.

1 GeV, while the radiation observed at 1 GHz is due to electrons with energy ten times higher.

The full analytical calculation gives a power spectrum of the radiation from a single electron in terms of the parameter x:

$$P(\nu)\mathrm{d}\nu = \sqrt{3}\frac{e^3 B \sin\alpha}{mc^2}F(x)\mathrm{d}\nu, \qquad (7.19)$$

where $F(x)$, which contains the shape of the spectrum, involves an integral of a modified Bessel function of order 5/3:

$$F(x) \equiv x \int_x^\infty K_{5/3}(\xi)\mathrm{d}\xi. \qquad (7.20)$$

The function F is plotted in Figure 7.3, showing the concentration of the synchrotron emission spectrum at a frequency of the order of γ^2 times the cyclotron frequency (i.e. γ^3 times the gyrofrequency).

Synchrotron radiation is polarized, and it is convenient to resolve the radiation into components perpendicular and parallel to the magnetic field. This involves another function, $G(x)$:

$$G(x) \equiv x \int_x^\infty K_{2/3}(\xi)\mathrm{d}\xi. \qquad (7.21)$$

The power spectra of the two polarized components are

$$P_{\perp,\parallel}(\nu) = \frac{\sqrt{3}}{2}\frac{e^3 B \sin\alpha}{mc^2}\begin{cases} F(x) + G(x) & (\perp B), \\ F(x) - G(x) & (\parallel B). \end{cases} \qquad (7.22)$$

The difference between $P_\perp$ and $P_\parallel$ implies that the synchrotron radiation is linearly polarized in the plane of motion of the electron, and the degree of linear polarization $\Pi(\nu)$ in the radiation is

$$\Pi(\nu) = \frac{P_\perp(\nu) - P_\parallel(\nu)}{P_\perp(\nu) + P_\parallel(\nu)} = \frac{G(x)}{F(x)}. \tag{7.23}$$

The degree of polarization is very little dependent on α and ν. In most circumstances the degree of linear polarization is about 70%–75%. Strong linear polarization, therefore, is the hallmark of synchrotron radiation.

The total power radiated by a relativistic electron moving at an angle α to the direction of the field B is

$$P = \frac{2}{3} \frac{e^4 \gamma^2 B^2 (\sin\alpha)^2}{m^2 c^3}. \tag{7.24}$$

An assemblage of electrons will have some distribution in pitch angle, so an average over all angles must be taken. If the electrons all have the same energy, but are distributed isotropically, the average of the factor $(\sin\alpha)^2 = 2/3$.

The loss of energy by the electron may be considered as a collision process, in which an electron with the classical Thomson cross-section σ_T encounters a magnetic field with energy density $U_B = B^2/(8\pi)$. The Thomson cross-section is $\sigma_T = (8\pi/3) r_0^2$ (r_0 is the classical electron radius $e^2/(mc^2)$). From Equation (7.24), with an isotropic electron distribution a highly relativistic electron radiates a power

$$P = \frac{4}{3} \gamma^2 c \sigma_T U_B. \tag{7.25}$$

From the energy-loss rate, Equation (7.25), the lifetime of a relativistic electron follows. For large γ the electron energy decays by half in a time

$$t_{1/2} = mc^2 \left(\frac{4}{3} \gamma c \sigma_T U_B \right)^{-1} \tag{7.26}$$

or in numerical terms

$$t_{1/2} = \frac{16.4}{B_\perp^2 \gamma} \text{ yr.} \tag{7.27}$$

Here $B_\perp$ is in gauss. Note that $B_\perp$ is the perpendicular component; if the field direction is randomly oriented for an assembly of electrons, the average lifetime is $25 B^{-2} \gamma^{-1}$ yr. Thus, a 10-GeV galactic cosmic-ray electron in a field of 3 microgauss has a lifetime of about 10^8 yr.

7.3 A power-law energy distribution

Up to this point, we have considered the radiation from a single relativistic electron. We now consider a relativistic gas with a density distribution in energy $N(E)$; we assume that

Table 7.1. *The function a(p) as a function of the energy spectral index p (Ginzburg and Syrovatskii, 1965)*

p	1	1.5	2	2.5	3	4	5
$a(p)$	0.283	0.147	0.103	0.0852	0.0742	0.0725	0.0922

it has an isotropic pitch-angle distribution. The energy distribution of cosmic rays, and apparently also of the radiating electrons in many other synchrotron sources, is in the form of a power law,

$$dN(E) = CE^{-p}\,dE, \tag{7.28}$$

where p is the spectral index and C is a constant. The spectrum emitted by a single electron has been given in Equation (7.19), which can be convolved with the energy distribution. Following Ginzburg and Syrovatsky (1969), with a power-law distribution and an isotropic pitch-angle distribution, the emissivity, i.e. the power emitted per unit volume per solid angle (see Section 7.1), can be written in closed form. A function $a(p)$ (given here in Table 7.1) is usually used to simplify the power-law index dependence. For typical cases that are met in radio astronomy, the function $a(p)$ is of the order of 0.1. The net result for the specific emissivity (radiation produced per unit volume per unit solid angle) for a power-law electron distribution with spectral index p and isotropic pitch-angle distribution is

$$j_\nu = \frac{e^3}{mc^2}\left(\frac{3e}{4\pi m^3 c^5}\right)^{(p-1)/2} CB^{(p+1)/2}\nu^{-(p-1)/2}a(p). \tag{7.29}$$

For an optically-thin source, the surface brightness follows on integrating along the line of sight (see Section 7.1 for this elementary solution of the equation of transfer). When numerical values are substituted, one obtains for uniform density and path length L a surface brightness

$$I(\nu) = 1.35 \times 10^{-22} CB^{(p+1)/2}\left(\frac{6.26 \times 10^{18}}{\nu}\right)^{(p-1)/2} a(p)L \ \text{erg cm}^{-2}\,\text{s}^{-1}\,\text{ster}^{-1}\,\text{Hz}^{-1}. \tag{7.30}$$

This shows that an ensemble of relativistic electrons with a power-law distribution in energy produces a synchrotron spectrum that is another power-law distribution:

$$dN(E) \approx E^{-p} \implies I(\nu) \approx \nu^{-(p-1)/2}. \tag{7.31}$$

Note that this result does not depend on a detailed integration of $F(x)$, but follows simply from the scaling described in Equation (7.19); it may be derived by assuming simply that the synchrotron emission from an electron with relativistic factor γ occurs only at frequency $\gamma^2\nu_{\text{cyc}}$.

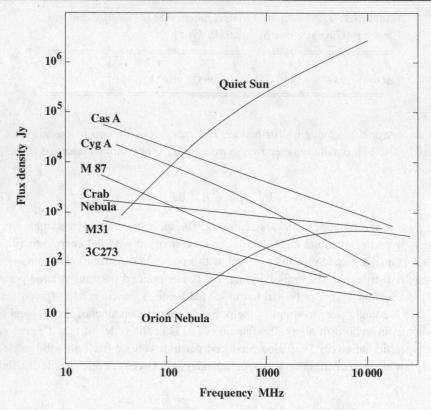

Figure 7.4. Spectra of synchrotron sources and two thermal sources (the Sun and the Orion Nebula).

In later chapters we will find many examples where the observed flux density $S(\nu)$ of continuum radiation follows a power law[1] $S(\nu) \propto \nu^\alpha$. Equation (7.31) shows that α may be related to the electron energy spectrum by the relation $\alpha = (1 - p)/2$. For most radio galaxies and quasars α lies in the range -2 to 0. Figure 7.4 shows examples, including for contrast the thermal spectra of the Sun and the Orion Nebula (Chapter 11).

When the relativistic electron gas is contained in an ordered magnetic field, the electrons in every energy range will produce linearly polarized radiation. The polarization analysis for a single electron leading to Equation (7.23) can now be extended for an energy power-law distribution of index p by integrating over energy and pitch angle. The resulting polarization, for an isotropic pitch-angle distribution, is

$$\Pi = \frac{p+1}{p+7/3}. \tag{7.32}$$

---

[1] Note the sign convention for the spectral index α; earlier work in radio astronomy, including the first edition of this text, uses $S(\nu) \propto \nu^{-\alpha}$, which is similar to the convention for the cosmic-ray power law. At long wavelengths most radio sources have synchrotron spectra that fall steeply with frequency; the earlier convention therefore gave the same sign for the cosmic-ray and radio spectral indices. In contrast, the positive convention is always used at millimetric and infrared wavelengths, at which spectra that increase with frequency are often observed, and we recommend its use throughout the radio range.

For a typical spectral index of -2.5, the resulting degree of linear polarization will be 0.72, close to the value for a single electron.

7.4 Synchrotron self-absorption

As we have seen in Section 7.1, the specific intensity of radiation from a source in thermodynamic equilibrium cannot exceed a limit set by the temperature of the source; at this limit emission and absorption are balanced along a line of sight. This must apply to synchrotron radiation, although in this case the source temperature must be related to the kinetic energy E of the radiating electrons, which are not necessarily in thermal equilibrium with their surroundings. The temperature is then a kinetic temperature T_k. As we have seen in Section 7.1, the emission must then be balanced by an inverse process of absorption, which is known as *synchrotron self-absorption*. The processes of radiation and self-absorption of synchrotron radiation are spread over a range of energies and frequencies, as seen in the spectral function of Figure 7.3, but it is nevertheless useful to assign a single energy to each frequency, as in our discussion of the power spectrum. If the radiating source is optically thick, $T_b = T_k$ at each frequency, the spectrum of optically thick synchrotron radiation is then found by equating the two temperatures:

$$T_k = E/k \approx 8 \times 10^3 B^{1/2} \nu^{1/2} k^{-1} \, \text{K}, \tag{7.33}$$

$$T_b = S\theta^{-2} c^2 (2k)^{-1} \nu^{-2} \, \text{K}. \tag{7.34}$$

Combining these gives a spectrum $S \propto \nu^{2.5}$, i.e. an index $\alpha = 2.5$ in place of the normal $\alpha = 2$ for optically thick radiation from a source in full thermodynamic equilibrium.

Synchrotron self-absorption is encountered notably in the compact quasar-like sources known as gigahertz-peaked sources (GPSs), which will be discussed in Chapter 13. An example of the spectrum of a GPS is shown in Figure 7.5.

7.5 Free–free radiation

Free–free, or bremsstrahlung, radiation is the broad-bandwidth radiation due to the acceleration of a free electron in the field of an ion; the electron is unbound before and after the collision. In the free–free collision, the time taken for emitting most of the radiation is so brief that, particularly in the radio spectrum, the acceleration can be approximated by a delta-function. A full calculation requires a summation over all directions, impact parameters and velocities. A summary of the principal results is given by Rybicki and Lightman.

For a completely ionized gas with electron density n_e and ion density n_i with charge number Z per ion, the specific emissivity for free–free emission, $j_{ff}(\nu)$, will be

$$j_{ff}(\nu) = \frac{1}{2\pi} mc^2 \sigma_T^{3/2} \left(\frac{mc^2}{kT} \right)^{1/2} Z^2 n_e n_i \exp\left(-\frac{h\nu}{kT} \right) \bar{g}_{ff}(T, Z, \nu). \tag{7.35}$$

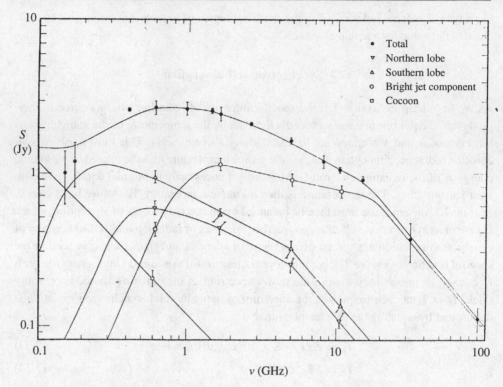

Figure 7.5. The radio spectrum of the gigahertz-peaked source 2352+495 (Polatidis *et al.* 1999). This compact source contains several components that have a spectral peak dividing the optically thin radiation at high frequencies from the low frequencies at which the source is optically thick, and the spectral index approaches 2.5.

Here, σ_T is the classical Thomson cross-section of the electron and $\bar{g}_{ff}(T, Z, \nu)$ is the velocity-averaged *Gaunt factor*, whose value varies between 4 and 5 over the radio range for temperatures of order 10^4 K. Numerically

$$j_{ff}(\nu) = 0.54 \times 10^{-38} Z^2 n_e n_i T^{-1/2} \exp\left(-\frac{h\nu}{kT}\right) \bar{g}_{ff} \, \text{erg cm}^{-3} \, \text{ster}^{-1}. \quad (7.36)$$

(For X-rays in the kilovolt range and for a million-volt plasma, the Gaunt factor is of the order of unity. A detailed plot can be found in Figure 5.3 of Rybicki and Lightman.) A useful practical form of this equation, at radio wavelengths, for temperature T_4 measured in units of 10^4 K and frequency in gigahertz, is

$$j_{ff}(\nu) \approx 3.15 \times 10^{-38} Z^2 n_e n_i T_4^{-1/2} \left[13.2 + \ln\left(\frac{T_4^{3/2}}{Z\nu_{GHz}}\right) \right] \text{erg cm}^{-3} \, \text{ster}^{-1}. \quad (7.37)$$

The exact form, Equation (7.35) above, has an exponential cut-off at the high-energy, short-wavelength end of the spectrum, expressing the fact that a colliding electron cannot emit more energy than it possesses. With this term the expression is as valid for radiation in the

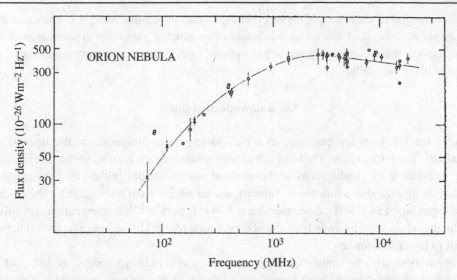

Figure 7.6. The spectrum of the observed radio emission from the Orion nebula M42, showing the effect of increasing optical thickness at lower radio frequencies (after Terzian and Parrish (1970)).

X-ray domain as it is at radio wavelengths. When X-ray spectral observations of a source show this exponential behaviour, the term *soft spectrum* is used, to distinguish it from the power-law *hard spectrum* of synchrotron radiation.

With the emission spectrum for free–free radiation defined, Kirchoff's law, Equation (7.4), can be used to derive the linear absorption coefficient, $\kappa_{\rm ff}$ for free–free radiation:

$$\kappa_{\rm ff} = j_{\rm ff}(\nu)\frac{c^2}{2kT\nu^2}. \tag{7.38}$$

Note that, whereas the free–free emission spectrum is nearly flat, the absorption coefficient increases inversely as the square of the frequency. Over a wide frequency range, an ionized gas cloud may be optically thick at low frequencies and optically thin at high frequencies. At low frequencies it will have a thermal spectrum corresponding to the electron temperature; a Rayleigh–Jeans spectrum, with the brightness increasing as the square of the frequency, will be observed. At a sufficiently high frequency, however, the plasma must become optically thin, and the spectrum will bend over to become approximately flat, diminishing slowly under the influence of the Gaunt factor. A spectrum of this sort is shown in Figure 7.6, which shows the theoretical and observed continuum spectrum of the Orion nebula, M42.

The optical depth τ_ν has been defined in Section 7.3 as the integral of the absorption coefficient along the line of sight. For ionized hydrogen in the radio domain, where $Z = 1$ and $n_{\rm e} = n_{\rm i}$, Mezger and Henderson (1967) have given the following useful expression, which is valid to $\approx 5\%$ for frequencies less than 10 GHz:

$$\tau_\nu \approx 8.235 \times 10^{-2} T_{\rm e}^{-1.35}\nu^{-2.1} \int n_{\rm e}^2 \, {\rm d}l. \tag{7.39}$$

The integral of the square of the electron density along the line of sight is known as the *emission measure*, and it is a commonly met observational parameter in plasmas such as H II regions. It is usually expressed in units of cm^{-6} pc, as here, since these are the units of choice observationally.

7.6 Radio spectral lines

Radio spectral lines are generated in a rich variety of environments in the interstellar medium. The intrinsic line widths of most radio transitions are narrow, so the line profiles are determined by combinations of thermal and bulk velocities in the gas. Line profiles are often described as a function of velocity, scaled simply from frequency; for the 21-cm hydrogen line at 1420 MHz, the conversion is 4.74 kHz per km s^{-1}. By convention, a positive velocity means velocity away from the observer, so a redshift in the optical convention is a shift to lower frequencies.

For an optically thin emitter, which is often the case in radio astronomy, the line profile $\phi(v)$ is determined by the local velocity distribution of the emitting atoms or molecules, since it is the Doppler shift arising from the component of velocity along the line of sight that will shape the line. The line profile is usually written as a normalized function such that its integral over all frequencies is unity. For a source in thermal equilibrium at kinetic temperature T the line profile can be written in terms of a thermal width parameter σ_t:

$$\phi(v_z) = \frac{1}{\sigma_t \sqrt{\pi}} \exp\left(-\frac{v_z}{\sigma_t}\right)^2. \tag{7.40}$$

The width parameter $\sigma_t = 0.83 T_{100} \sqrt{m_H}$, where m_H is the mass of the emitter in terms of the hydrogen mass and $T_{100} = T/100$ K.

The thermal velocity distribution within the volume under observation might not determine the line profile, since there may be much larger turbulent velocities present, and these may, but need not, have a Gaussian distribution. There may also be a systematic flow that results in an asymmetric or complex line shape, or there may be several different components, centred on different velocities.

The specific emissivity j_v of an incoherent source (introduced in Section 7.1) determines the amount of energy dE emitted into a solid angle dΩ, in a direction $\hat{n}$, in a frequency interval dv, per time interval dt, from a volume element dV:

$$j_v = \frac{dE}{d\Omega \, dv \, dt \, dV}. \tag{7.41}$$

The specific emissivity depends upon a variety of pseudo-temperature parameters. If the gas is reasonably close to kinetic equilibrium, the velocity distribution $f(v_z)$ will depend upon T_k, but the transition rates will depend upon the state temperature, discussed in Section 1.2 and defined by Equation (1.1). If the transition is between the ith and jth levels of a system (the ith state being lower), one may have a state temperature T_i that describes the population of the ith state compared with the ground state, and another state temperature T_{ij} that describes the relative populations of the two states involved in the transition. Fortunately, one is often observing lines that terminate in the ground state, and there is then only one state temperature, T_s. If the system partition function[2] is $\mathcal{Z}$, the statistical weight of the upper (radiating) state is g_i, and it lies at an energy ϵ_i above the ground state and the transition probability is A_{ij}, the specific emissivity is

$$j_\nu = \left[\frac{g_i}{\mathcal{Z}} \exp\left(-\frac{h\nu_{ij}}{kT_s}\right) \right] \left(\frac{h\nu_{ij}}{4\pi}\right) A_{ij} n\phi(\nu). \tag{7.42}$$

The first set of brackets is simply the fraction of the population that is in the upper state, and the velocity distribution function $f(v_z)$ has been transformed to frequency units, so that $n\phi(\nu)$ is the number density of emitters per frequency interval. For a transition to the ground state, provided that the emitted photon has an energy that is small compared with the state temperature and no higher levels are significantly populated, Equation (7.42) simplifies to

$$j_\nu = n \left(\frac{g_1}{g_1 + g_0} \right) \left(\frac{h\nu}{4\pi}\right) A_{ij} \phi(\nu). \tag{7.43}$$

These conditions are met for the 21-cm hydrogen line, for which $g_1 = 3$ for the upper $F = 1$ state and $g_0 = 0$ for the ground $f = 0$ state. For a discrete hydrogen cloud that is optically thin, the resulting brightness temperature follows from Equation (7.5), one of the elementary solutions to the radiative transfer equation:

$$T_b = \frac{3}{32\pi} \left(\frac{hc\lambda}{k}\right) A_{10} N_\nu. \tag{7.44}$$

where N_ν is the surface density of hydrogen atoms per hertz along the line of sight (i.e. the integral of $n\phi(\nu)$ along the ray path). Note that the result is independent of state temperature for this approximation, since $h\nu \ll kT$.

When the medium is not optically thin, the optical depth τ_ν, defined by Equation (7.7), must be calculated from the absorption coefficient κ_ν. This follows from Kirchoff's law, Equation (7.4), so that, with state temperature T_s,

$$\tau_\nu = \int \kappa_\nu \, dz = \frac{\lambda^2}{2k} \int \frac{j_\nu}{T_s} \, dz. \tag{7.45}$$

The emissivity can be taken from Equation (7.43) for any two-state system that satisfies its requirements. For the special case of the 21-cm hydrogen line, $g_1/(g_1 + g_0) = 3/4$, as noted above. The more general expression, (7.42), is needed for most molecular systems in thermal equilibrium. This is seldom the case in the interstellar medium, where the state distribution frequently cannot be described in terms of a single state temperature. In such cases, due caution must be exercised.

---

[2] The partition function $\mathcal{Z}$ normalizes the probability for each state; it is the summed relative probability of finding a system in a particular state, taken over all posssible states of the system. Thus, if a system has degeneracy g_i and energy ϵ_i above the ground state, $(g_i/\mathcal{Z})\exp[\epsilon_i/(kT)]$ is the properly normalized probability of finding the system in that state.

For the 21-cm line, when a single state temperature holds along the ray path, the optical depth is related to the population of hydrogen atoms by

$$\tau_v = \frac{N_v}{1.835 \times 10^{18} T_s}.$$

(7.46)

In this expression N_v is the number of atoms per square cm, per km s^{-1}, along the line of sight. The approximation is good for estimation purposes, but in the interstellar medium the temperature is not uniform; if the density and/or the state temperature varies along the line of sight (in H atoms per cubic cm, per km s^{-1}) at velocity v is ρ_v, the integral of ρ_v / T_s must be taken along the line of sight.

7.7 Masers

The maser phenomenon was first discovered in radio astronomy during the period 1965–1967, when studies of the 18-cm hydroxyl lines showed that, when the line was seen in emission near an H II region, the specific intensity was far too great to be generated by a thermal source. Under normal conditions, the ordinary expectation is that the linear absorption coefficient κ_v will be a positive quantity, giving the rate of attenuation of the specific intensity along a ray path. The probability of induced absorption is exactly the same as the probability of induced emission and, since the energy-level populations are greater for the lower-lying states in any system, the induced absorptions predominate. If a non-equilibrium disturbance is introduced, this need no longer be the case, since a population inversion (a negative state temperature) can occur and the induced emission transitions will be greater than the absorptions. The specific intensity will grow along the ray path and the 'absorption' coefficient will be negative. Such a system is known as a *maser* (an acronym for microwave amplification by stimulated emission of radiation; the equivalent for light is a *laser*).

Maser action requires a 'pump' to maintain the population of the upper energy level above the equilibrium thermal level. The pump may be a source of radiation, or the mechanism may be collisional, but, in order to satisfy the second law of thermodynamics, the equivalent temperature of the pumping process must be higher than the temperature corresponding to the energy difference of the energy-level pair that serves as the maser.

A simplified maser model can be constructed that demonstrates the principal parameters that control maser behaviour. Assume that the system has two energy levels, with populations n_1 and n_2 in the lower and upper states. The photon energy corresponding to the energy difference will be $h\nu_{12}$, the probability of a spontaneous transition from the upper to the lower state will be the Einstein coefficient A, and the probability of stimulated transition will be the Einstein coefficient B. The equation of radiative transfer, given in its basic form in Equation (7.3), can now be written, for the line's centre frequency, in the form

$$\frac{dI}{ds} = h\nu_{12}[B(N_2 - N_1)I + A].$$

(7.47)

For most molecular transitions, the spontaneous transition rate is small, and it will be negligible if the maser action is strong. If we postulate a pumping mechanism that populates the upper level at a rate R, and if there are collisions that induce transitions at a rate C between the two states (with no production of photons), in this approximation the equilibrium populations of states 1 and 2 will be determined by equating the rate of transition from the upper and lower states (setting $A = 0$):

$$N_1(BI + C) \approx N_2(BI + C) - R. \tag{7.48}$$

Solving this expression for the population difference and substituting into Equation (7.47) gives

$$\frac{dI}{ds} \approx h\nu_{12}\frac{BRI}{C + BI}. \tag{7.49}$$

When the collision term C dominates the induced radiation term BI, the solution of this equation gives an exponentially growing specific intensity along a ray path s:

$$I = I_0 e^{\alpha_{12}s}, \tag{7.50}$$

where the negative absorption coefficient α_{12} is

$$\alpha_{12} = \frac{h\nu_{12}BR}{C}. \tag{7.51}$$

In this regime, the maser is said to be unsaturated. If the collision rate C is high, it drives the state populations towards equality and the net gain is small; if the pump rate R is high, the gain is correspondingly larger.

If the specific intensity grows to the point that the transitions are overwhelmingly induced by the radiation field, the transition rate is determined completely by the net induced radiative transitions, and hence by the pump rate; the right-hand side of Equation (7.49) then becomes constant. The specific intensity grows linearly and the maser is *saturated*:

$$I = \frac{h\nu_{12}R}{C}s. \tag{7.52}$$

In this saturated condition, the transitions from the upper state are wholly induced by the radiation field, and the maser output is controlled by the rate at which the upper state is populated by the pump.

In the astrophysical context, the behaviour of maser sources is far more complex, as will appear in Chapter 11. The real-world cases, in addition to being saturated or unsaturated, can also have a partially saturated condition, and in a given masering cloud different regions may be in different states of saturation. There may also be a high degree of polarization in maser line emission, due to Zeeman splitting of spectral lines in a magnetic field. The orientation of the magnetic field determines the polarization, which can be circular, elliptical or linear. The review by Elitzur (1992) may be consulted for further details.

7.8 Propagation through ionized gas

The upper atmosphere, at heights above 100 km, is partially ionized by solar ultraviolet radiation, forming the ionosphere. This has a major effect on the propagation of radio waves; at low frequencies, below about 10 MHz, there may even be total reflection at vertical incidence. The Earth's magnetic field has a large influence on propagation through the ionosphere, causing birefringence, which may have complex effects on the polarization of reflected and refracted radio waves. Fortunately for the radio astronomer, the refractivity of the ionosphere decreases as ν^{-2}, and at the higher frequencies used in most observations the refractivity may be described comparatively simply. The same is true for the interstellar medium, where the very small refractivity can provide a useful probe of the density of the ionized gas and of the strength of the interstellar magnetic field.

The refractive index n is most simply expressed in terms of the resonance frequency ν_p of the plasma:

$$n = \left(1 - \nu_p^2/\nu^2\right)^{1/2}, \tag{7.53}$$

where

$$\nu_p^2 = [1/(2\pi\epsilon_0)]^2(Ne^2/m) \qquad \text{(mks units)} \tag{7.54}$$

or

$$\nu_p^2 = [1/(2\pi)]^2[Ne^2/(mc)] \qquad \text{(Gaussian units).} \tag{7.55}$$

To a good approximation $\nu_p = 9N^{1/2}$ Hz, where N is in m^{-3}; for interstellar space the electron density is usually quoted in cm^{-3}, giving the same numerical result in kilohertz. A typical interstellar electron density is 0.03 cm^{-3}, corresponding to a plasma frequency of 1.56 kHz; the refractivity $1 - n$ for a 100-MHz radio wave is then 1.2×10^{-10}.

Since n is less than unity the velocity of the wave c/n is greater than c; this is the *phase velocity* of an infinite monochromatic wave. Energy (and information) is carried at the *group velocity*, which is the velocity of a wave with finite bandwidth. The frequency components of the wave group travel with different velocities, because of the *dispersion* in refractive index which appears in Equation (7.53). The dispersion is such that the product of the group and the phase velocities is c^2; the group velocity is thus cn, which is less than c, giving a propagation delay. Equations (7.53) and (7.54) lead to a time delay of

$$\tau = e^2c/(2\pi m\nu^2)\int n_e\, ds, \tag{7.56}$$

where the integral of n_e along the line of sight is called the *dispersion measure* (this is discussed more completely in Chapter 12). For example, a pulse observed at a frequency of 100 MHz from a pulsar at a distance of 100 light years, travelling through interstellar ionized gas with a density of 0.03 cm^{-3}, will arrive 3.4 s later than the same pulse observed at a much higher frequency, say 1000 MHz.

The group velocity is

$$v_g = cn = c\left[1 - v^2/(nu_p^2)\right]^{1/2} \approx c\left(1 - \frac{1}{2}v^2/(nu_p^2) - \frac{1}{8}v^4/(nu_p^4)\cdots\right). \quad (7.57)$$

In the solar corona the plasma density is high enough for the refractive index to approach zero for radio frequencies up to and beyond 100 MHz. Since the power flux along a ray path is given by the Poynting vector, $|S| = |\mathcal{E}|^2/Z = \sqrt{\epsilon/\mu}|\mathcal{E}|^2$ is constant along the ray, so, when ϵ tends to zero, the field $\mathcal{E}$ becomes very large. This greatly increases the absorption, and the plasma may become opaque. The brightness temperature of the Sun at low frequencies is dominated by this effect. For example, at 100 MHz the brightness temperature is approximately 10^6 K over an area that is much larger than the visible disc; this is the temperature of the outer parts of the corona. At progressively higher radio frequencies the corona and then the chromosphere become transparent, and the brightness temperature falls correspondingly towards the photospheric temperature (see Chapter 11).

7.9 Faraday rotation

When a wave passes through a plasma, the electrons respond to the electric field by oscillating and re-radiating at the wave frequency; this is the origin of the refractive index in Equation (7.53). The response of the electrons is affected by any static magnetic field, which forces the electrons into curved paths. If the incident wave is circularly polarized, the electron response will be a circular motion whose amplitude depends on the direction of the static field. The addition of a static magnetic field B in the direction of propagation therefore splits the refractive index into two components, and the plasma becomes *birefringent*; for the two handednesses of circular polarization the refractive index becomes

$$n^2 = 1 - \frac{v_p^2}{v(v \pm v_B)}, \quad (7.58)$$

where v_B is the gyrofrequency

$$v_B = \frac{eB}{2\pi m} \text{ (mks units)}; \qquad \frac{eB}{2\pi mc} \text{ (Gaussian units)}. \quad (7.59)$$

The electron gyrofrequency in a field of 1 Tesla is 2.8×10^{10} Hz. Interstellar magnetic fields are usually quoted in microgauss, so it is useful to remember that the gyrofrequency is 2.8 Hz per microgauss. (The same factor occurs in the Zeeman frequency splitting of the 21-cm hydrogen line.)

The analysis of propagation along the direction of the magnetic field is the 'transverse' case, with the right-hand and left-hand circularly polarized waves being the normal modes. The general formalism applies equally well at large angles to the field, taking the component $B\cos\theta$ of the field along the wave normal. (This 'quasi-longitudinal' approximation is generally applicable in astrophysical circumstances, but might not be applicable when the wave frequency is comparable to the plasma frequency, as may occur for low radio

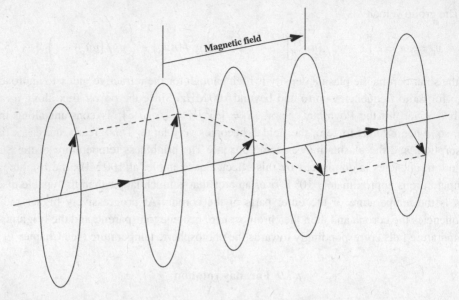

Figure 7.7. Faraday rotation. Propagation along a magnetic field in a plasma rotates the plane of polarization of a radio wave.

frequencies in the ionosphere.) In all cases, the ionized medium is birefringent when the proper normal modes are identified (see for example Ginzberg (1970)).

A plane polarized wave may be regarded as the sum of two circularly polarized components of opposite handedness. In a plasma with a static magnetic field these two components propagate with different phase velocities; when they are added after travelling some distance, their relative phase has altered and the plane of polarization of the resultant has rotated. This is Faraday rotation (Figure 7.7), which was first observed by Faraday in plane-polarized light in a glass block between the poles of an electromagnet.

In radio observations the rotation along a ray path follows from Equation (7.58) above:

$$\theta_R = R\lambda^2 \tag{7.60}$$

where R is the *rotation measure* of that path, usually expressed in units of rad m^{-2}.

Using the conventional astrophysical units, measuring distances in parsecs, N in cm^{-3} and B in microgauss, the rotation measure is

$$R = 0.81 \int BN \cos\theta \, dl. \tag{7.61}$$

For an astronomical source, such as a pulsar, which radiates with the same plane of polarization over a wide frequency range, Faraday rotation may be measured from the variation of position angle with radio frequency. The difference in position angle between adjacent frequencies ν and $\nu + \delta\nu$ MHz is

$$\Delta\theta_R = -2R(300/\nu)^2 \, \delta\nu/\nu. \tag{7.62}$$

Measurement on two frequencies alone may be insufficient, since a whole number of turns may be missed, measuring only the fractional number of turns.

7.10 Scintillation

The plasmas of the ionosphere, of interplanetary space and of the interstellar medium all contain random irregularities. Propagation through a medium with random fluctuations in refractive index initially corrugates the wavefront, and then leads to the amplitude fluctuations which are familiar in the optical domain as the twinkling of stars. Scintillation is observed only when the source has a sufficiently small angular diameter. In the radio domain the most important example is due to the propagation through interstellar space of radio pulses from pulsars.

A simple presentation of the theory by Scheuer (1968) in terms of random phase changes imposed on a plane wavefront by irregularities with typical dimension a and with differences in refractive index due to fluctuations ΔN_e, contained in a slab with thickness D, leads to phase perturbations $\Delta\phi$ across the wavefront:

$$\Delta\phi = D^{1/2}a^{1/2}r_e\lambda\,\Delta N_e. \tag{7.63}$$

The angle of scattering is given by

$$\theta_{\text{scat}} = [\Delta\phi/(2\pi)]\lambda/a = (1/2)(D/a)^{1/2}r_e\,\Delta N_e\,\lambda^2. \tag{7.64}$$

This is the apparent angular size of a distant point source observed through the screen.

As the wave progresses beyond the screen to distances L greater than $L\theta_{\text{scat}}$ the rays cross and an interference pattern develops (Figure 7.8). The ray paths will differ by $\approx (1/2)\,\theta_{\text{scat}}^2 L$, which will usually be many wavelengths. Interference between rays will therefore depend on the wavelength; the resultant amplitude will vary over a wavelength difference $\delta\lambda$ given by

$$\delta\lambda/\lambda = 2\lambda/(\theta_{\text{scat}}^2 L). \tag{7.65}$$

The model may be extended to the practical situation where the screen fills the whole line of sight over a distance L by setting $L = D$; only a small numerical factor is involved in the results for scattering angle and interference. The frequency bandwidth B_{scat} then becomes

$$B_{\text{scat}} = \frac{8\pi^2 ac}{D^2(\Delta N_e)^2 r_e^2\lambda^4}. \tag{7.66}$$

The lateral scale S of the interference pattern is the same as the scale a of the density irregularities provided that $\Delta\phi \ll 1$. This is the case of weak scattering. When $\Delta\phi \gg 1$, that is, for strong scattering, the lateral scale S becomes

$$S = a/\Delta\phi = \lambda/(2\pi\theta_{\text{scat}}). \tag{7.67}$$

The lateral structure moves past the observer due to the combined velocities of the source and the observer relative to the screen, and the source is seen to scintillate.

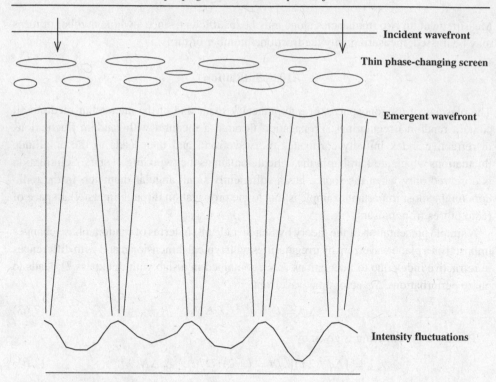

Figure 7.8. Scintillation: fluctuations in refractive index corrugate a plane wavefront, which then develops amplitude irregularities.

In the important case of scintillation in the turbulent interstellar medium, there is a very wide spectrum of random size and amplitude of density irregularities to be considered. Two distinct regimes of scintillation may be observed, corresponding to large and small scales of turbulence (see a review by Rickett (1990)). On the large scale, scintillation may be regarded as the effect of individual ionized clouds, which act as prisms or lenses; this is referred to as refractive scintillation, which is responsible for some of the slow changes in observed intensity of pulsars (Chapter 12) and quasars (Chapter 13). The more usual scintillation effects are understood as random diffraction.

A wavefront with randomly distributed fluctuations of phase can be treated by diffraction theory; the wave spreads in angle as it propagates, with an angular spectrum that is determined from the phase distribution by a Fourier transform. (The treatment is closely related to the diffraction theory of an antenna aperture, discussed in Chapter 4.) A lateral scale a of phase structure $\Delta\phi$ corresponds to a scattered wave at angle $[\Delta\phi/(2\pi)](\lambda/a)$. If the total phase irregularities are less than a radian, there is an unscattered plane-wave component; in this case, known as weak scattering, the appearance of a point source will be as an unscattered point surrounded by a scattered halo. For strong scattering there is no unscattered wave, and the total amplitude of the randomly scintillating wave varies according to a Rayleigh distribution.

The effects of scintillation can be used as a tool for determining the angular sizes of radio sources; we note two examples here. First, the development of VLBI was encouraged by the observation of scintillations of compact radio sources in the solar wind, which indicated that there were compact sources as small as a milliarcsecond in size. Second, the ionized interstellar medium will cause scintillations in radio sources such as pulsars whose angular size is of the order of a microarcsecond or smaller; notably the radio afterglow of the gamma-ray bursters shows scintillations that damp out as the radio source expands, showing that the angular diameter of the afterglow was less than 3 µarcsec during the first month of its expansion (Frail *et al.* 1997).

7.11 Propagation in the Earth's atmosphere

The refractive index of the neutral, dry atmosphere is almost the same at optical and radio wavelengths. The effect of water vapour is, however, more than 20 times greater at radio wavelengths; this is due to the permanent dipole moment of water molecules. The refractive index n of air at radio wavelengths and at temperatures encountered in the atmosphere is conveniently quoted in terms of a refractivity $N = 10^6(n - 1)$. For most purposes this is given adequately by the Smith and Weintraub (1953) formula

$$N = 77.6T^{-1}(P_d + 4810P_v T^{-1}), \tag{7.68}$$

where P_d is the partial pressure of the dry air and P_v is the partial pressure of the water vapour, both in millibars (1 mbar = 100 Pa; 1 atm = 1013 mbar). At frequencies above 100 GHz the component due to water vapour is slightly greater because of the effect of the very broad infrared transitions (Hill and Clifford 1981).

Refraction in the atmosphere affects the direction of arrival of a wavefront at a radio telescope in much the same way as for an optical telescope, except that radio observations are more often made close to the horizon. If the atmosphere can be treated as a flat slab, with refractive index n at ground level, the apparent increase Δz in elevation at zenith angle z is given simply by

$$\Delta z = (n - 1)\tan z. \tag{7.69}$$

This simple approximation must be elaborated near the horizon by the results of ray tracing, taking account of the Earth's curvature. It is useful to remember, however, that $n - 1$ is approximately 3×10^{-4}, so the refraction near the horizon is approximately 20 arcmin.

The refractivity of the atmosphere is also important in interferometric observations because of the extra path length which is introduced. A plane-parallel slab atmosphere affects all telescope elements of an interferometer array in the same way, so the only important changes in path length are those due to differences in the atmosphere above the individual telescope sites; this is important for VLBI, where there may be large differences in water-vapour content at the widely spaced sites. In VLBI there may also be important differences in path length due to the Earth's curvature, where the elevation of a source above the horizon may differ greatly among the various contributing sites. Differences in

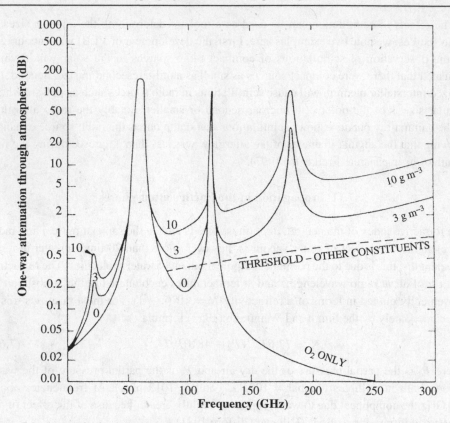

Figure 7.9. Atmospheric absorption as a function of frequency. The total attenuation at the zenith is shown for the oxygen content alone and for surface water-vapour densities of $3 \, \mathrm{g \, m^{-3}}$ and $10 \, \mathrm{g \, m^{-3}}$ (scale height 2 km).

path length may be estimated from measurements of partial pressures at the telescope sites, but, since the excess path lengths may be large (more than 10 m for zenith distances greater than 80°), correction for refraction may be a serious limitation. The effects on path length are obviously more serious at shorter wavelengths; for millimetre-wavelength arrays such as ALMA they may prove to be the limiting factor in the overall performance.

At radio frequencies below about 10 GHz atmospheric absorption is small and often negligible; at metre wavelengths even the rain might not matter much, except for the effects the liquid water might have on the electrical components of the receiver. The attenuation at higher frequencies arises from molecular resonances of oxygen, ozone and water vapour. These resonances occur within the radio band from 22 GHz upwards, but their pressure-broadened absorption bands also affect atmospheric transmission at lower frequencies. Figure 7.9 from Smith (1982) shows the attenuation at the zenith for frequencies up to 300 GHz for a standard atmosphere with low and moderate water-vapour content[3].

---

[3] The water-vapour content is often quoted as millimetres of precipitable water vapour; for a scale height of 2 km a density of $1 \, \mathrm{g \, m^{-3}}$ at the surface is equivalent to 2 mm of precipitable water vapour.

Atmospheric absorption has a second effect on radio-astronomical observations. Simple thermodynamic arguments show that not only is a signal from above the atmosphere reduced, but also there is an added component of thermal radiation from the atmosphere (Section 3.5). If the antenna temperature would in free space have been T_a, but is attenuated by a factor α in a medium with temperature T_0, the antenna temperature becomes

$$\alpha T_a + (1 - \alpha)T_0. \tag{7.70}$$

Low-noise receiving systems may operate with noise temperatures much lower than atmospheric temperatures, so that absorption by only 1 dB may double the total system temperature; this is a far more serious effect than the attenuation of the signal itself. The effects of water vapour are sufficiently serious for millimetre-wave observations to drive site choice to otherwise inconvenient locations. In Antarctica, the South Pole is probably the observing site with the lowest average water-vapour content through the year (about 1 mm precipitable water); Dome C is higher and probably drier, but lacks the logistical support. The ALMA site, at its best, may be comparable in quality. Mauna Kea, in Hawaii, is an excellent site with greater accessibility, but is not as dry on average. Other locations, on lower mountain peaks, are distinctly inferior, but have the advantage of easy accessibility.

Further reading

Astrophysical Formulae, Lang K. R., Berlin, Springer-Verlag, 1974.

Classical Electrodynamics, Jackson J. D., New York, Wiley, 3rd edn, 1998.

Classical Theory of Fields, Landau L. D. and Lifshitz E. M., Amsterdam, Elsevier, 4th edn, 1980.

Radiative Processes in Astrophysics, Rybicki G. B. and Lightman A. P., New York, Wiley, 1979.

The Propagation of Electromagnetic Waves in Plasmas, Ginzburg V. L., Oxford, Pergamon Press, 2nd edn, 1970.

8

The local Universe

The Universe that we observe had its beginning in the *big bang*, the cosmic fireball. After an initial phase of element formation, this settled into a primeval plasma of protons and electrons, helium nuclei and a trace of lithium nuclei. When the primeval plasma became optically thin at an age of about 370 000 years (the *era of decoupling*), electromagnetic radiation could reach the Earth, and appears as the cosmic microwave background (CMB), discovered by Penzias and Wilson (1965) and observable only in the radio and far-infrared parts of the spectrum. Observations of the CMB and their implications are described in Chapter 14; a discussion of anisotropies in the CMB follows in Chapter 15. Here, we discuss the local aftermath, our Milky Way Galaxy and its surroundings.

8.1 Stars and galaxies

Our home Galaxy, the Milky Way system, has stars as its most visible component; but radio astronomy is usually more concerned with the interstellar medium (ISM). The ISM is composed principally of hydrogen and helium, with a trace of heavier elements, having a mass fraction of about 1%. These occur in both atomic and molecular form, including larger aggregates referred to generically as dust. In addition there is a third, more tenuous component, the high-energy medium, composed mostly of energetic particles, principally protons and electrons, with energies extending well beyond 10^{18} electron-volts (cosmic rays in the energy range 10^{19}–10^{20} electron-volts cannot be contained by the Galactic magnetic field, and may come from extragalactic sources). These pervade the Galaxy and are contained within it by a magnetic field of a few microgauss; they are detected on Earth as cosmic rays. Finally, it has become clear that our Galaxy, and galaxies in general, are embedded in a much larger halo of *dark matter*, detectable only through its gravitational influence on the stars and the ISM. Radio observations contribute to the study of all these components: the evidence for dark matter is discussed in Chapters 10 and 14.

Our Galaxy is only one of the enormous number of galaxies that make up the Universe. The galaxies come in many forms and sizes, and these can be classified, to a first approximation, by their optical appearance; the main division is between spirals and ellipticals. The most commonly used classification system is that of Edwin Hubble, which was refined by Allan Sandage. Prototypes of this system are collected in Sandage's *Hubble Atlas of*

Galaxies, a striking collection of photographs, mostly taken with the Palomar 200-inch telescope (Sandage 1961). A larger collection of prototypes, accompanied by a more detailed classification scheme, can be found in *The Carnegie Atlas of Galaxies* (Sandage and Berdke 1994). The alternative classification system of Gérard de Vaucouleurs (1964) can also be consulted.

There is a different classification, originally by stellar population, that recognizes that stars differ widely in age. The oldest stars are designated Population II, and are characterized by a low metal (i.e. carbon and all heavier elements) abundance compared with that of our Sun. These are the oldest stars in galaxies, having formed early in the process of galaxy formation. When these stars reach the ends of their lives, the more massive stars explode as supernovae, and heavy elements are generated by thermonuclear reactions. Younger stars are formed from the ISM, now enriched in 'metals' injected into the ISM by supernovae, and these evolve in similar fashion, further enriching the ISM. The youngest stars, with a high metal abundance, are designated Population I. The dust and gas of the ISM, in addition to the youngest stars, are characteristic of 'extreme Population I', while the stars in most globular clusters and in the Galactic stellar halo are 'extreme Population II'. Most stars are intermediate; the Sun, for example, belongs to the intermediate population. The system of galaxy types, distinguishing spirals from ellipticals, is interlinked with the population concept; spiral arms are associated with Population I and elliptical galaxies with Population II.

There is unknown territory between the era of decoupling, when the age of the Universe was approximately 370 000 years, a factor of about 200 greater in redshift than the earliest galaxies that can be observed (a bit beyond redshift 6 at the time of writing). These earliest galaxies contain 'metals', albeit at a much lower concentration than that found in young stars. The problem is that there is no known way for elements heavier than lithium to be made in the primeval fireball. The existence of heavy metals in the Population II stars is attributed to a hypothetical Population III, a class of stars so far unobserved, that started the process of heavy-element formation. These must have come into existence during the 'Dark Ages' when the primeval gas was neutral. These (or quasars from the same era) eventually reionized the medium (the era of reionization), a phenomenon that may be observable by low-frequency radio telescopes now being built (Chapter 17).

Elliptical galaxies, which contain little gas or dust, are nearly featureless, with stars of extreme Population II, and are principally described by their intrinsic ellipticity. The E0 ellipticals are nearly spherical, and increasingly elliptical galaxies are designated E1 to E7; the most elliptical have axial ratios of about 3 : 1. The most massive elliptical galaxies, the giant ellipticals, are a special class, designated cD galaxies.

Spiral galaxies are distinguished by a flat disc, with a variously developed system of spiral arms, and have a central bulge of stars that may have a bar-like structure linked to the spiral arms. The dust and gas of the ISM, and the young stars forming from the ISM at present, are strongly concentrated towards the plane of the disc, typically with an axial ratio of about 100 : 1, and are designated extreme Population I. The stellar disc, composed of stars of intermediate population, is a flattened ellipsoid, but thicker, with an aspect ratio of around 10 : 1. The galaxies are designated S0 if they have no discernable spiral arms, and Sa, Sb, Sc for those with increasingly open spiral structure; the designation S becomes SB for barred spirals. The central bulge is most prominent in Sa spirals and least in Sc spirals; the spiral arms are clearly delineated in Sa and Sb spirals, but are rough, loosely wound and irregular in the Sc galaxies. In spiral galaxies, the oldest stars are found in a less conspicuous Population II halo surrounding the galaxy, together with the globular clusters, which are found in an approximately spherical distribution around the central region. Our Galaxy is classified as SBb, on the basis of H I spectral line observations that we present in Chapter 10. Star formation is proceeding in all classes of spiral galaxies, but some of these have unusually high rates of star formation. These are known as *star-burst galaxies*, and are more common at large redshifts.

There are galaxies that have no discernable spiral structure but are rich in young stars, gas and dust. These are given the catch-all title of irregular, Irr. In addition, there are two more recognized classes: the first of these are the dwarf galaxies, such sparse collections of stars that they are barely discernable among the foreground stars; there are also peculiar galaxies that defy classification. Examples of these have been collected by Halton Arp in his *Atlas of Peculiar Galaxies* (Arp 1966).

Our Milky Way system has two satellite galaxies, the Large Magellanic Cloud (LMC) and the Small Magellanic Cloud (SMC), both classified as Irr, having masses of roughly 10^9 and 10^8 solar masses. Both are gravitationally bound to the Milky Way, and are at present at distances of roughly 45 kiloparsecs (kpc)[1], six times the Sun's distance from the Galactic centre. Our Galaxy, in turn, is a member of the Local Group, a gravitationally bound system of 30 or so galaxies, all closer than a megaparsec, the largest of which are the Milky Way system and the well-known Andromeda spiral, M31. The Local Group is a relatively small cluster, but is a member of a much larger group called the Local Supercluster that contains both field galaxies and clusters of galaxies, several richer than the Local Group by orders of magnitude. Its extent is not well defined, but it extends to more than 40 megaparsecs.

A wide variety of radio observations both of our Milky Way system and of external galaxies are treated in later chapters. Here is a sampling of the topics.

- Galactic rotation is revealed with high precision by H I observations, revealing the dynamics of spiral galaxies and the influence of the dark-matter halo.
- For external galaxies, the overall velocity width of the total hydrogen emission, when combined with a galaxy's infrared flux, gives one of the best indicators of a galaxy's absolute luminosity (Tully and Fisher 1977).
- Molecular emission gives a large range of physical parameters, including the molecular hydrogen density, which is otherwise a difficult quantity to measure. There are now hundreds of molecular varieties that have been detected in the radio spectrum, and these give a new tool with which to study interstellar chemistry and the processes that lead to star formation.
- In addition to these examples of thermal emission, there are galaxies that are extraordinarily bright at radio wavelengths, whose radiation is generated by non-thermal processes, primarily synchrotron

[1] The astronomical unit of distance, the parsec, is based on the measurement of distance by parallax due to the Earth's orbital motion: 1 parsec (pc) $= 3.08 \times 10^{16}$ m $= 3.26$ light-years.

Figure 8.1. A panoramic view of the Milky Way as seen optically (Lund Observatory).

radiation from relativistic electrons in magnetic fields, which is often associated with relativistic jets originating in the region around the massive black hole at the nucleus of the galaxy. This non-thermal radio emission can be associated with either ellipticals or spirals, and is not easily correlated with optical appearance.

8.2 Aspects of the Milky Way

An air-brush map prepared at the Lund Observatory, showing the optical appearance of the entire sky in Galactic coordinates[2] is presented in Figure 8.1. The equal-area Aitoff projection distorts the constellations, but Ursa Major, Orion and several other bright-star constellations can be identified easily. The dark lanes along the Milky Way give evidence of the interstellar dust, and there is a marginal indication of the central bulge of stars. Because of the extinction caused by the dust, most of the visible stars are within a kiloparsec of the Sun, only a fraction of the distance of 8 kpc to the Galactic centre. The Large and Small Magellanic Clouds can be seen as diffuse patches in the lower right-hand quadrant, and the Andromeda galaxy, M31, is the small elliptical patch below the Milky Way, near $\ell = 130°$, $\ell = -20°$. There is little indication in this map of the three-dimensional structure of the Galaxy.

The mid-infrared sky, mapped at 60 μm by the infrared observatory IRAS, is shown in Figure 8.2. Here, the stars are relatively faint except for a few luminous supergiants and dust-shrouded young stars; the dust component of the ISM dominates. Its extreme

[2] The Galactic coordinate system, longitude ℓ, and latitude ℓ, has been defined by the Intenational Astronomical Union, and is based on radio H I measurements. The definition as given in Allen's *Astronomical Quantities* approximates the plane of the Galaxy, with the origin ($\ell = 0°$, $\ell = 0°$) chosen close to the actual centre, the central massive black hole.

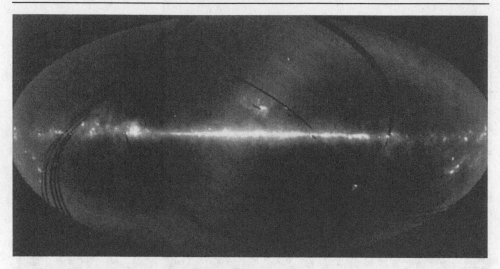

Figure 8.2. The sky in the infrared at 60 μm (C. Beichman).

Population I character is evident from its strong concentration towards the Galactic plane. Molecules such as CO, CN, OH and NH_3, detectable at radio wavelengths, are associated with the dust, but in a simple way; regions of high density favour the formation of larger molecules which will be discussed in the following chapter. The dust particles scatter radiation, causing attenuation of starlight that is so severe that, as noted above, few of the stars in the Lund map along the Milky Way are at a distance greater than a parsec or so. At sufficiently long wavelengths, in the Rayleigh-scattering regime, the absorption will vary as λ^{-4}. Since the dust particles are distributed over a wide range of sizes, from a few hundreds of micrometres to several micrometres, the actual extinction due to interstellar dust varies approximately as λ^{-1} over the visible and near-infrared regimes. The dust particles are also thermal emitters, and it is their thermal radiation that appears in the 60-μm emission of Figure 8.2. Because the greatest fraction of the dust consists of particles far smaller than the 60-μm observing wavelength, the ISM is sufficiently transparent to allow much of the radiation to reach the observer.

At radio wavelengths the sky has an entirely different appearance, as shown in the 408-MHz map of Figure 8.3. It is clear that this radiation is less concentrated towards the Galactic plane than the thermal components seen in Figure 8.2 (the extreme Population I), but it does nevertheless peak on the plane of the Milky Way. This radiation originates neither in stars nor in the ISM, but comes primarily from synchrotron radiation emitted by cosmic-ray electrons moving in the large-scale magnetic field of the Milky Way, although there is a thermal component at low Galactic latitudes. The properties of this high-energy component are considered in this chapter.

In the following chapter we discuss the thermal component of the ISM. This includes radio spectroscopy of the atoms and molecules that inhabit the ISM, and the H II regions that mark the sites of star formation; the two are closely related. The ISM clouds are partially

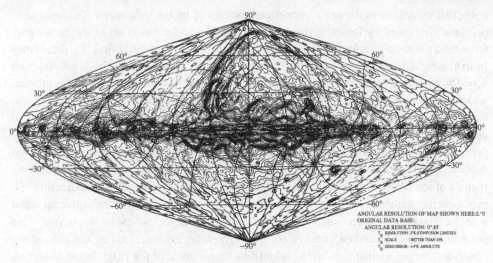

Figure 8.3. An all-sky map at 408 MHz (Haslam *et al.* 1982).

ionized, and are revealed through radio dispersion and scintillation effects. Chapter 10 examines the dynamical information that is gained through spectroscopy, especially by studies of the 21-cm hydrogen line, and its implications for the large-scale structure of the Milky Way system. These dynamical observations, supplemented by observations of gravitationally lensed quasars and galaxies, allow the study of the unseen dark-matter (DM) component. This does not show its presence directly, but the sum of all the evidence (discussed in Chapters 10 and 14), indicates that the DM contribution to the mass density of the Universe is about five times greater than the contribution of the baryonic matter that we observe directly.

8.3 Measurement of sky brightness temperature

The production of the 408-MHz all-sky map (Figure 8.3) provides a good example of the data-reduction techniques that must be addressed in single-aperture radio astronomy. Observations from three steerable paraboloids in Europe and Australia were combined so as to cover the whole sky. The individual maps had to be convolved with the antenna pattern of the smallest of the telescopes and merged into a common map. In the overlap regions, differences between the maps had to be resolved; for example, the lowest level of background brightness is particularly difficult to measure, and the temperature scales of the different maps must be reconciled. The actual measurements of antenna temperature were made with Dicke-switch radiometers (Chapter 3) that measured the effective noise temperature at the radiometer terminals.

In order to derive the brightness temperature of the sky from the observed antenna temperature, a number of corrections must be applied. There are several sources of extraneous noise that contribute to the noise temperature measured by the radiometer: the noise temperature T_f, caused by losses in the telescope feed and its connection to the radiometer terminals; the atmospheric contribution, T_{atm}; and the sidelobe contribution, T_{sl}, that comes from the radiation received in the sidelobes of the antenna pattern. The last correction actually has three components: radiation received in the far sidelobes from the ground, which will depend upon the elevation angle and the mean temperature of the ground; the sidelobe contribution from the atmosphere, which will depend upon the zenith angle and upon both the water-vapour content and temperature profile of the atmosphere; and finally the contribution picked up in the sidelobes from the whole sky. The latter contribution depends upon the map itself, and can be surprisingly large, particularly when observing regions of low surface brightness. This can be understood by considering, from Section 4.4, that a perfect antenna observing a patch of sky at temperature T_b will only give an equal antenna temperature if the patch fills the beam, and if the sidelobes are negligible. This is never the case. There is a quantity, the *beam efficiency* (η_b), defined by the ratio of the observed antenna temperature to the brightness temperature of the patch being observed, but that assumes that the patch is the same size as the primary beam. This is not a totally accurate definition, but in practice it can be taken as a heuristic definition. If the sky varies slowly over the angle subtended by the primary beam, the brightness temperature, corrected for sidelobe contributions, is approximately

$$T = \left(T_a - \int_x G(\theta, \phi) T(\theta, \phi) d\Omega \right) / \eta_b, \qquad (8.1)$$

where the integral is taken over the visible hemisphere, with the exception of the main beam, and the gain is normalized with respect to unity in the direction of the main beam. Note that the integration over the ground hemisphere to obtain the radiation contribution of the ground is a separate procedure. Except at low elevations, this contribution comes from spillover, the radiation that enters the feed from outside the paraboloidal main aperture. This correction, which depends upon the zenith angle, is determined separately. It is a tedious iterative procedure to carry out the calculation exactly, and in practice approximations are used.

The sidelobe effects are minimal in telescopes designed with an offset feed, such as the GBT described in Section 4.8. This avoids blockage of the aperture by the feed, and eliminates the wide-angle sidelobes induced by scattering from the blocking structures. For centre-fed telescopes, the cross-section of the feed structure should be made as small as possible. The importance of aperture blockage can be appreciated by considering the effect on a centre-fed paraboloid with an area A. If the aperture is blocked by a fraction α, the on-axis gain is reduced by 2α (provided that α is much less than unity and that the geometrical-optics approximation is valid, which is usually the case) and this loss in gain must appear in the elevated gain in the sidelobes, as explained in Section 4.4. As an illustration, the typical Cassegrain telescope has a secondary whose diameter is about 10% that of the main aperture, or 1% of the area. The feed support has approximately the same blocking cross-section, therefore the reduction in gain, and hence the elevation of the

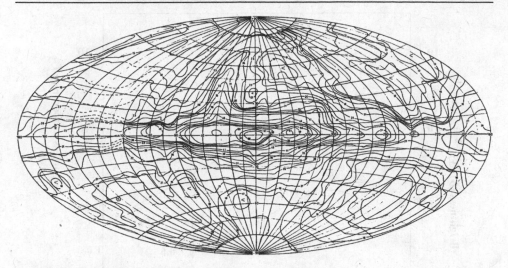

Figure 8.4. An all-sky map at 30 MHz. The contours of brightness temperature are labelled in units of 1000 K (Cane 1978).

sidelobes, will be 4%. These sidelobes will not be uniformly distributed over the sky; the feed-support sidelobes will have the form of narrow spokes across the sky, subtending a relatively small solid angle, and thus will have a surprisingly high sensitivity.

The relative importance of the various corrections depends upon the observing frequency. For instruments to study the CMB at wavelengths of 1 cm and shorter, the feed noise (several kelvins), atmospheric contribution (5–20 K, unless the instrument is in space) and ground contribution can be large compared with the brightness temperature of the sky, 2.74 K. The ground contribution can be minimized by properly shielding the telescope, but the telescope gain pattern must be carefully measured, to compensate for radiation from the Galactic plane into the sidelobes. For the 408-MHz map of Figure 8.3, the lowest sky brightness temperatures are about 10–20 K, so the sidelobe contributions from ground radiation and from the brighter regions near the Galactic plane may be considerable. Feed noise and the atmospheric contribution (if it is not raining) are negligible. At this frequency and higher, the minimum sky brightness, near the Galactic poles, has to be measured separately, using a telescope of smaller size but with accurately known ground-radiation contributions. The contribution of 2.74 K from the CMB must also be subtracted. For example, in a careful measurement at 2000 MHz Bersanelli *et al.* (1994) found that the Galactic contribution at the South Galactic Pole is 0.29 K.

A contrasting case is presented by the 30-MHz sky map of Figure 8.4 (Cane 1978). The minimum brightness temperature at the poles is close to 20 000 K, ranging up to 100 000 K, so feed noise, ground radiation and atmospheric contributions are negligible. On the other hand, the beam efficiency has to be well-calibrated, sidelobe corrections may be large, and the relatively poor angular resolution of a single-aperture telescope at this long wavelength introduces smoothing problems.

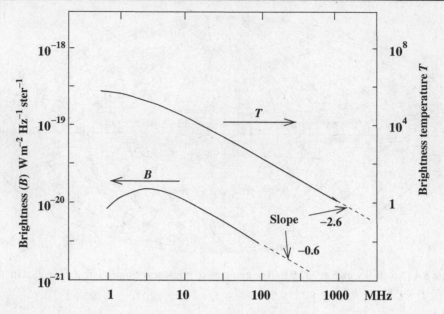

Figure 8.5. Brightness temperature and specific intensity of the Galaxy at high Galactic latitudes, after Cane (1979).

Although telescopes with an angular resolution of about 1° are adequate to delineate the main large-scale features of the continuum radio emission from the Galaxy, much more is revealed when greater resolution is available. A survey by the Effelsberg 100-m telescope at 1.4 GHz has mapped large sections of the sky with resolution of 9 arcmin, including measuring polarization. A first section of the survey was published by Uyaniker *et al.* (1999). It is salutary to contemplate the magnitude of the task: there are about 40 000 square degrees in the sky, so a full survey at this resolution would involve 2 000 000 individual telescope pointings.

8.4 The spectrum of the Galactic continuum

Jansky's first observations posed a severe challenge to our understanding of the Galaxy. He showed clearly that the radiation at 20 MHz had to be associated with the Galaxy, and could not be coming from stars such as the Sun. The historical review in Appendix 3 gives references that illustrate the imaginative approach that was needed to understand the new regime. The 1947 paper by Townes, which includes the first use of the term 'brightness temperature', is an illuminating example: Townes concluded that Jansky's observed brightness temperature, 150 000 K, supported by the early work of Reber and of Hey, was too high to be explained by a reasonable thermal origin. The critical nature of the problem is illustrated in Figure 8.5, which plots the average brightness temperature at high Galactic latitudes as a function of frequency. At frequencies above 1 GHz the temperature is nearly constant at 2.74 K over a wide frequency range: this is the CMB, which is the relict radiation from the early Universe (Chapter 14). The brightness temperature of the high-latitude Galactic component is below the CMB at frequencies above 1 GHz, but it has a very steep spectrum and rises to about 1.6×10^7 K at 1 MHz.

The maps of Figures 8.3 and 8.4 show considerable structure in this high-latitude Galactic emission. The structure becomes increasingly difficult to resolve at longer wavelengths, but it is nevertheless useful to follow the overall average spectrum down to the lowest possible frequencies. Figure 8.5 from Cane (1979) shows the spectrum in terms of specific intensity I_ν (related to brightness temperature by the equation $I_\nu = 2kT/\lambda^2$; see Chapter 2). Above 10 MHz the spectrum is well fitted by a power law of the form

$$I_\nu = 9 \times 10^{-21} (\nu/10_{MHz})^{-0.55} \, \text{W m}^{-2} \, \text{Hz}^{-1} \, \text{ster}^{-1} \tag{8.2}$$

or, in terms of temperature,

$$T_b = 3 \times 10^5 (\nu/10_{MHz})^{-2.55} \, \text{K}. \tag{8.3}$$

The *spectral index*[3] α is then -0.55 in this spectral region. Below 3 MHz the high-latitude spectrum falls steeply; this is obviously an interesting, although badly observed, region of the spectrum. Unfortunately, the terrestrial ionosphere makes it very difficult to observe at such low frequencies, while satellite observations from above the atmosphere necessarily have very low angular resolution and are not much help in sorting out the reason for the dramatic turnover in the spectrum.

The character of the observed Galactic radiation at lower Galactic latitudes is complicated by the presence of discrete sources, both Galactic and extragalactic; these must be subtracted from the maps to obtain the underlying continuum. At very low frequencies the angular resolution even of the largest arrays is still several degrees, and separation of discrete sources from the background is an uncertain process. Note also that at high Galactic latitudes the continuum background becomes fainter than the CMB at frequencies above 1 GHz. Despite these difficulties, the mean spectral index of the Galactic synchrotron radiation between 100 MHz and 1 GHz is realistically determined to be -0.6 ± 0.2. The spectrum is distinctly curved, changing smoothly from a spectral index of -0.4 at lower frequencies to -1.0 above 1 GHz (Reich and Reich 1988; Davies *et al.* 2006).

Among the discrete Galactic sources that are superimposed upon the Galactic continuum there are many compact sources such as stars (Chapter 11) and pulsars (Chapter 12); these contribute little to the measured sky brightness. Supernova remnants and H II regions, on the other hand, show up prominently in the continuum maps, mainly at low Galactic latitudes. The supernova remnants exhibit non-thermal spectra, and typically have high surface brightnesses and spectra that follow an inverse power law with frequency, with spectral indices not greatly different from those of the Galactic background. The surface brightness can be high; the youngest supernova remnants have surface brightnesses in

---

[3] Note that in many earlier texts the opposite sign is used, so that $I_\nu \propto \nu^{-\alpha}$. See the footnote to Section 7.3.

excess of 10^{-18} W m^{-2} Hz^{-1} ster^{-1} at 100 MHz, corresponding to a brightness temperature of 300 000 K.

The H II regions are strong thermal sources: they are ionized gas clouds, heated and ionized by the ultraviolet radiation from young stars. Their temperature is limited to below 10 000 K by the efficient cooling effect of forbidden-line radiation from ionized oxygen. The radio spectrum of an H II region is that of a plasma, which may be optically thick or thin (Chapter 7). The Orion Nebula, M42, is a typical example of a moderate-sized H II region; its low-frequency flux density is nearly proportional to ν^2, following the Rayleigh–Jeans law closely (Figure 7.6). This means that the brightness temperature is constant (at about 9000 K), the plasma is optically thick, and the observed brightness temperature is a measure of the kinetic temperature. At higher frequencies, the spectrum is nearly flat, resembling the classical optically-thin bremsstrahlung spectrum discussed in Chapter 7. If the spectrum could be expanded, it would show discrete spectral lines, which are the hydrogen-atom-recombination lines.

At low frequencies, as seen in the 30-MHz map of Figure 8.4, optically thick H II regions obscure the more distant bright non-thermal sources and reduce the brightness temperature on the plane. This can be seen in more detail in a survey of the northern sky at 22 MHz by Rogers *et al.* (1999). The more widespread fall in the spectrum below 3 MHz, seen in Figure 8.5, may be due to absorption in more local ionized hydrogen subtending a larger solid angle; this is examined further in Section 8.10 below. Delineation of the sources of non-thermal radiation in the disc is a difficult problem due to this absorption at low frequencies and the comparative brightness of the thermal sources at high frequencies.

8.5 Synchrotron radiation: emissivity

Interpreting the synchrotron radiation of the Galaxy in terms of electron energies and magnetic field strength requires a three-dimensional model. In other spiral galaxies we see a thin disc containing the spiral arms with, in many cases, an extended halo that is apparent when the galaxy is seen edge on. We make the same distinction between disc and halo for the radio emission from our Galaxy, following Beuermann *et al.* (1985), who distinguished between a thin disc, containing a mixture of ionized and neutral hydrogen with supernova remnants, and a thick disc, which merges smoothly with the high-latitude radiation. Radiation from the thick disc, or halo, comprises 90% of the total Galactic emission at 408 MHz, which is 9×10^{21} W Hz^{-1}. The thick disc has a full equivalent width of 2.3 kpc in the vicinity of the Sun (Galactic radius $R = 8$ kpc), increasing to 6.3 kpc beyond $R = 12$ kpc. The observed brightness temperature T_b at 408 MHz can now be assigned to an emissivity $B_t = 7$ K kpc^{-1} in the thick disc at $R = 8.5$ kpc, increasing to 31.5 K kpc^{-1} at $R = 4$ kpc (Reich and Reich 1986). We now follow the analysis of synchrotron radiation in Chapter 7 to obtain the flux of cosmic-ray electrons and the strength of the magnetic field.

There are some reasonably direct measurements of the strength of the general Galactic magnetic field (Section 8.8), but these are mainly confined to the ionized regions of the thin disc. Again, there are measurements of the flux of electrons at cosmic-ray energies arriving at the Earth, but this might not be representative of the flux in the thick disc. The only measure for both the magnetic field and the electron flux in the thick disc is from the synchrotron emissivity itself. The radio emissivity depends on both quantities, but they can be separated if it can be assumed that the energy density of cosmic rays is in equilibrium with the energy density of the magnetic field. This assumption of *equipartition* is made in the interpretation of other radio sources, and we briefly outline the analysis. We require the condition that the total energy contained in the field and in the particles is a minimum. Note that the assumption of equipartition and minimum total energy is a common hypothesis, but not supported by rigorous theory. It may well be a valid assumption if quasi-equilibrium conditions hold.

From Equation (7.13) the total power P emitted from N electrons with relativistic energy γ in a field B varies as

$$P \propto N\gamma^2 B^2. \tag{8.4}$$

We use the approximation that radiation at frequency ν originates from electrons with relativistic energy γ, where

$$\nu \propto \gamma^2 B. \tag{8.5}$$

Equipartition requires a minimum in the total energy density E_t, which is the sum of the cosmic-ray and magnetic-field energies:

$$E_t = E_{cr} + E_B. \tag{8.6}$$

Since $E_B = B^2/(8\pi)$ and E_{cr} is proportional to $N\gamma$, the minimum is obtained by differentiating,

$$E_t = c_{cr}N\gamma + c_B B^2, \tag{8.7}$$

where c_{cr} and c_B are constants, giving the equilibrium value

$$B \propto P^{-3/7}\nu^{-1/7}. \tag{8.8}$$

The constant c_{cr} includes the ratio of the total cosmic-ray energy (including protons) to the electron energy; this ratio is usually taken to be ≈ 100. The constant is also a function of the energy-spectral index of the cosmic-ray particles, since Equation (8.5) is a simplification depending on the spectrum (see Chapter 7; detailed calculations can be found in the survey paper by Heiles (1995)). Characterizing the volume emissivity of the Galaxy in terms of brightness temperature at 400 MHz as 7.3 K kpc^{-1}, Heiles finds a magnetic field strength of 7.4 microgauss. Note that from Equation (8.8) this is a moment average $\langle B_{t,min}^{7/2}\rangle^{2/7}$, which is weighted towards higher values; we note later that there is in fact a range of field strengths from about 2 to 10 microgauss.

It is gratifying to find that the value of the field found in this way agrees well with other measurements, which we discuss in the next section. The assumption of equilibrium does not, however, apply in some other galaxies, where the magnetic field has been found from Faraday-rotation measurements to be an order of magnitude higher than the equilibrium

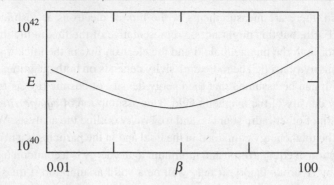

Figure 8.6. The total energy in magnetic field and particles versus β, the ratio of magnetic to particle energy for a synchrotron source of radio luminosity 10^{31} ergs s^{-1} of size 100 by 20 mas at a distance of 1 kpc. A radio spectral index of -0.6 has been assumed.

value. The calculation of the total energy density may nevertheless be useful, since this does not depend critically on exact equipartition. Spencer (1996) shows in Figure 8.6 that in a typical discrete source the total energy varies by less than an order of magnitude when the ratio of magnetic to particle energy is between 0.01 and 100.

8.6 The energy spectrum of cosmic rays

The electrons responsible for the Galactic synchrotron radiation mainly have energies in the range 1–10 GeV. They can be observed directly as cosmic-ray electrons from balloon-borne spectrometers, which largely avoid the losses in the Earth's atmosphere. At low energies the flux can be reduced by the deviation in the solar magnetic field (Moskalenko and Strong 1998), but the energy spectrum of cosmic rays can be extended to extremely high energies by observations of extensive cosmic-ray showers produced by nuclear collisions in the high atmosphere. The most abundant high-energy particles are protons. Helium nuclei have a flux that is about an order of magnitude less, while cosmic-ray electrons have a flux that is about two orders of magnitude lower than the proton flux. Heavier nuclei are also present, but with still lower fluxes.

The mechanism by which high-energy nuclei and electrons are accelerated is partly understood, starting with the work of Fermi (1949a, 1949b). Recognizing that the ISM is permeated by magnetic fields, and that the medium is not uniform, but clumped into clouds, Fermi showed that charged particles could be accelerated by colliding with these magnetized clouds. In the collision, the cloud acts as a body of enormous mass, and its random motion gives it an enormous effective temperature. By the same second-order collisional process as that by which gases come to equilibrium, the particles increase in energy until the magnetic field cannot prevent them from passing through the cloud.

The shock wave of supernovae expanding into the magnetized ISM can also accelerate particles, this time as a first-order process. A survey of present-day cosmic-ray techniques

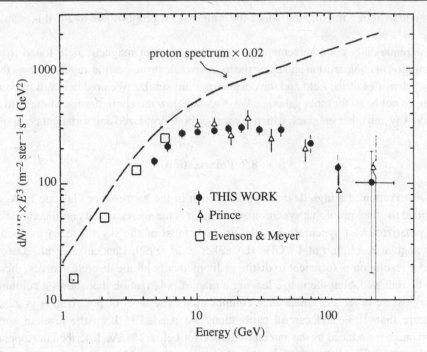

Figure 8.7. The observed energy spectrum of cosmic-ray electrons (Rockstroh and Webber 1978). The differential spectrum dN/dE has been multiplied by E^3 in this plot, since the population falls steeply with energy. The proton spectrum, scaled down by a factor of 50, is shown for comparison.

and knowledge, together with references, is given by Longair in *High Energy Astrophysics* (1994).

The observed spectrum of cosmic-ray electrons in the range accessible by balloons is shown in Figure 8.7. Above an energy of about 1 GeV the flux of particles reaching the Earth is reliably measured, with little influence from the magnetic field entrained in the solar wind. Between 5 and 50 GeV Golden *et al.* (1984) found that the flux of electrons follows a simple power-law spectrum $S = S_0(E/E_0)^{-\beta}$, with index $\beta = 3.15 \pm 0.2$. Tang (1984) extended the energy range to 300 GeV, finding that the spectrum steepens at higher energies, with the spectral index increasing to 3.5 above 40 GeV. The spectral index β for electrons with energies of about 10^{9-10} eV is estimated at 2.5 (see also Grimani *et al.* (2002)).

The theory of synchrotron radiation, outlined in Chapter 7, shows that radiation from cosmic-ray electrons with such an energy power law is expected to have a power-law spectrum, with index $\alpha = (1 - \beta)/2$, that is, -0.75. This agrees well with the observed spectral index near the Galactic pole (-0.4 at low and -1.0 at high radio frequencies: see Section 8.4 above). The curvature of the radio spectrum indicates that the spectral index of the electron energy spectrum continues to fall at lower energies, where no direct measurements are possible. The volume emissivity of the Galaxy also agrees well with the measured intensity of energetic electrons. Rockstroh and Webber (1978) used the value of 7.3 K kpc^{-1} for the radio emissivity at 400 MHz, and found that this required a Galactic magnetic field of 7 microgauss. Since the emissivity is proportional to B^2, this is an r.m.s. value $\langle B^2 \rangle^{1/2}$.

The remarkably good agreement between the values of magnetic field found from the assumption of equipartition and using the measured electron–cosmic-ray flux shows that the energy densities of the field and the cosmic rays are similar. We note below, however, that this need not be so for other galaxies. We discuss below the configuration of the field in the Milky Way and other galaxies, which comprises both organized and turbulent components.

8.7 Polarization

The observation of a high degree of polarization in the metre-wave Galactic background provided the final proof that synchrotron radiation is the source of the continuum radiation. The polarization of synchrotron radiation from most of the sky has been mapped with high angular resolution at 1.4 GHz (Uyaniker *et al.* 1999; Landecker *et al.* 2006). The angular resolution is sufficient to distinguish practically all the discrete sources; these can then be removed from the maps, leaving a map of polarization that shows a combination of large-scale order and small-scale complication. The level of polarization is generally no more than 10%, and careful calibrations are needed to keep the level of spurious polarization introduced by the antenna and receiver below 1%. As described in Appendix 4, the characteristics of the telescope must be evaluated as a Mueller matrix; an example is the survey by Wolleben *et al.* (2006), using a 26-m reflector to survey the northern sky at wavelength 1.4 m.

From Chapter 7, the degree of linear polarization of synchrotron radiation in a well-ordered field should be up to 70%. Such a high degree of polarization would normally be reduced by the superposition along a line of sight through the Galaxy of sources with different polarization position angles; in several directions, however, the magnetic field is sufficiently well ordered for the polarization to exceed 10%. An excellent example, shown in Figure 8.8, is centred on $\ell = 140°$, $b = 10°$ (Brouw and Spoelstra 1976). The two maps show observations of the same region at 465 MHz and at 1411 MHz; the highly polarized region is more disordered at the lower frequency, and the position angles of the vectors have opened out like a fan. This is due to Faraday rotation in the ionized interstellar medium; at the centre of the polarized region the line of sight is perpendicular to the magnetic field, and on either side there is a component along the line of sight, rotating the vectors in opposite directions.

At the edges of this region the degree of polarization is reduced because of the Faraday rotation occurring within the source; this is often a more important reason for the low degree of polarization in a synchrotron source than the topology of the field itself.

8.8 Faraday rotation: the Galactic magnetic field

There is detailed structure in the polarized Galactic radiation, even at high Galactic latitudes and away from the Galactic centre. The magnetic field apparently has an ordered and a

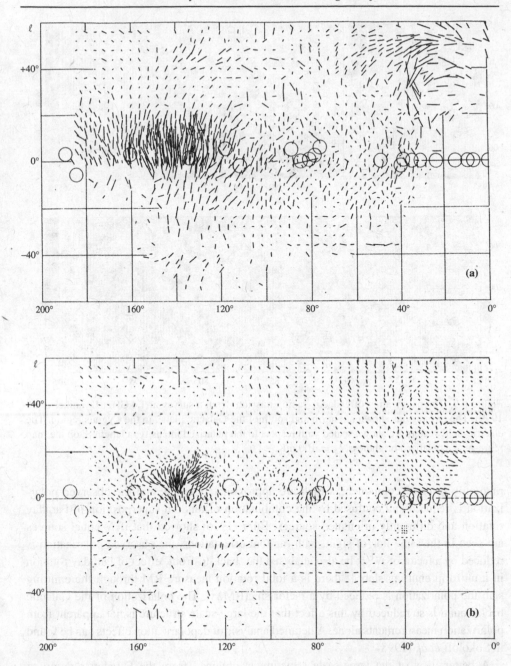

Figure 8.8. Polarization in a region of well-ordered magnetic field. The two maps show the region centred on $\ell = 140°$, $\ell = 10°$, observed at 465 MHz (a) and at 1411 MHz (b) (Brouw and Spoelstra 1976).

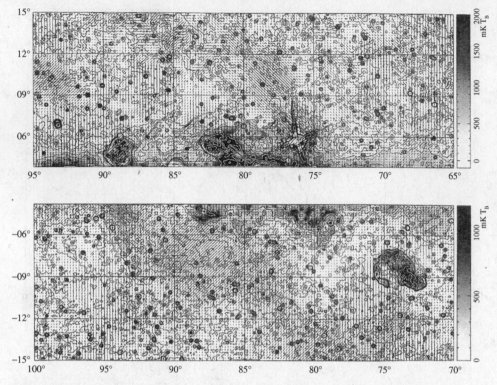

Figure 8.9. The polarization of Galactic radio emission, from a survey at 1.4 GHz (Uyaniker *et al.* 1999). The two regions shown are north and south of the Galactic plane, in the Cygnus region. The electric-field vectors are shown scaled with respect to the polarized intensity; a length 8′ on the map represents a polarized brightness $T_B = 100$ mK.

randomly oriented component, which are approximately equal, and the observed polarization is reduced and confused by the combination of random superposition of Faraday rotation and differently orientated sources. Burn (1966) showed that if N equal sources are seen in the same line of sight, with random orientations, the observed polarization is reduced by a factor of $N^{1/2}$; he also analysed the 'back-to-front' effect of Faraday rotation in a uniform emitting slab. If there is a total rotation measure RM through the emitting slab the polarization is reduced by a factor sinc(RMλ^2). The polarization of the Galactic background is so reduced by this effect that the large-scale structure is not apparent from polarization measurements alone. A detailed analysis of depolarization effects can be found in Sokoloff *et al.* (1998).

A better view of the large-scale structure is obtained from the Faraday rotation of polarized radiation from extragalactic sources. Faraday rotation is proportional to the product of electron density and the magnetic-field component along the line of sight; it is also dispersive, so it can be measured as a difference in polarization position angle at

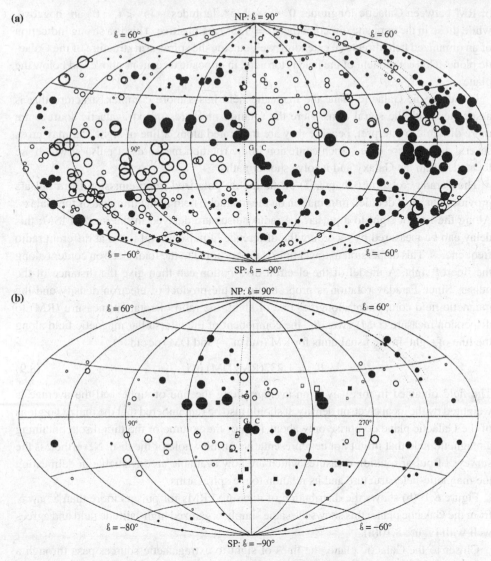

Figure 8.10. Rotation measures of (a) extragalactic sources (note the division between positive and negative values indicated by the dotted line at $\ell = 8°$) and (b) pulsars at high Galactic latitudes $|\delta| > 8°$. The area of the symbols is proportional to $|RM|$, ranging from 5 to 150 rad m^{-2}. Filled and open circles represent positive and negative RM values, respectively (Han *et al.* 1999).

adjacent frequencies (Chapter 7). It obviously includes rotation within the extragalactic source itself, but an average over several sources can be used to find the Galactic component in a region of sky. The distribution of rotation measures of extragalactic sources is shown in Figure 8.10(a), where positive and negative values are distinguished by open and filled circles. A remarkable degree of organization emerges: for example, almost all values of RM between Galactic longitudes $0° < \ell < 135°$, latitudes $-45° < \ell < 0°$ are negative, while those in the opposite direction (near $\ell = 135°$) are positive. This is a strong indication of an organized field direction, extending well outside the spiral-arm structure of the Galactic plane. This component of the field is toroidal, in opposite directions above and below the plane.

Closer to the Galactic plane the magnetic field has a more complex structure that is associated with the spiral arms. Here the rotation measures of extragalactic sources are more difficult to interpret, because they are integrated along a line of sight that may cross several arms. There is less such confusion in the rotation measures of pulsars, which are located within the Galaxy and involve shorter paths.

The Faraday rotation (Chapter 7) of the highly polarized radio emission from pulsars provides our most detailed information on the magnetic field in the thin disc of the Galaxy. Along the line of sight to a pulsar, the radio pulses are delayed in the ionized ISM; this delay can be measured by recording the difference in pulse arrival times at different radio frequencies. This dispersion in travel time is proportional to the total electron content along the line of sight. A model of the electron distribution can then give the distance of the pulsar. Since Faraday rotation is proportional to the product of electron density and the magnetic-field component along the line of sight, the ratio of rotation measure (RM) to dispersion measure (DM) gives B_L, the component of the interstellar magnetic field along the line of sight. In the usual units for RM (rad m^{-2}) and DM (pc cm^{-3}),

$$B_L = 1.232(\text{RM/DM})\,\mu\text{G}. \tag{8.9}$$

The field obtained in this way is an average along the line of sight, and the average is weighted by the local electron density. It should also be remembered that the ionized regions of the Galactic plane comprise only about 10% of the volume of the thin disc, containing a magnetic field that might not be representative of the whole of the disc. Nevertheless the surveys of pulsar rotation measures which are now available are invaluable in delineating the magnetic-field structure and its relation to the spiral arms.

Figure 8.10(b) shows the distribution of measured RMs for pulsars more than 8° away from the Galactic plane; this shows the same simply organized high-latitude field and agrees well with Figure 8.10(a).

Closer to the Galactic plane, the lines of sight to extragalactic sources pass through a system of spiral arms, in which the magnetic field has a more complicated structure than the toroidal field above and below the Galactic plane. Here the dispersion measures of pulsars are invaluable because of their shorter lines of sight. Figure 8.11 (Han 2007) shows a compilation of rotation measures of low-latitude pulsars, projected onto the Galactic plane, with rotation measures from extragalactic sources around the periphery. The diagram is centred on the Galactic centre, and the Sun is located towards the top; Galactic longitudes are shown. The locations of three spiral arms are shown; the evidence for these depends mainly on the pulsar rotation measures. On the basis of a similar compilation, Brown *et al.* (2007) show that the magnetic field follows the spiral arms in the inner Galaxy, clockwise

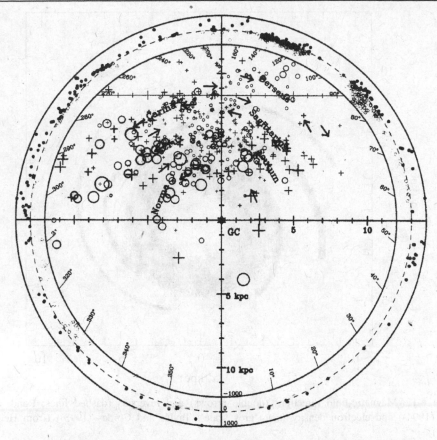

Figure 8.11. Rotation measures of pulsars and extragalactic sources with Galactic latitude $|\delta| < 8°$. The size of the symbols is proportional to $RM^{1/2}$, + are positive and ○ negative RM. (Han 2007).

in the Sagittarius–Carina arm and anticlockwise in the Scutum–Crux arm. Similar reversals have been proposed for the outer arms; Figure 8.12 is a sketch of the arms as delineated by electron density, with magnetic-field directions shown by arrows (Heiles 1995).

In addition to the rotation measures described above, evidence for the magnitude and structure of the Galactic magnetic field is obtained from the intensity and polarization of the Galactic synchrotron radiation, and, within some discrete sources, from Zeeman splitting of maser line radiation. Our location within the Galactic plane requires a synthesis of all the evidence to proceed via a model, involving a toroidal field out of the plane and a spiral-arm field within it (Sun *et al.* 2008). The connection between these two systems is not understood.

There are both large- and small-scale deviations from this pattern. On small scales within the arms there is a random component of approximately equal magnitude to the organized field; this is the same random field as that seen in the thick disc as structure in

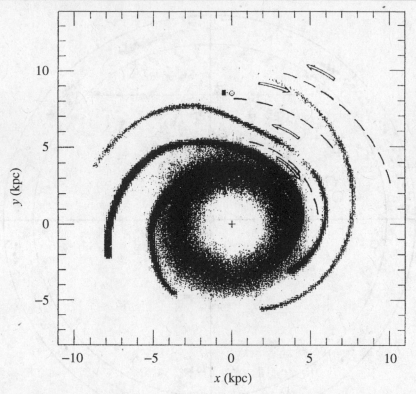

Figure 8.12. Magnetic-field directions (double arrows) and reversals (dashed lines; Rand and Lyne (1994)) and electron density $\langle n_e \rangle$ ('grey scale'; Taylor and Cordes (1993)) (from Heiles (1995)).

the polarization of the synchrotron radiation. Between the arms, where the large-scale field is small, the random field predominates.

Most estimates of the strength of the Galactic magnetic field are in the range 3–8 µG, comprising roughly equal turbulent and organized components. The organized component is larger towards the Galactic centre, falling to 2 µG near the Sun. The origin of the field is much debated. A 'primordial' field, built in from the early Universe, may play a role, but that is still a conjecture, unsupported by observations or rigorous theory. Possibly, the large-scale field is derived from stellar magnetic fields via supernova explosions. The development of the organized field would then be due to a dynamo action of the large-scale differential velocities in the Galaxy; the turbulent field would then be degrading to large-scale organization, the reverse of many other turbulent phenomena (see Beck *et al.* (1996) for a review). On larger scales it seems that major dynamical events both in the Galactic centre and, at larger Galactic radii, in supernova explosions, may drive the field into quite different configurations. This is most readily seen in other galaxies, where we have a distant perspective rather than the confusing view from inside.

Figure 8.13. The orientation of the large-scale magnetic field in M51, from observations of the polarization of radio emission at high radio frequencies (Neininger 1992).

In many other galaxies the overall organization of the magnetic field is evident from the polarization of synchrotron radio emission. Figure 8.13 shows the orientation of the large-scale magnetic field in M51, where the field lines follow the spiral structure; in contrast the field lines in NGC 4631 (Figure 8.14) show a radial pattern, which is believed to be due to an outward flow of ionized interstellar gas (Golla and Hummel 1994). Comprehensive reviews of magnetic fields in galaxies have been published by Kronberg (1994) and by Wielebinski and Krause (1993); see also *Cosmic Magnetic Fields*, edited by Beck and Wielebinski (2005).

8.9 Loops and spurs

The structure which can be seen away from the plane of the Galaxy, especially at low radio frequencies as in Figure 8.3, has no counterpart in stars or any other visible object. The most obvious feature in this high-latitude structure is the North Galactic Spur, which leaves the plane at $\ell = 30°$ and reaches to above $b = 30°$. It is best described as part of an approximately circular loop, with diameter $116°$, centred on $\ell = 329°$, $b = 17°$. This is also known as Loop I; two similar features are designated Loops II and III. All are believed to

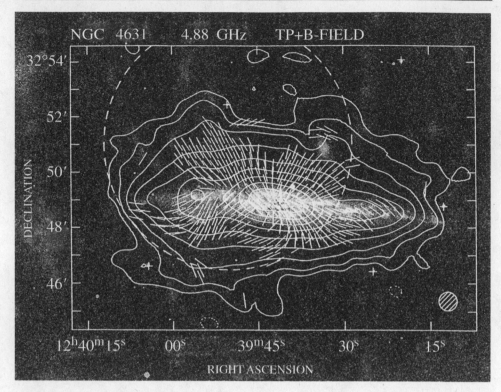

Figure 8.14. The radial magnetic field around NGC 4631 (Golla and Hummel 1994).

be the outer shells of supernova remnants, and possibly the superposition of several such shells originating in an area of active star formation.

The radio emission from Loop I has a steep spectrum characteristic of synchrotron radiation. It is strongly linearly polarized; as expected, the degree of polarization is greater at higher frequencies where the effects of depolarization are least; see the WMAP observations at 22 GHz (Page *et al.* 2007). The magnetic-field direction is mainly along the line of the Loop, indicating that the field has been swept up by an expanding shell. There is, however, hydrogen-line emission from the outer part of the shell, indicating an enhanced density of cool gas, and also enhanced X-ray emission from within the shell, indicating a very-high-temperature region. This suggests that hot gas within a bubble is compressing the cold ISM as it expands (Heiles *et al.* 1980), and that this region already has an abnormally high density. The hydrogen-line radiation shows Zeeman splitting, giving a value of 5.5 μG for the magnetic field (Verschuur 1989). The effects of this field can also be seen in Faraday rotation of radio sources beyond the Loop.

Loop I is now regarded as a *superbubble*, which is the effect of a series of supernova explosions in an active region; the prominent shell is due to the most recent of these explosions (Wolleben 2007). The distance of the shell is probably about 100–200 pc. It is not seen below the plane; this is presumably due to the location of its supernova progenitor above the plane, and to a sufficient concentration of mass in the plane to halt expansion in this direction.

These Loops are evidently local and comparatively short-lived features of the Galaxy. As we shall see in Chapter 11, supernova explosions may be an important source of energy in the ionized gas which extends more than one kiloparsec from the plane, and the effects of an individual explosion may last as long as 10^6 yr. The only direct evidence of the supernova itself would be the presence of a neutron star in the right place and with the right age. Such evidence is given by the gamma-ray source Geminga. This is one of the three brightest gamma-ray sources in the sky, the other two being the Crab and Vela pulsars. Almost 20 years after the discovery of Geminga, it was found to be a strong X-ray source that is pulsating with a period of 237 ms; the same periodicity was then found in the gamma-ray observations. Although no radio pulsar can be seen, Geminga is evidently a neutron star. Its age is about 300 000 years, as determined from its slowdown rate. It is visible as a very faint object with a very large proper motion; even at a distance of only 100 pc its speed must be over $100 \, \mathrm{km \, s^{-1}}$. Its position and age are at least consistent with its identification as the collapsed centre of the explosion which produced Loop I.

8.10 The Local Bubble

In the next chapter we shall frequently refer to the inhomogeneity and variety of the ISM. The loops and spurs of the preceding section are obvious examples. Less obvious, although probably very similar, is a bubble-like structure, which actually contains the Sun. This 'Local Bubble' is a region with dimensions of order 100 pc, with a comparatively low density of hydrogen, about $5 \times 10^{-3} \, \mathrm{cm^{-3}}$, and a high temperature, about 10^6 K. The product of density and temperature is a pressure; this is reasonably in equilibrium with the surrounding regions.

Above Galactic latitude $60°$, the column density of neutral hydrogen drops more rapidly than it should for a plane-parallel atmosphere, and this may be evidence for most of the high-latitude gas being ionized (see Dickey and Lockman (1990) for a summary of the evidence). In particular, there is a high-latitude 'hole' in the ISM, located in the constellation Ursa Major (RA 10 h 40 m, Dec $+58°$) where the local hydrogen falls below a column density of $4 \times 10^{19} \, \mathrm{cm^{-2}}$. This region has been studied in detail by Jahoda *et al.* (1990).

The best evidence for the Local Bubble comes from combining soft-X-ray data with measurements of the Lyα absorption in nearby O and B stars. The X-ray emission comes from the hot, ionized gas, while the absorption depends on the line-of-sight density of neutral hydrogen. The bubble boundary is defined by a column density of 10^{19} hydrogen atoms $\mathrm{cm^{-2}}$ (Paresce 1984). The temperature is given by the X-ray brightness. Its origin is unknown; it may be formed by outflowing hot gas from a group of young stars, or it may be the result of a single supernova explosion. If it is a supernova remnant, its age would be about 10^5 yr, and it would have occurred within 100 pc of the Sun. We have already remarked that the X-ray pulsar Geminga might have originated in a supernova that created Loop I; it is equally possible instead that this was the origin of the Local Bubble (Gehrels and Wan Chen 1993; Bignami *et al.* 1993). Only a general conclusion can be drawn: the complex structure of the ISM, as seen in the radio maps of Figures 8.3 and 8.4, is at least partly formed by the transient effects of supernova explosions.

The extent and outline of the Local Bubble has been described in the review by Cox and Reynolds (1987). A large new body of evidence from satellites observing in the extreme ultraviolet will provide far more extensive data on such local structures.

8.11 Other galaxies

Although radio astronomy has revealed our Milky Way Galaxy to be the seat of very energetic phenomena, ranging from the cosmic rays generating the bright radio background to the black hole at the Galactic centre (Chapter 10), the Galaxy is not in the same class as the radio galaxies and quasars which we discuss in a later chapter. Nearly all of these are elliptical galaxies; they are such powerful emitters that they can be detected at very great distances, and consequently they dominate the now extensive catalogues of extra-galactic radio sources. Several comparatively nearby spirals are, however, accessible to radio observation (Dressel and Condon 1978). Many of these, like our Galaxy, have total radio luminosities of about 10^{37} erg s^{-1}; other normal spirals including the Andromeda Nebula, M31, have emissivities up to ten times greater. For some of the nearer galaxies it is possible to estimate the flux of high-energy electrons from observations of the gamma-ray flux, which results from electron–proton collisions (Chi and Wolfendale 1993), allowing an estimate of the magnetic field from the equipartition argument in Section 8.5 above. Comparison with more direct measurements from Faraday rotation suggests that equipartition might not apply, since the measured fields are found to be double those expected from equipartition.

In Chapter 13 we describe the much more powerful Seyfert and star-burst galaxies, which have luminosities up to about 10^{40} erg s^{-1}. The luminosities of quasars and radio galaxies are again much larger, ranging up to 10^{45} erg s^{-1}.

Further reading

Atlas of Peculiar Galaxies, Arp H., *Astrophys. J. Suppl.* **14**, 1, 1966.
The Structure and Evolution of Normal Galaxies, ed. Fall S. M. and Lynden-Bell D., Cambridge, Cambridge University Press, 1981.
Cosmic Magnetic Fields, ed. Beck R. and Wielebinski R., Lecture Notes in Physics, No. 664, Berlin, Springer-Verlag, 2005.
Galactic and Extragalactic Magnetic Fields, IAU Symposium 140, ed. Beck R., Kronberg P. P. and Wielebinski R., Dordrecht, Kluwer, 1990.
High Energy Astrophysics, Longair M. S. (2 volumes), Cambridge, Cambridge University Press, 2nd edn, 1994.

Radio Astrophysics. Nonthermal Processes in Galactic and Extragalactic Sources,
Pacholczyk A. G., San Francisco, CA, Freeman, 1970.
Reference Catalogue of Bright Galaxies, de Vaucouleurs G. and de Vaucouleurs A.,
Houston, Texas University Press, 1964.
The Carnegie Atlas of Galaxies, Washington, DC, Carnegie Institute of Washington, 1994.
The Hubble Atlas of Galaxies, Sandage A., Washington, DC, Carnegie Institute of
Washington, 1961.

9

The interstellar medium

Between the stars, the thin mixtures of gas, dust and high-energy particles that make up the interstellar medium have a strong influence on observations despite the rarefied density. At optical wavelengths, the dust absorbs starlight; at infrared wavelengths the dust emits thermal radiation; at sub-millimetre and millimetre wavelengths the gas molecules emit a rich array of spectral lines; at 21 cm the strong hyperfine line of interstellar atomic hydrogen is a tracer of galactic dynamics; at centimetre and metre wavelengths, there is thermal bremsstrahlung from the ionized hydrogen component; as the hydrogen atoms recombine, the high-quantum-number transitions give rise to recombination lines throughout the radio spectrum. In this chapter, the emphasis will be on the physical properties of the gaseous component as revealed by spectroscopic studies, and on the electron component as revealed by its effect on radio propagation. The following chapter will address the application of atomic and molecular radio spectroscopy to large-scale aspects of galactic structure. By general convention, 'interstellar medium' is usually abbreviated to ISM.

9.1 Atoms and molecules

Some atomic species, notably neutral hydrogen, are observable by virtue of transitions between high-quantum-number orbital states. These spectral lines, which are a continuation of the Balmer and other series in the visible spectrum, are known as *recombination lines*. They occur when hydrogen atoms are ionized by ultraviolet photons and recombine with the electron, initially in orbits with high quantum number, cascading down to lower orbital energies. The wavelength of a transition between quantum levels n_1 and n_2 is given by the Rydberg formula familiar in the low-quantum-number regime of optical physics,

$$\frac{1}{\lambda} = R_H \left(\frac{1}{n_1^2} - \frac{1}{n_2^2} \right), \tag{9.1}$$

where the Rydberg constant R_H is specific to hydrogen but does not differ greatly for electrons recombining with other atomic species, given large n. The first such line to be discovered (Höglund and Mezger 1965) was the transition $110 \rightarrow 109$, designated 109α, at 5009 MHz. Radio recombination lines have also been discovered in several other atomic

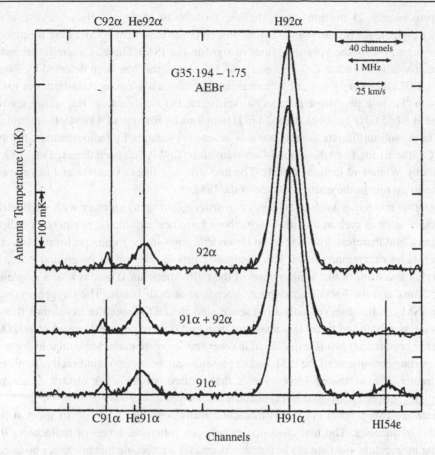

Figure 9.1. Radio recombination lines from the Galactic H II region G35.194-1.75. The observations of two adjacent hydrogen lines in the series, 91α and 92α, are shown together with their sum (Quireza *et al.* 2006. arXiv:astro-ph/0603133).

species, notably helium and carbon. Figure 9.1 shows an example of a spectrum from a survey by Quireza *et al.* (2006), in which lines near 8.5 GHz from carbon, helium and hydrogen can be seen. This example shows the similarity between adjacent lines in the series and between lines from different atomic species. The hydrogen H154ϵ line is a transition in the series with $\delta n = 4$.

Apart from the high-n recombination lines, radio spectral lines of atoms arise from the low-energy transitions between fine- and hyperfine-structure states. The hyperfine transition in the ground state of the hydrogen atom is the quintessential example: observations of the 21-cm (1420.406-MHz) photon emitted when the atom flips between the parallel ($F = 1$) and antiparallel ($F = 0$) configurations of the electron and proton have provided a powerful tool for studying galactic dynamics (Chapter 10). The lifetime of an atom in the upper state is approximately 11 million years; the low probability of decay arises because it is a magnetic-dipole transition, but the low hydrogen density ($\sim 1\,\mathrm{cm^{-3}}$) and long lifetime are compensated for by the very long lines of sight in the ISM. There is a similar transition in deuterium between the $F = 3/2$ and $F = 1/2$ states; this has been detected by Rogers *et al.* (2007). The singly ionized helium atom also has an analogous transition, not for the common ^{4}He isotope, whose nucleus has spin zero, but for the rarer ^{3}He, spin-$\frac{1}{2}$ species; its line at 8.665 GHz has been detected in H II regions by Rood *et al.* (1984). Fine-structure lines in the sub-millimetre domain are now accessible to study by radio methods; the 493-GHz C I line arising from the ground-state transition $^3P_1$–$^3P_0$ has been detected in the Orion Nebula by White and Padman (1991). The fine-structure lines of carbon and oxygen play an important role in the energy balance of the ISM.

Diatomic molecules such as molecular hydrogen, nitrogen, together with symmetrical molecules such as carbon dioxide and methane have neither electric nor magnetic dipole moments, and therefore give rise to no observable lines in the radio spectrum. Molecular oxygen is an exception, since it has a magnetic dipole moment in its ground state, and has a rich spectrum in the 5-mm region of the radio spectrum. There is also a single line at 2.52 mm, and the Earth's atmosphere is opaque at both bands. The oxygen molecule in the ISM can be observed only from space, and the ODIN satellite has shown that the molecule is a thousand times less abundant than had been expected (Larsson *et al.* 2007), probably because oxygen is sequestered in other stable compounds. Molecular hydrogen is an important constituent of the ISM, and its presence can be observed indirectly. To observe H_2 directly, it is necessary to observe in the infrared, where weak electric-quadrupole transitions between rotational levels occur.

There are two general classes of molecular transitions that give rise to spectral lines at radio frequencies. The first class occurs between rotational levels of molecules. Total angular momentum must always be conserved, and for a molecule that possesses no angular momentum other than from rotation, the laws of quantum mechanics require that the square of the angular momentum must be $J(J + 1)\hbar^2$, where $\hbar = h/(2\pi)$ and J, the angular-momentum quantum number, must take on integral values, including 0. The energy levels E_J of a rigid-rotor linear molecule must be equal to the square of the angular momentum divided by twice I, the moment of inertia, so, for rotational state J,

$$E_J = \frac{1}{2}J(J + 1)\hbar^2/I. \tag{9.2}$$

Defining the rotational constant $B_J = \frac{1}{2}\hbar^2/I$, the spectrum of a rigid-rotor linear molecule forms a ladder of linearly increasing frequencies; the frequency of the transition from $J + 1$ to J is $\nu = B_J(2J + 1)$. This is only an approximation, for molecules are not rigid, and act as harmonic oscillators, following quantum rules. The energy, expressed as the Hamiltonian in quantum mechanics, must include the vibrational state, the electronic state and the hyperfine effects of the nuclear spin, in addition to the rotational state. Fortunately these are well separated in energy, so interactions can be accounted for by using perturbation techniques (with a few notable exceptions). Thus the stretching of a molecule changes the frequency by an amount that can be expressed as a power series in J (see Rohlfs and Wilson, *The Tools of Radio Astronomy*, p. 367).

Many of the molecules of interest in radio astronomy have non-linear structures, and in general, since they have three principal moments of inertia, their spectra require a more complex description. For symmetric-top molecules, like methyl cyanide, CH_3CN, the two moments of inertia perpendicular to the symmetry axis are equal, and the Hamiltonian has a simple form, depending only on the total angular momentum and on the component of the angular momentum along the figure axis, as in classical mechanics. (The ammonia molecule NH_3 is a symmetrical rotor, but it is a special case, considered below.)

The general asymmetric-rotor molecules offer no such simplification. The water molecule H_2O is shaped like a boomerang, and has an important transition at 22.234 GHz. The total angular momentum, J, is a good quantum number, and for every J there are $2J + 1$ substates, expressed in the form J_{ab} where each a, b varies from 0 to J (there is only one J_{00} level). The important 22-GHz transition for the water molecule is designated 6_{16}–5_{23}. Most of the transitions occur in the infrared, because of the small moments of inertia, but by chance this transition has relatively low energy. The 3_{13}–2_{20} transition occurs at 183 GHz, and there are several in the 300–400-Hz sub-millimetre range. The rotational energy-level diagram for H_2O is shown in Figure 9.2. Note that there are two 'families' of energy levels, designated ortho- and para-hydrogen. This distinction arises because the two hydrogen atoms in the water molecule can be interchanged by a half-rotation about the symmetry axis, and the protons that form the nuclei are fermions. Thus they must be in an odd state with respect to interchange; the 22-GHz transition is in the ortho family, which by definition is an even state (proton spins both up), so the odd character is given by having a rotational wave function that is odd. The ground state, 0_{00}, has to be odd with respect to the interchange since the rotational wave function is even; hence the nuclear spin state must be odd (total nuclear spin 0, with the antiparallel spins in an odd linear combination). This symmetry characteristic is met in many molecules, such as formaldehyde H_2CO. When there is threefold symmetry, such as for a molecule with a methyl group CH_3, three families are necessary to satisfy the Fermi statistics.

In the second class of molecular transitions, a degeneracy is broken by the molecular rotation. For example, the oxygen molecule O_2 has a spin of 1 (contrary to what one might naively expect, the electrons do not pair each other off completely, so it is in a triplet state), and hence has a net magnetic moment. If the molecule could not rotate, its energy would be the same no matter which way along the molecular axis the magnetic dipole was pointing. This degeneracy is lifted when molecule rotation is taken into account. The actual frequency of the transition will depend upon the rotational state of the molecule. Thus, there is a rich cluster of lines in the region around 55 GHz.

A related and simpler example is given by the hydroxyl radical OH, which exhibits the phenomenon known as Λ-doubling. For a linear molecule, the angular momentum of the electron has to be quantized along the molecular axis. The diatomic OH molecule has an odd number of electrons, and the net angular momentum gives orbital quantum number Λ, which is 1, while the total electron spin is $\frac{1}{2}$. Thus, the ground electronic state exhibits

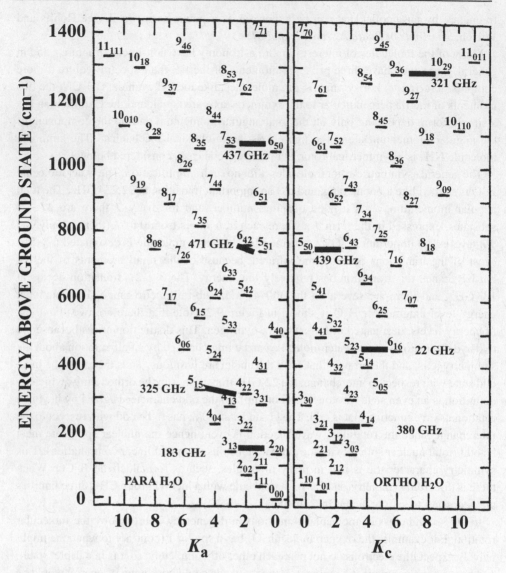

Figure 9.2. Energy-level diagrams for ortho- and para-H_2O. The marked transitions are in the radio range, and are observable as masers (Rohlfs and Wilson, *Tools of Radio Astronomy*, p. 374).

fine-structure splitting, with a $^2\Pi_{3/2}$ ground state and a $^2\Pi_{1/2}$ state lying about $180\,\mathrm{cm}^{-1}$ higher. In the absence of rotation, the projection of the total angular momentum along the molecular axis, whether $3/2$ or $1/2$, would give degenerate energies no matter what the direction might be. However, just as for O_2, molecular rotation breaks the degeneracy into a pair of states of opposite parity, and electric-dipole transitions are allowed, with an energy difference corresponding to a frequency of about $1.6\,\mathrm{GHz}$ for the ground state, with the

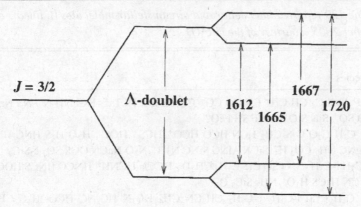

Figure 9.3. Energy levels of OH, showing Λ-doubling. F is the total angular momentum, including hyperfine splitting (Wilson *et al.* 1990).

splitting increasing for higher rotational quantum numbers. In addition, the proton has a magnetic dipole moment, so there is hyperfine splitting as well. There are then four possible states, leading to four lines at 1612, 1665, 1667 and 1720 MHz. The energy diagram for the Λ-doubling of both the $^2\Pi_{3/2}$ and the $^2\Pi_{1/2}$ states is shown in Figure 9.3.

Predictions of spectral lines at radio frequencies (other than the 21-cm hydrogen line) were given by Townes (1957) and by Shklovsky (1960), although absorption lines from interstellar CN, CH and CH$^+$ had been observed optically. In 1963 Weinreb *et al.* discovered emission from the Λ-doublets of OH at 1665 and 1667 MHz. Lines from H_2O, NH_3 and H_2CO (formaldehyde) were discovered shortly afterwards, and, when the Kitt Peak 11-m telescope of the NRAO began operations in 1970, a veritable flood of new discoveries came: at the time of writing (2008), approximately 200 molecular species have been detected in the ISM, and the list is lengthening rapidly. The $J = 1$–0 line of CO at 115 GHz has been a particularly useful tool, especially when the observations have been coupled with measurements of higher rotational states and isotope species.

The ammonia molecule NH_3 presents another type of degeneracy-lifting. The molecule is pyramidal in form, and the nitrogen atom can tunnel through the triangle of hydrogen atoms, so its ground-state vibration will consist of its continually inverting, in what has been described as an oil-can mode. As a result, the ground-state vibrational state can be regarded as a linear superposition of two states, one odd and one even, and, because the rotation lifts the degeneracy, there is an electric-dipole transition between the two, at about 25 GHz. The energy depends upon the rotational state, so, just as in the case of OH, there is a rich set of lines. A somewhat different case is presented by methanol, CH_3OH. The OH group sits on top of the methyl pyramid, and the H-atom angles off to the side. There is thus an allowed internal rotation, influenced by the threefold potential of the methyl group. See *Microwave Spectroscopy* by Townes and Schawlow (1955) for a more detailed description of the allowed states.

Table 9.1. *The 129 reported interstellar and circumstellar molecules (l, linear; c, cyclic) as of November 2005 (courtesy of the NRAO)*

Number of atoms	Species
2	AlF AlCl C_2 CH CH^+ CN CO CO^+ CP CS CSi HCl H_2 KCl NH NO NS NaCl OH PN SO SO^+ SiN SiO SiS HF SH FeO?
3	C_3 C_2H C_2O C_2S CH_2 HCN HCO HCO^+ HCS^+ HOC^+ H_2O H_2S HNC HNO MgCN MgNC NH_2 CO_2 H_3^+ SiCN AlNC SiNC N_2H^+ N_2O NaCN OCS SO_2 c-SiC_2
4	c-C_3H l-C_3H C_3N C_3O C_3S C_2H_2 CH_2D^+ HCCN $HCNH^+$ HNCO HNCS $HOCO^+$ H_2CO H_2CN H_2CS H_3O^+ NH_3 SiC_3 C_4
5	C_5 C_4H C_4Si l-C_3H_2 c-C_3H_2 CH_2CN CH_4 HC_3N HC_2NC HCOOH H_2CHN H_2C_2O H_2NCN HNC_3 SiH_4 H_2COH^+
6	C_5H C_5O C_2H_4 CH_3CN CH_3NC CH_3OH CH_3SH HC_3NH^+ l-H_2C_4 HC_2CHO NH_2CHO C_5N HC_4N
7	C_6H CH_2CHCN CH_3C_2H HC_5N $HCOCH_3$ NH_2CH_3 c-C_2H_4O CH_2CHOH
8	CH_3C_3N $HCOOCH_3$ $CH_3COOH?$ C_7H H_2C_6 CH_2OHCHO CH_2CHCHO
9	CH_3C_4H CH_3CH_2CN $(CH_3)_2O$ CH_3CH_2OH HC_7N C_8H
10	$CH_3C_5N?$ $(CH_3)_2CO$ $NH_2CH_2COOH?$ $CH_3CH_2CHO?$
11	HC_9N
12	$CH_3OC_2H_5$
13	$HC_{11}N$

The number of molecules in the ISM identified through their radio spectral lines is continuously increasing; 129 species are listed in Table 9.1. For many of these several lines have been detected, and in the stronger lines it is often possible to detect lines from isotopic species, such as ^{13}CO or $C^{18}O$. Some of the simpler molecules with hydrogen are observed in deuterated form, such as HDO, DCN and the deuterated forms of ammonia, especially NDH_2. (See also the *Cologne Database*, Müller *et al.* (2005).) The less-abundant isotopic species can be especially valuable when, as for CO, the lines of the principal species are optically thick; also, when there is a question of thermodynamic equilibrium, the lines of the isotopic species can provide a cross-check on local thermodynamic equilibrium.

Many of the species listed in Table 9.1 are found only in the densest of the 'giant molecular clouds', such as the Orion Nebula complex or in the uniquely important cloud Sgr B2, near the centre of the Galaxy. The interstellar chemical processes through which they form are complex, being only partially understood, but this is an active area of research. A pair of single atoms, for example, can only rarely combine to form a diatomic molecule, because a two-body process cannot conserve both energy and momentum. A catalyst that has many internal degrees of freedom is needed, such as a much larger molecule or a dust grain. Various routes have been proposed for the basic pathways (see *IAU Symposium No. 231*, 2005). Molecular hydrogen, for example, might start as a proton (from cosmic-ray ionization), picking up a neutral hydrogen atom by colliding with a dust grain to form the hydrogen molecular ion H_2^+. This can collide with a hydrogen atom in a collision that has a large cross-section because of the many degrees of freedom of the triatomic ion; for example:

$$H_2^+ + H \rightarrow H_3^+.$$

A subsequent collision with an electron gives a multi-body reaction that goes easily:

$$H_3^+ + e \rightarrow H_2 + H.$$

A collision with a neutral H or with neutral H_2 would also work. This is one example of how, given a way of creating H_3^+, further reactions are possible, with the formation of OH being a particularly useful first step. H_3^+ is a species whose line radiation at 2 μm has been detected in the atmosphere of Jupiter (Drossart *et al.* 1989), and it is likely to be detected in the ISM eventually. Unfortunately, the ion has no lines at radio frequencies.

It will be observed that the molecular species are predominantly organic. Although some are remarkably large, there is as yet no indication that any precursors of life, such as the amino acids, are to be found in the ISM. Most of the larger molecules are long chains rather than ring structures: the only known rings are SiC_2 and the transient species C_3H_2. There are larger molecules in the ISM that exhibit ring structures, in the form of polycyclic aromatic hydrocarbons (PAHs); these cause absorption in the infrared, and probably play an important role in interstellar chemistry.

9.2 Kinetic, radiation and state temperatures

The interstellar gas is far from being a homogeneous medium filling the space between the stars; neither is it in a simple state of thermal equilibrium. Temperatures of the various components are often quoted, and it is necessary to distinguish three different meanings. Radiation temperature, as discussed in Chapter 2, relates to the flux of radiation in a particular wavelength band; it is the brightness temperature which may be measured by a radiometer. Kinetic temperature is related to the random velocities of the atoms or molecules in a particular region of the ISM; the various species may have different kinetic temperatures. State temperature refers to the excitation of an atom or molecule, as in Section 7.6. The state temperature may refer to any form of excitation; at the low energies usually encountered in the ISM it is most often encountered as a rotational temperature in a molecule.

The rotational levels in the molecules of the ISM are populated by collisions with hydrogen molecules, which are the dominant constituent in dense regions. De-excitation by radiation occurs rapidly, within hours only, so the population of the excited level need not be in equilibrium with the kinetic temperature. Consequently, even where the CO or CS lines are saturated, the brightness temperature is lower than the kinetic temperature. In order to maintain a spin temperature that approximates the kinetic temperature, the molecules must be in a region of high enough density that the mean interval between collisions is shorter than or of the order of magnitude of the radiation lifetime. This requires an H_2 density of $100 \, \text{cm}^{-3}$ for CO, and a density of $10^4 \, \text{cm}^{-3}$ for CS.

One might have expected that the CO radiation from distant galaxies would be far feebler than that from nearby galaxies, but this is not the case. Observing at constant frequency, the redshift brings spectral lines from higher J-values into the receiving band, and these, in the rest frame, become more intense as J increases. The statistical weight increases, and the radiation probability increases, and these two effects compensate for the factor of $(1 + z)^{-2} z^{-2}$ that diminishes the flux. The result is that the total flux from a galaxy is approximately independent of redshift. The significance of this effect is explored in Chapter 16.

The ISM is in a dynamical state of constant change, driven by ultraviolet heating where stars are forming, by supersonic expansion of shells of supernova explosions, and subject to dynamical instabilities on scales ranging from a small fraction of a parsec to thousands of parsecs, dimensions comparable to those of the entire Galaxy. Wide variations of local densities are found, from below 0.01 to more than 100 H atoms cm^{-3}. Despite these complications, which pose interesting and fundamental problems at the present time, the physical conditions can be measured in a local, quasi-adiabatic sense, and approximations that allow simplified models to be constructed can be made. Under nearly all conditions, the ISM is either in a state that is nearly completely ionized, or it is in a neutral state (with relatively few electrons and ions present).

The ionized component of the ISM has at least two forms. The widespread *warm ionized medium* (WIM), with a density of about 0.025 electrons cm^{-3} is at a temperature close to $10\,000$ K. The *hot interstellar medium* exhibits much higher temperatures, in excess of $500\,000$ K. There are also relatively dense regions of ionized gas, usually associated with a hot star or stars responsible for the ionization. These are known as H II regions; here the density may be from 10 to 10^6 electrons cm^{-3}. An important task for understanding the ionized component (and the neutral component as well) is to identify the heating and cooling mechanisms that lead to the observed state.

The neutral ISM also occurs in several temperature regimes. The main component is the *cool* neutral ISM close to the Galactic plane, which is traced by the 21-cm hydrogen line, and is at a temperature of about 100 K (within a factor of 2). The mean density of neutral hydrogen atoms is here about $0.4 \, \text{cm}^{-3}$. The *cold* neutral medium is encountered more locally; it is at a temperature of a few tens of degrees Kelvin, and is characteristically the medium that is studied by molecular-line emission. The boundary between the cool and cold neutral ISM regimes is not a sharp one, but the distinction can still be useful. There is also a *warm* neutral medium, which can exhibit temperatures from a few hundred Kelvins to 1000 K (or somewhat higher). Despite the wide range of conditions, there must be a reasonably close pressure equilibrium between adjacent regions. There will be exceptions, of course, when transient changes, marked by supersonic velocities such as those generated by turbulence or by supernova explosions, lead to timescales that are short compared with the travel time for sound waves. One can start with the equilibrium approximations, however, bearing in mind that the final description will be more complicated.

The various temperature regimes of the ISM can be looked on as different phases of the medium, responding to the heating and cooling processes that are acting (Cox 2005). For greater detail concerning the physics of line formation, and the complexities of the interactions between H_2 and molecules such as CO, reference can be made to Rohlfs and Wilson.

9.3 The 21-cm spectral line of neutral hydrogen

The 21-cm spectral line of neutral hydrogen is emitted by the most common component of the ISM and is a potent probe of the ISM. It not only provides the basic information on the dynamics of galactic rotation (Chapter 10) but also allows us, at least in principle, to determine the distribution of neutral hydrogen both in density and in temperature. We recall the basic results from Chapter 7: if a radio telescope observes the 21-cm H I line in a cloud of neutral hydrogen atoms with total column density N_H and uniform spin temperature T_s (which normally equals the local gas-kinetic temperature) then

- for an *optically thin* cloud, the brightness temperature T_b integrated across the whole emission line measures N_H, and the line profile gives the distribution of N_H with velocity v;
- the optical depth τ, as observed in an absorption line for a source beyond the cloud, measures N_H/T_s; and
- for an *optically thick* cloud, the observed brightness temperature equals the spin temperature T_s.

The intrinsic line width is always negligible, and the observed profile is determined by a combination of thermal and bulk velocities. Usually the width is determined by the bulk velocity, and it is convenient to calibrate the receiver bandwidth Δv in terms of velocity v in units of km s^{-1} (1 kHz $\equiv 4.74$ km s^{-1}). If the width is determined only by temperature, the full line width at half maximum (FWHM) $\delta v_{1/2} \approx T^{1/2}$ km s^{-1}. When the line is optically thin, and when the line profile (measured in brightness temperature as a function of velocity and in km s^{-1}) is integrated, Equation (7.43) leads to

$$N_H = 1.8 \times 10^{18} \int T_b \, dv \text{ cm}^{-2}. \qquad (9.3)$$

When the line profile is Gaussian, of FWHM Δv, Equation (7.46) becomes

$$\tau = 5.2 \times 10^{-19} N_H (\Delta v \, T_s)^{-1}. \qquad (9.4)$$

A typical emission-line profile (Figure 9.4) contains several components with different velocities, corresponding to separate components of the spiral-arm structure of the Galaxy (Chapter 10). When the line structure is simple, as in Figure 9.4, broad and narrow components appear superposed; these originate from the 'cool' and 'cold' components, respectively, and their widths are determined mainly by the temperatures of these different regions.

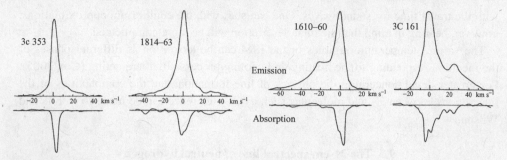

Figure 9.4. Typical H I emission and absorption line profiles towards extragalactic radio sources (Radhakrishnan *et al.* 1972). The absorption on the line of sight to the extragalactic source occurs mainly in colder H I regions; the emission is all H I in the telescope beam.

The cold components may have considerable optical depth; if the cold regions in which they originate are situated in front of the cool regions, there may be absorption of the cool component within the narrower line of the cold component.

Figure 9.4 shows absorption-line profiles for a continuum source outside the Galaxy, with the emission spectrum for an adjacent region. The H I clouds include both cool and cold components. The cold components show clear absorption, but the warmer 'cool' component, which has a broader velocity spread, is virtually unabsorbed because the higher temperature lowers the absorption coefficient. A close comparison between these spectra depends on the assumption that the emission spectrum in the direction of the source is the same as that from adjacent regions; this is often far from true. A better observation can be made using a pulsar as the distant source, since a pulsar conveniently turns on and off through the pulse cycle and can therefore be distinguished from other sources. Figure 9.5 shows the emission and absorption spectra in the direction of the pulsar PSR 1904+06; here there is no absorption at velocities corresponding to hydrogen beyond the pulsar, but the emission is from the whole line of sight through the Galaxy. In both these examples the absorption lines are from cooler gas and are narrower than the emission lines. The wide components, which we have already identified with the 'warm' components, do not show up in absorption because of the inverse relation between optical depth and temperature in Equation (9.4).

Using the integrated brightness to measure the total neutral hydrogen content in a line of sight, and extending this to measure the total neutral hydrogen content of the Galaxy, obviously depends on the assumption in Equation (9.4) that the optical depth is much less than unity. This is not true for the 'cold' component, where the brightness temperature often approaches the value of T_s deduced from measurements of absorption lines. A suitable allowance for self-absorption depends on the detailed geometry of the clouds, which may be difficult to determine because the clouds do not fill the beam of the observing telescope; the total content of the 'cold' clouds through long lines towards the centre of the Galaxy cannot, therefore, be measured accurately. It turns out, however, that at least half of the total

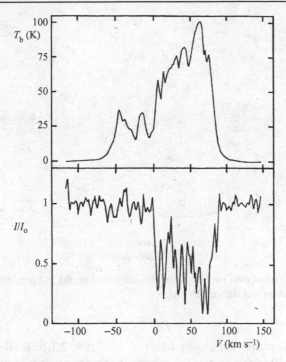

Figure 9.5. The H I emission spectrum in the direction of pulsar PSR 1904+06 and the pulsar absorption spectrum (Clifton *et al.* 1988).

hydrogen content is in the 'warm' components, where there is little or no self-absorption and Equation (9.3) applies without correction.

An outstanding survey of Galactic neutral hydrogen, showing its distribution in intensity and velocity, has been completed by Hartmann and Burton (1995). In addition to the maps presented in this survey, there is an authoritative discussion of the effects of out-of-beam radiation received in the sidelobes of the radio telescope (see Sections 3.4 and 8.3). The corrections can be significant, since observations of weak emission can be contaminated by the strong Milky Way emission, attenuated by the sidelobes but subtending such a large solid angle that the resulting signal may be comparable to the in-beam signal.

Figure 9.6 shows the distribution over the northern sky of the integrated brightness $\int T_b \, dv$ as measured by Hartmann and Burton. The contour intervals are in units of $K \, km \, s^{-1}$; for optically thin hydrogen the column density in the plane of N_H is then $0.182 \times 10^{19} \int T_b \, dv \, cm^{-2}$. Making allowance for self-absorption in the Galactic plane, the integration over this map gives a total mass of neutral hydrogen in the Galaxy between 3 and 5×10^9 solar masses (Fich and Tremaine 1991). The major part of the neutral-hydrogen distribution is clearly closely confined to the plane of the Galaxy; it can be fitted to a model with a density that is nearly independent of the Galacto-centric distance R between $R = 4$ kpc and $R = 20$ kpc, in which the density falls exponentially with vertical

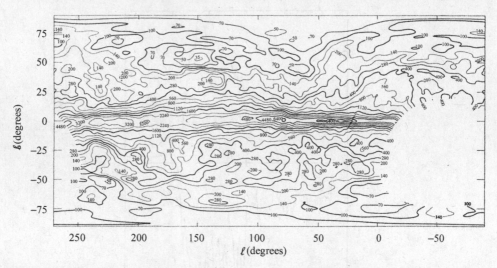

Figure 9.6. The distribution over the Galaxy of the integrated brightness temperature $\int T_b \, d\nu$ of H I emission (from Hartmann and Burton (1995)).

distance $|z|$ from the plane. The density falls to e^{-1} at $|z| \approx 250\,pc$ for the 'warm' component and at $|z| \simeq 130\,pc$ for the 'cold' component. The density on the plane is $0.7\,cm^{-3}$, and the total projected density on the plane is 3×10^{20} atoms cm^{-2}.

It must be emphasized that this is a very much smoothed-out representation of the distribution. The 'cold' component is the more non-uniform; reference to Figure 9.5 shows that the absorption is concentrated in discrete clouds that fill only part of the line of sight. There are on average about four such clouds per kiloparsec along a line of sight at low Galactic latitude, and the clouds fill less than one-tenth of the volume of the disc. Even this description may be misleading; more detailed structure can be seen in the absorption along a single line of sight to an extragalactic radio source, and this structure may be different along two adjacent lines of sight. Diamond *et al.* (1989) found that absorption can differ along the lines of sight to the two separate lobes of a single distant radio galaxy, showing that the scale in our Galaxy is as small as 25 AU.

At high Galactic latitudes there are discrete emission components forming long streaks in the sky, more like cirrus than cumulus clouds. Most of these complex structures have high velocities, and appear to be completely distinct and detached from the hydrogen in the Galactic plane. Figure 9.7 shows hydrogen structures with local velocities greater than $90\,km\,s^{-1}$. Most of these clouds are falling in towards the plane; some form a continuous stream that seems to have been pulled out of the Magellanic Clouds by the gravitation of the Galaxy (see Section 10.4 for a discussion of the dynamics).

Since hardly any absorption is observed in the 'warm' component, the only estimate of temperature is from the line width. This gives an upper limit to the kinetic temperature, which cannot exceed 8000 K, which, as it happens, is easily accounted for theoretically.

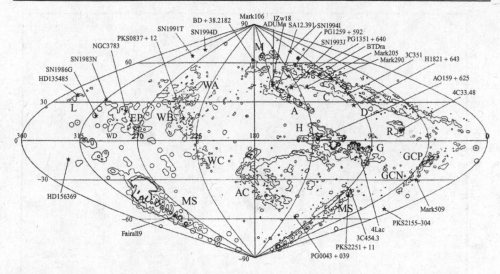

Figure 9.7. High-velocity clouds of H I, with velocities relative to the local standard of rest $|v_{lsr}| > 90\,\mathrm{km\,s}^{-1}$. Contours of brightness temperature are shown at 0.04, 0.5 and 1.5 K. Many of the complexes are named. The asterisks show extragalactic sources whose spectra at 21 cm show absorption by high-velocity clouds (Wakker and van Woerden 1997).

Temperatures between 40 and 140 K are found for individual clouds in the 'cold' component; as we see later, there is no simple mechanism of heating or cooling that would tend to give a particular value of temperature in this range.

9.4 H II regions and supernova remnants

The classic H II regions are discrete ionized clouds surrounding very hot O-type stars. A typical cloud might be approximately spherical (a *Strömgren sphere*) several parsecs across, bounded by a well-defined ionization front, inside which there is an equilibrium balance between ionization by ultraviolet light from the star and recombination. Hydrogen atoms spend some hundreds of years in the ionized state and some months in the neutral state. They are maintained at a temperature of 8000–10 000 K, above which the region loses energy very rapidly through the excitation of oxygen ions (see Section 9.5). Lower-energy photons penetrate beyond the limit of the H II region, and can excite or ionize atoms and molecules in a *photodissociation region* (PDR) surrounding the Strömgren sphere. The PDR contains a rich mixture of atomic and molecular species, as in the example of Figure 9.1.

The major part of the radio spectrum observed from H II regions is the continuum from free–free emission, in which the electron is unbound before and after a collision with an ion (Section 7.5).

Figure 9.8 shows the free–free emission from a group of H II regions known as W3. The radio spectrum from such regions follows the classic pattern of optically-thin

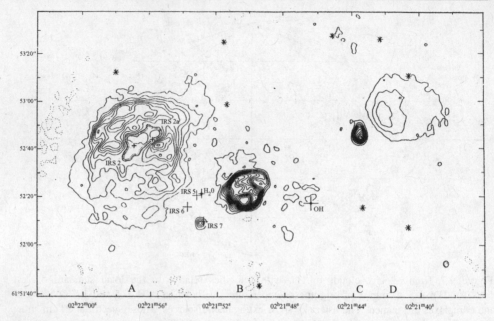

Figure 9.8. The free–free emission at 5 GHz from the W3 group of H II regions. The contour interval is 255 K (Harris and Wynn-Williams 1976).

bremsstrahlung (spectral index −0.1) above a frequency of the order of 1 GHz; at lower frequencies the spectrum progressively approaches the optically-thick case in which the spectral index becomes +2.0 (see Section 7.5). At millimetre and sub-millimetre wavelengths (Ladd *et al.* 1993) the H II regions are less conspicuous, while the 20-μm emission is concentrated on regions of star formation (Wynn-Williams *et al.* 1972).

The ratio of radio recombination-line emission to free–free emission provides a useful measurement of temperature, since it depends on the ionization ratio in the H II region. The emissivity of the line radiation, like the continuum free–free radiation, is proportional to the square of the electron density; the ratio depends on the radio frequency ν, quantum number n and temperature T_e as

$$\text{line/continuum} \propto \nu^{2.1} n^{-1} T_e^{-1.15}. \tag{9.5}$$

Typical temperatures determined from this ratio are between 8000 K and 10 000 K. The interpretation of this ratio is, of course, less certain when conditions within the H II region are not uniform; this is certainly the case for the high-density 'compact' H II regions, such as W3. The ratio of line to continuum radiation is also more difficult to interpret for diffuse sources away from the Galactic plane; here the background radiation is primarily synchrotron, and the ratio of line to continuum is less meaningful.

Large H II regions are detectable optically in the plane of the Galaxy out to distances of several kiloparsecs. Georgelin and Georgelin (1976) showed that they follow the spiral-arm pattern much as can be seen in other galaxies. This delineation of the spiral structure can be extended through the whole Galaxy using radio recombination lines, since the Galaxy is transparent at radio wavelengths; Wilson *et al.* (1970) mapped all the bright H II regions of the Milky Way in 109α recombination-line emission.

The total electron content of the spiral arms, including these discrete concentrations, is an important parameter in interpreting the dispersion measures along the lines of sight to pulsars (Chapter 12). The 'warm' ionized components are responsible for the majority of the electrons which cause dispersion in pulse-arrival times from pulsars. The distances of many pulsars are known either from direct parallax measurements or from H I absorption (which places them in relation to the spiral arms). The dispersion measure of a pulsar gives the integrated electron content along the line of sight, allowing a model of the electron density in at least the nearer half of the Galaxy to be constructed. Lyne *et al.* (1985) set out such a model for use in determining the distances of the majority of pulsars, for which only the dispersion measures are available. Their model (Equation (9.6)) has two axisymmetric components: an extensive thick disc with scale height 1000 pc, and a thinner plane component, both of which have a central density depending on Galacto-centric distance R:

$$n_e = \frac{2}{1 + R/10} \left(0.025e^{-|z|/1000} + 0.015e^{-|z|/70}\right). \tag{9.6}$$

A more detailed model by Taylor and Cordes (1993), which includes the spiral-arm structure as deduced from the location of H II regions, allows the distances of most pulsars to be measured to accuracies of 25% or better from their dispersion measures. It includes a more reasonable model of the z-variation of density for both components of Equation (9.6), in which the distribution with z follows a $\mathrm{sech}^2 z/h$ law. Figure 9.9 shows the profiles of their three components as a function of R and z. Their 'inner-galaxy' component is represented here as an annulus with maximum density at radius 3.5 kpc; it is, however, equally possible to represent this component by a filled-centre disc.

9.5 Heating and cooling mechanisms

Absorption of starlight in the ultraviolet, observed over recent years with satellites such as IUE, shows up as prominent resonance lines of heavy elements. Some, such as O VI, are highly ionized, indicating a temperature of at least 2×10^5 K. Assuming a normal abundance ratio of hydrogen to heavier elements, the optical depth of the absorption indicates a hydrogen column density of 2×10^{18} cm^{-2} perpendicular to the Galactic plane. This hot ionized component is also detected in observations of low-energy X-rays; these do not reach us from further than about 100 pc from the Sun, but it is reasonable to suppose that the hot component is widespread in the Galaxy. The hot ionized component has a low density, in accordance with the general pressure equilibrium among the four components of the ISM; consequently it does not contribute significantly to the total mass of the Galaxy. It may, however, contribute significantly to the dispersion-measure model of Equation (9.6),

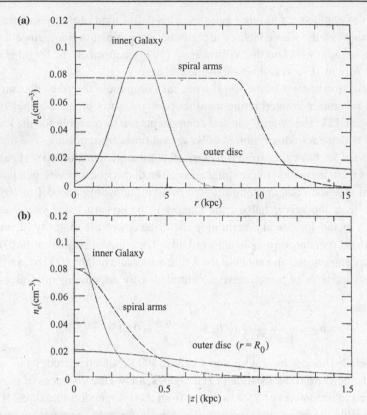

Figure 9.9. Three components of the distribution of electron density n_e as a function (a) of R and (b) of z. This model of the electron distribution is used to determine the distances of pulsars from their dispersion measures (Taylor and Cordes 1993).

which requires a column density of 5×10^{19} cm^{-2} on the plane at small Galacto-centric radii.

The surprising variety of components of the ISM is at least matched by the variety of sources of energy and cooling processes involved in the maintenance of their various temperatures. In a classic H II region (the Strömgren sphere) surrounding a hot O-type star, the cooling process can be observed directly: the visible light from the nebula contains prominent emission lines from minor constituents of the gas, most notably oxygen. These lines are from forbidden transitions, but the levels are so easily excited when the gas temperature reaches 8000–10 000 K, and the timescales are so long, that most of the energy is lost in this way. Both doubly and singly ionized oxygen are involved; the 372.6-nm and 372.9-nm lines from O II are the 'nebulium' lines, so-called from the time when their origin was a mystery.

There can also be little doubt about the origin of the hot ionized component of the ISM. Temperatures of the order of 10^5–10^6 K, found over large regions of space, can only originate in supernova explosions. The expanding bubble of hot gas from a supernova may

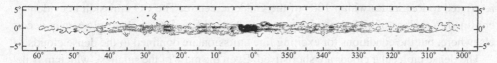

Figure 9.10. The distribution of the brightness temperature of the 2.6-mm carbon monoxide line (Dame and Thaddeus 1985).

reach a diameter of 50 pc, and it is expected to cool only after 10^7 yr. At the rate of one supernova per 30 yr there would be 3×10^5 bubbles in existence at any time, occupying a large part of the Galactic disc. This provides not only an explanation for the hot component but also a source of energy for major dynamical processes in the halo of the Galaxy.

At temperatures below about 600 K the main cooling process involves a different minor constituent, ionized carbon. Here the excitation involves fine-structure levels, which radiate at the infrared wavelength of 157.7 μm. Warming may also involve carbon, which will be ionized by starlight. Carbon alone does not provide a full explanation: the population of C I is insufficient to maintain the temperature even of the 'cold' H I regions. Yet another minor constituent, interstellar dust, is involved in heating, certainly for the 'cold' and possibly for the 'warm' regions. The process is photoelectric emission, in which electrons with energy of about 1 eV are emitted, lose their energy by multiple collisions, and eventually recombine on the dust grains. The considerable amount of energy involved, about 10^{42} erg s^{-1}, is derived from ultraviolet light from the stars.

Apart from the excitation by supernovae, all the ionizing processes are dominated by Lyman-α ultraviolet radiation from hot stars. This also plays a part in the thermalization of the spin temperature of neutral hydrogen, which we have assumed to equal the gas-kinetic temperature. Collisions between photons and hydrogen atoms are more effective in coupling spin to kinetic energy at low temperatures; at 8000 K this mechanism is relatively ineffective. The Lyman 'photons' are scattered many times by the atoms, and the line profile acquires a bias in shape, that is, a 'colour', which couples the relative population of the spin levels to the kinetic temperature of the atoms.

9.6 Dense molecular clouds

So far in this chapter we have discussed the diffuse components of the ISM, which are unrelated to stars except for the H II regions in the immediate vicinity of very hot O-type stars. We now turn to the smaller and denser clouds, which are probably the seat of star formation. They are of particular interest in radio astronomy, because the molecular species which form preferentially at high density are mainly observable only via spectral lines that occur at millimetre wavelengths.

The contrast between the sources of H I and molecular-line emission is clearly shown in Figure 9.10, which shows the distribution over the Galaxy of the brightness temperature of the 2.6-mm line of carbon monoxide. These maps show that CO is found in small discrete clouds, concentrated in the plane of the Galaxy; further, that these clouds are mainly restricted to longitudes within $100°$ of the Galactic centre. The CO emission is evidently confined to regions of high density; as we note later, the molecules themselves may occur elsewhere, but the conditions for exciting line emission are confined to the dense clouds.

The dense molecular clouds are short-lived. Even the mass of the smallest, which may be $10 M_{\odot}$ or less, is more than the Jeans mass, indicating that they are in a state of gravitational collapse. These have no source of heating apart from cosmic rays, and they are cooled by exciting the first rotational level of carbon monoxide. Their equilibrium temperature is only about $10\,\mathrm{K}$. The larger clouds, known as the giant molecular clouds, have masses up to $1000 M_{\odot}$: these usually contain hot stars, and are complex structures with a range of densities and temperatures. Within such clouds there may be both gas condensing to form stars and gas flowing out from stars at various stages of evolution.

9.7 Interstellar scintillation

Interstellar scintillation, due to the scattering of radio waves by irregularities in the ionized ISM, is regarded by pulsar observers as a nuisance in the search for pulsars and in finding their intrinsic characteristics. Its observation is, however, a powerful way of investigating the structure of the ISM. There are many observations of fluctuations of intensity, as a function of both time and radio frequency. There are also observations of pulse broadening in pulsars, and of the scattered angular diameters of molecular maser sources and quasars. The theory of propagation in an irregular medium characterized by a generalized structure function is well understood, but the inverse problem of relating the observations to the actual structure of the ISM is difficult, not least because the structure itself is complex, comprising both an easily characterized random component and large-scale structures due to discrete objects and events such as H II regions and expanding supernova ionization fronts. We follow a discussion by Lambert and Rickett (2000), who give many useful references to observation and theory.

A useful model of the random structure as a function of wavenumber q is a combination of a power law and an exponential:

$$P_{N_{\mathrm{e}}}(q) = \frac{C_{N_{\mathrm{e}}}^2(z)}{(q^2 + k_0^2)^{\beta}} \exp\left(-\frac{q^2}{4k_{\mathrm{i}}^2}\right). \tag{9.7}$$

Here the power law has exponent β, and the spectrum of irregularities is limited by inner and outer scales k_{i}^{-1} and k_0^{-1}, respectively. It turns out that these limiting scales are many orders of magnitude apart, so that the index β is a sufficient characterization over some ten orders of magnitude in q. If random fluctuations are caused by an energy input on a large scale, followed by an inertial cascade to smaller scales, the spectrum will be a *Kolmogorov* distribution, with $\beta = 11/3$; this proves to be a good match to many observations. A different spectrum, with $\beta = 4$, would be expected for irregularities related only to a set of discrete objects. Although these two indices are surprisingly nearly the same, Lambert and Rickett point out that there are observable differences between the types of scintillation they cause; and they show that the two patterns of irregularity co-exist.

Two regimes of scintillation effects can be distinguished. *Diffractive scintillation*, which may be considered in terms of waves diffracted at a phase-changing two- or three-dimensional screen, is a useful way of understanding the effects of the smaller scales of irregularity, say 10^7–10^8 m. Simple scaling laws follow; for example the frequency structure of scintillation has a bandwidth proportional to ν^4, where ν is the observing frequency, the pulse lengthening is proportional to ν^{-4}, and the angular size of a point source varies as ν^{-2}.

The larger scales of irregularity, say 10^{11}–10^{12} m, may be considered as the source of *refractive scintillation* (Lambert and Rickett 1999), in which rays reach the observer by a number of different paths and interfere on arrival. The scaling laws are similar to those of diffractive scintillation, but the effects are on a longer timescale. The two regimes may appear together, with the effect that a source may scintillate on a short timescale, say some minutes, while its average flux density varies on a much longer timescale, say several days.

Refractive scintillation is also the cause of random fluctuations at frequencies of several gigahertz in many compact quasars, on a timescale of hours or days (Lovell *et al.* 2003). These sources have angular diameters that, like those of the pulsars, are small enough for the scintillation pattern to be observed (incidentally, this indicates that the apparent brightness temperature of these quasars may exceed 10^{12} K). The rate of scintillation has been observed to depend on the phase of the Earth's orbital motion, allowing the transverse velocity of the scintillation pattern to be deduced from a vector subtraction of the Earth's orbital velocity (Dennett-Thorpe and de Bruyn 2003).

A related approach to probing the ISM is to monitor the small variations in dispersion measure observable in the precise timing of millisecond pulsars (You *et al.* 2007). As observed at frequencies around 1 GHz, these fluctuations are related to refractive scintillation, and may help in evaluating the outer scale of sizes of irregularities.

9.8 Supernova remnants (SNRs)

All stars evolve and eventually die, leaving small, condensed fossil remains. The death of stars with masses greater than $\sim 8M_\odot$ is catastrophic and spectacular: the core of the star becomes a white dwarf whose mass reaches the limit of stability, which then collapses within a second to form a neutron star or a black hole. A similar collapse can occur when a white dwarf in a binary system accretes mass from its companion star (Section 12.15). The energy released in the collapse drives infalling matter outwards in a violent explosion that disrupts the surrounding remnants of the star, producing the visible supernova which can outshine its entire parent galaxy for several weeks (see Burrows (2000) for a brief review).

A supernova explosion is essentially a phenomenon for the visible spectrum and for neutrino astronomy; radio astronomy takes over progressively as the expanding cloud of debris becomes more diffuse. Radio emission has, however, been observed from recent supernovae, notably SN 1987A in the Large Magellanic Cloud and SN 1993J in the galaxy M81. At the earliest stage the emission is probably from the expanding supernova debris, but it is the subsequent interaction with the surrounding ISM that produces the shell-like

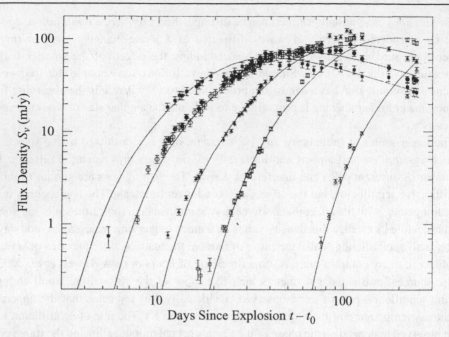

Figure 9.11. The growth of radio emission from SN 1993J in the galaxy M81, over the first 200 days. From left to right the curves show the flux density at 1.3, 2, 3.6, 6 and 20 cm (van Dyk *et al.* 1994).

supernova remnants (SNRs) which are the long-lasting memorials to the death of the star. The evolution over the first 200 days of the radio emission from SN 1993J in the galaxy M81 is shown in Figure 9.11. Short wavelengths are observed first, and the spectrum evolves rapidly towards longer wavelengths. This is synchrotron radiation from an expanding shell round the supernova. The expansion was measured by Bartel *et al.* (1994), showing a linear growth to a diameter of 250 μarcsec over 250 days.

The evolution of a young SNR at a later stage has been seen in the star-burst galaxy M82 (see Chapter 13), which contains at least 40 discrete radio sources that have been identified as SNRs (Muxlow *et al.* 1994). Figure 9.12 shows the expansion of one of these, which is now about 50 milliarcseconds across, over 11 years. The angular velocity of this expansion is about one arcminute per year, corresponding to a velocity of 10^4 km s^{-1} and a birth date in the early 1960s (see also Beswick *et al.* (2006)).

The supernova remnant Cas A, which is about 300 years old and is the youngest SNR in the Galaxy, is the brightest radio object in the sky, but was not noticed optically until radio observations drew attention to it (Figure 9.13). It contains filaments moving radially outwards at 6000 km s^{-1}; others have been nearly stopped by collision with the surrounding ISM. The spherical shell shows where the ISM has been compressed by the supernova. The synchrotron radio emission from this and similar shells exhibits linear polarization neatly arranged round the circumference, showing where the interstellar magnetic field has been swept up and compressed; a good example is 3C10, shown in Figure 9.14.

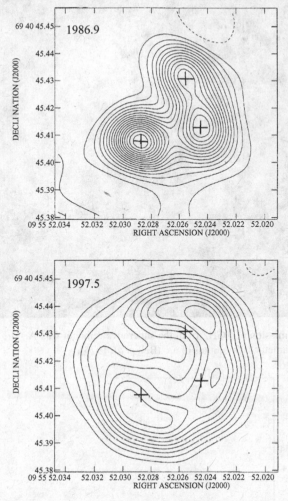

Figure 9.12. The expansion of the supernova remnant 43.31+592 in the starburst galaxy M82. The two contour maps were made in 1986 and 1997 using the EVN at wavelength 18 cm (Pedlar *et al.* 1999).

The ages of the observable radio remnants extend at least to 100 000 yr (see Raymond (1984) for a review). If they all behaved identically, we would have a complete picture of the development and fading of their radio emission, and of its relation to the physical conditions in the expanding cloud. The 300-year-old remnant Cas A is still expanding rapidly, but its radio emission is decreasing; the same has been found for the 400-year-old remnants SNR 1604 (Kepler) and SNR 1574 (Tycho). Two 1000-year-old remnants are also observed to be expanding: the Crab Nebula (1054) and SNR 1006 (Figure 9.15). The remains of much older SNRs can also be observed: IC 443 (Figure 9.16), the Cygnus Loop (Dickel and Willis 1980) and the Vela Nebula. These are probably between 10 000 and

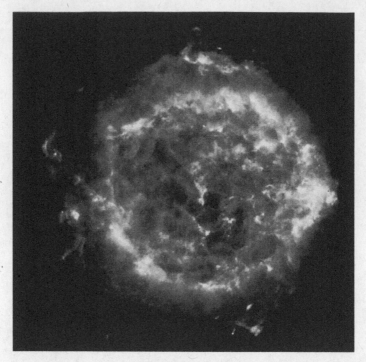

Figure 9.13. Radio emission at 5 GHz from the supernova remnant Cas A. The contour interval is 480 K. The overall diameter is 5 minutes of arc (Bell *et al.* 1975).

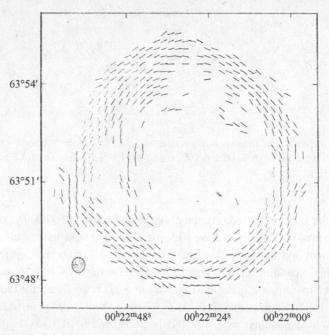

Figure 9.14. Polarized radio emission on the circumference of the supernova remnant 3C10. The map shows the intrinsic polarization-angle distribution as derived from observations at 6 and 21 cm, correcting for Faraday rotation (Duin and Strom 1975).

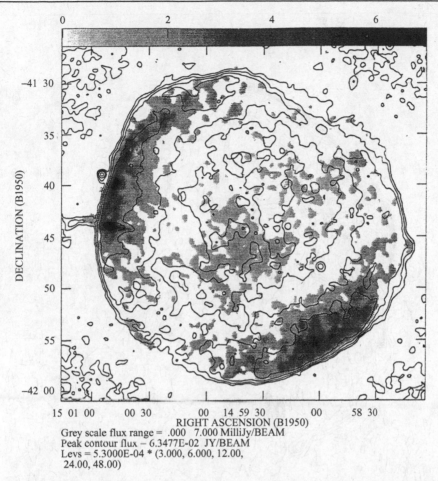

Grey scale flux range = .000   7.000 MilliJy/BEAM
Peak contour flux − 6.3477E-02 JY/BEAM
Levs = 5.3000E-04 * (3.000, 6.000, 12.00,
24.00, 48.00)

Figure 9.15. Radio emission at 1370 MHz from the supernova remnant SNR 1006. The grey scale shows the polarized component (Reynolds and Gilmore 1986).

100 000 yr old. The North Polar Spur (Chapter 8) is also probably part of an old SNR shell. An invaluable working catalogue of SNRs is maintained by D. A. Green at http://www.mrao.cam.ac.uk/surveys/snrs/.

The common feature of all these various SNRs must be the catastrophic release of energetic particles, with a total energy of order 10^{51} erg, expanding into the ISM with a velocity of the order of 6000 km s^{-1}. Although the mean free path is high, this blast wave is coupled to the interstellar gas by the ambient magnetic field, and the gas is swept up with it. Up to the point where the swept-up mass becomes comparable to the mass of the ejecta, the expansion is a simple unhindered explosion; after this point, the shape and speed of the expansion depend on the density and distribution of the interstellar gas and its magnetic field. Parts of Cas A are still expanding at 6000 km s^{-1}; the older SNRs appear to have been slowed down by interstellar gas. The shapes of IC 443, the Vela Nebula and the Cygnus

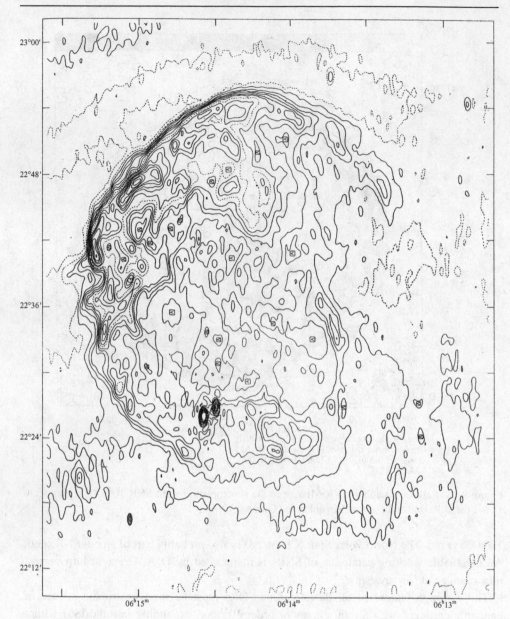

Figure 9.16. Radio emission at wavelength 21 cm from the supernova remnant IC 443 (Duin and van der Laan 1975).

Loop seem to be determined mainly by collisions with interstellar clouds; these remaining wisps are moving only at speeds of some tens of $km\,s^{-1}$. The effect on the clouds is best seen at millimetre and sub-millimetre wavelengths: for example, van Dishoek *et al.* (1993) have shown that the outer parts of the IC 443 shell contain shocked molecular gas emitting

Figure 9.17. The Crab Nebula (FORS Team, 8.2-m VLT, ESO).

both lines, such as CO, and radio continuum. There is, however, a distinctive feature of the Crab Nebula (Figure 9.8), which is shared by a minority of others. Over the wide spectrum through which it can be observed, the emission comes from a filled sphere rather than an expanding shell; such SNRs are known as *plerions*. The Crab Nebula appears as a tangled web of filaments, with some optical line emission but mainly synchrotron radiation. The continuum follows a power law, with unusually low spectral index, extending to X-rays. The lifetime of electrons sufficiently energetic to radiate X-rays is only a few years (see Chapter 8); there must therefore be an energy supply within the nebula. This is, of course, the Crab Pulsar, which we discuss in Chapter 12.

Although the Crab Nebula is expanding at a lower rate than is the Cas A SNR, an extrapolation of its expansion back to its origin in AD 1054 shows that the expansion has not been slowed by sweeping up the surrounding interstellar gas and dust; on the contrary, it has accelerated. Again, the magnetic field required for its synchrotron radiation, about 100 microgauss, is hard to account for either as a remnant of the original stellar magnetic field or as a swept-up interstellar field. The pulsar provides the explanation for all these distinctive features. Only young SNRs can be expected to show these characteristics, but it is unknown what proportion of young SNRs will show the same evidence of an active pulsar. The 400-year-old Tycho and Kepler SNRs are already behaving very like the old nebulae such as IC 443 and the Cygnus Loop.

Further reading

Atlas of Galactic Neutral Hydrogen, Hartmann D. and Burton W. B., Cambridge, Cambridge University Press, 1997.

Astrochemistry, IAU Symposium No. 231, ed. Lis D. C., Blake G. A. and Herbst E., Cambridge, Cambridge University Press, 2006.

Interstellar Processes, ed. Hollenbach D. J. and Thronson H. A., Dordrecht, Reidel, 1987.

Microwave Spectroscopy, Townes C. H. and Schawlow A. L., New York, McGraw-Hill, 1955.

Physical Processes in the Interstellar Medium, Spitzer L., New York, Wiley, 1978.

Radiative Processes in Astrophysics, Rybicki G. B. and Lightman A. P., New York, Wiley, 1979.

The Chemically Controlled Cosmos: Astronomical Molecules from the Big Bang to Exploding Stars. Hartquist T. W. and Williams D. A., Cambridge, Cambridge University Press, 1995.

The Cologne Database for Molecular Spectroscopy, Müller H. S. P., Schlöder F., Stutzki J. and Winnewisser G., *J. Mol. Struct.* **742**, 215 (2005).

The Interstellar Disc–Halo Connection in Galaxies, IAU Symposium No. 144, ed. Bloemen H., Dordrecht, Kluwer, 1991.

The Physics of the Interstellar Medium, Dyson J. E. and Williams D. A., Bristol, Institute of Physics, 1997.

The Tools of Radio Astronomy, Rohlfs K. and Wilson T. L., Berlin, Springer-Verlag, 4th edn, 2000.

Unsolved Problems of the Milky Way, ed. Blitz L. and Teuben P., IAU Symposium No. 169, Dordrecht, Kluwer, 1996.

10

Galactic dynamics

The dynamical structure of our galaxy, the Milky Way system, is most simply described as a set of nested ellipsoids. The largest, approximately spherical halo, composed largely of dark matter, extends well beyond the stars of the Milky Way and is not yet understood. Within this there is a smaller, nearly spherical ellipsoid of old stars in orbits plunging through the plane of the Milky Way, appearing as the Population II high-velocity stars to Earth-bound observers, for our Sun is in an approximately circular orbit within the plane. The system of globular clusters, not as extended in distance from the centre, can be regarded as part of this old system. Within that, there is a quasi-ellipsoid of stars, denser near the Galactic centre, with an axial ratio of the order of 3 or 4 to 1; this includes the prominent bulge population of stars congregating about the Galactic centre in the central few kiloparsecs of the system. Finally, the Galactic plane is outlined by a thin disc of dust, gas and the youngest stars (Population I), stretching out to two or three times the Sun's distance from the centre, but with a thickness only about 1/100th the diameter of the disc. Much of Galactic radio astronomy concentrates on this extreme Population I region of the Milky Way. The stellar orbits are governed by the gravitational field of all the matter in the Milky Way system, with the Population I disc component in nearly circular motion, while the oldest stars (Population II) in the nearly spherical component execute highly eccentric orbits, passing through the innermost parts of the Galactic structure.

In the classification system described in Chapter 8, we are in a barred spiral, probably of type SBb. The spiral structure and the central bar may be regarded as perturbations, both in the tangential and in the radial direction. Stars are an imperfect tracer of the dynamics, since obscuration by dust limits the distance at which they can be observed, and the clumpy nature of the dust distribution biases the selection. In contrast, the spectral line radiation from neutral hydrogen and other common molecules in the ISM serves as an excellent tracer, giving both the velocity and the density of the interstellar medium. Other galaxies may also be studied in this way, with the advantage that their various different attitudes to the line of sight overcome the difficulty of delineating the structure of our own Galaxy from our position within it. This chapter describes the dynamics of our Galaxy and other spiral galaxies in terms of a circular motion of a disc on which the spiral structure is imposed. The radio observations provide a tracer of the total mass distribution; we will also note the

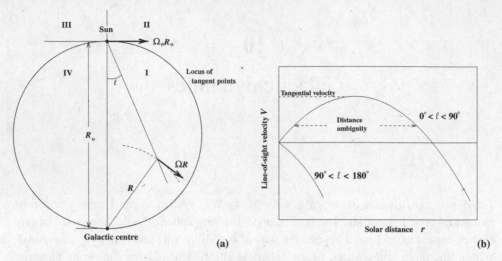

Figure 10.1. Line-of-sight velocities for a model Galaxy with circular orbits: (a) the geometry and (b) line-of-sight velocities, showing the region of ambiguity.

difficulty of accounting for this mass in terms of the mass of the stars and ISM, which leads to the concept of dark matter and its distribution throughout the Galaxy.

10.1 The circular approximation

As observers in the solar system, we are located within the disc of the Galaxy, slightly above the Galactic plane. If the gas were moving in circular orbits, the observed line-of-sight velocities would be related to the angular velocity through the geometry of Figure 10.1. The distance of the Sun from the Galactic centre, R_o, is the fundamental distance parameter, while the angular velocity $\Omega(R)$ is a function of radial distance R from the centre, with Ω_o being the angular velocity at the Sun.

Conversion of the observed radial velocity to the observer's *local standard of rest* (LSR) requires several corrections to be made to the data. The Earth–Moon system is rotating about a barycentre (with a velocity usually too small to be important); the Earth rotates on its axis (necessitating a daily correction of about $1 \, \mathrm{km \, s^{-1}}$); the Earth is in orbit about the Sun (velocity about $30 \, \mathrm{km \, s^{-1}}$); and the Sun itself is moving with respect to the average of the stars in its vicinity. The motion of the Sun with respect to this local standard of rest was originally assumed to be $20 \, \mathrm{km \, s^{-1}}$ in the direction $\alpha = 18^{\mathrm{h}}$, $\delta = +20°$, round numbers being used to remind one that the value was not well known. More accurate determinations have been carried out, and the value seems to depend upon the population that serves as the reference; a currently popular value is $16.5 \, \mathrm{km \, s^{-1}}$ towards $\ell = 53°$, $\ell = 25°$ (Binney and Tremaine 1987). The careful observer should understand what convention is being followed.

Once the observed radial velocities have been reduced to radial velocities V with respect to the LSR, the fundamental quantities involved in the circular approximation are the distance from the Sun to the centre, $R_\circ$, the angular velocity of the solar neighbourhood, $\Omega_\circ$, and the linear velocity of the solar neighbourhood, $\Theta_\circ = \Omega_\circ R_\circ$. We require the line of sight velocity V_r as a function of distance r in any direction, given the tangential velocity ΩR as a function of distance from the Galactic centre.

Figure 10.1 shows the geometry for a source in the Galactic plane. Assuming that all corrections have been made to the LSR, the measurements are being made in a rotating coordinate system, on a radiating hydrogen cloud at Galactic longitude ℓ and distance r from the observer. With respect to the Galactic centre, the cloud is situated at radius R and azimuthal angle θ. The corrected velocity V_r with respect to the LSR is then

$$V_r = \Theta(R)\cos(90° - \ell - \theta) - \Theta_\circ \cos(90° - \ell). \tag{10.1}$$

From this, after a little trigonometry, we obtain the fundamental equation

$$V_r = R_\circ[\Omega(R) - \Omega_\circ]\sin \ell \cos \ell. \tag{10.2}$$

Here the extra factor $\cos \ell$ allows for sources away from the plane, i.e. at Galactic latitude ℓ. If all the velocities really are circular round the Galactic centre, this equation allows the velocities in the profiles of Figures 10.6 and 10.7, shown later, to be interpreted as a distribution of angular velocity with distance. At the tangential points which lie on the circle through the Sun and the Galactic centre (Figure 10.1(a)), the maximum velocity will determine the rotation curve $\Omega(R)$. One can see immediately that within the solar circle and within $0° < \ell < 90°$ and $270° < \ell < 360°$ (the regions denoted I and IV in the diagram) there is an ambiguity in assigning a distance to a given radial velocity. The ambiguity can often be resolved from angular size alone: the more distant clouds appear smaller on average. Distances from pulsar absorption lines (see Section 9.4) or identification with associated known objects such as H II regions or molecular clouds can also be used to resolve the ambiguity. Outside the circle in Figure 10.1(a), in the regions denoted II and III, the relation Equation (10.2) is unique; given a rotation curve, the distance to the hydrogen cloud can be determined, or conversely, if the distance is known, the rotation law can be established. By convention, the rate of change of r determines the sign of the radial velocity, so that the velocity relative to the Sun will be positive in regions I and III, and negative in regions II and IV.

The current state of knowledge of the rotation curve of our Galaxy, the Milky Way system, is summarized in Figure 10.2. The inner part of the rotation curve is derived from H I and CO data, using the tangent points as in Figure 10.1(b). The curve beyond the solar circle uses optical data from observation of H II regions and open clusters; the scatter is probably caused by measurement uncertainties and will improve with time. Inside $R = 4$ kpc there are systematic differences between the velocities observed in regions I and IV, due to non-circular motions; Figure 10.2 shows averages. The simplest fit to the observations is a straight horizontal line from $R = 4$ kpc, that is, a constant velocity equal to the rotational velocity at R_0. This remarkable result is not peculiar to the Milky Way; it is a common

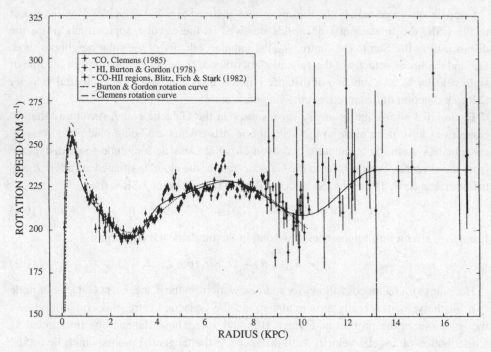

Figure 10.2. The rotation curve of the Galaxy, from observations of CO and H I emissions inside the solar radius, and from CO complexes associated with H II regions of known distances (Combes 1991).

phenomenon, observed in other spiral galaxies both in the radio observations (Wevers *et al.* 1986) and in observations of optical emission lines (Rubin *et al.* 1985). In a Keplerian system, with a point mass at the centre, the tangential velocity would diminish as $R^{-1/2}$; a model of the Milky Way in which the mass is the sum of the observable luminous matter, the gas and the dust gives a slower fall with radius, but there is no match possible to the nearly constant velocity from 4 kpc outwards, shown in Figure 10.2.

If the visible Milky Way resided in a singular isothermal sphere of matter, a constant rotational velocity, independent of radius, would be expected. Since the same phenomenon is observed in many other galaxies, this led to the conjecture that galaxies are often embedded in spherical mass distributions, larger than the visible galaxy, although the matter is not luminous. There is other evidence that *dark matter* is common: clusters of galaxies show dynamical behaviour that demands far more mass than can be accounted for by the observed luminous matter. Furthermore, gravitational lensing (Chapter 16) gives evidence that the observed multiple images require the presence of a large amount of dark matter in many cases. This led to the conclusion that the Universe contains a large quantity of dark matter that cannot be detected except by the influence of its gravitational field. It cannot be neutral hydrogen or ionized hydrogen, but it might be condensed objects, too small to become luminous stars, commonly referred to as massive compact halo objects (MACHOs). Two searches for such objects, the MACHO and OGLE projects, using gravitational lensing of background stars, have failed so far to show their existence if they are in the size range from the mass of Jupiter to the mass of a brown dwarf (0.001–0.05 solar masses). Another option, at the fundamental-particle scale, conjectures that the dark matter consists of weakly interacting massive particles (WIMPs). This has encouraged a number of searches for such particles, but with negative results so far. Nevertheless, the evidence for our Milky Way, and many other massive galaxies, is so strong, and other evidence for dark matter is so compelling, that its existence has to be granted. See Chapters 14 and 15 for a further discussion.

The size of the halo of the Milky Way is not known, but it has to be limited by tidal interactions with other galaxies; Fich and Tremaine (1991) suggest a nominal halo radius of 35 kpc, and a total mass of the Galaxy, including the halo, of

$$M_{\rm G} = 3.9 \times 10^{11} \left(\frac{\Theta}{220\,{\rm km\,s^{-1}}} \right)^2 \left(\frac{r_{\rm max}}{35\,{\rm kpc}} \right) M_{\odot}. \tag{10.3}$$

There are two commonly used parameters that are derivable from Galactic rotation, the Oort constants A and B, defined by

$$A \equiv -\frac{1}{2} R_{\circ} \left(\frac{{\rm d}\Omega}{{\rm d}R} \right)_{R_{\circ}}, \tag{10.4}$$

$$B \equiv -\frac{1}{2} \left[\frac{\Theta_{\circ}}{R_{\circ}} + \left(\frac{{\rm d}\Theta}{{\rm d}R} \right)_{R_{\circ}} \right]. \tag{10.5}$$

Both quantities refer to circular motion near the solar circle; the Oort A is proportional to the logarithmic derivative of the angular velocity $\Omega(R)$, while the Oort B is most easily interpreted by invoking its relation to the angular velocity of the Galaxy:

$$\Omega_{\circ} = A - B. \tag{10.6}$$

Oort originally introduced these constants in the context of the stellar motions in the solar vicinity, but they are also relevant to radio observations of the Milky Way. For example, the product $AR_{\circ}$ is related directly to the H I observations; by examining the trigonometric relationships for the velocity V observed at the circle of maximum velocity in Figure 10.1, and by use of the defining relationships for the Oort constants, the following equation is obtained if $R \approx R_{\circ}$:

$$AR_{\circ} \approx \lim_{R \to R_{\circ}} \frac{V}{2(1 - |\sin \ell|)\sin \ell}. \tag{10.7}$$

A detailed discussion of this and other relationships in the circular approximation can be found in the discussion by Burton (1988).

The Earth's quasi-circular Galactic orbit causes an apparent motion of the radio source Sgr A* at the Galactic centre with respect to a set of compact extragalactic radio sources; this provides an important method of measuring the distance $R_{\circ}$ from the Sun to the centre, as discussed in Section 10.5.

CO on Hα

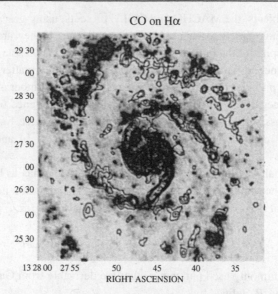

RIGHT ASCENSION

Figure 10.3. The spiral structure of M51, seen in CO line emission superposed on Hα emission (Rand and Kulkarni 1990). The contour intervals are 4.5, 7.8, 13.5, 22.5, 31.5 and 40.5 Jy km s^{-1} beam^{-1}.

The velocities of individual stars need not be exactly zero in their local rotating frame, and they will slowly oscillate about an equilibrium position. Given a Galactic gravitational potential having rotational symmetry, one can ask for the motion that would be executed by a test particle, displaced from its equilibrium radius. A linear approximation (see, for example, Binney and Tremaine (1987)) shows that the particle will execute an epicyclic motion of frequency κ:

$$\kappa = \sqrt{-4B(A - B)}. \tag{10.8}$$

In the case of our Milky Way, where a test particle in the solar neighbourhood completes a complete revolution in 230 million years, it will execute about 1.3 epicyclic oscillations during each revolution. A large section of the Milky Way moving coherently in this way forms a density wave that is seen as a spiral structure, which we now discuss.

10.2 Spiral structure

The beautiful patterns of spiral galaxies have radio counterparts; these can be seen both in the radio continuum and in the spectral lines from hydrogen and common molecules such as CO. This should be expected, because the optical spiral arms are outlined by star-forming regions, and the stars have formed from the ISM. This is illustrated in Figure 10.3, which shows the CO distribution in M51, constructed by aperture-synthesis measurements with the Owens Valley Millimeter Interferometer, superimposed on Hα emission that traces the

H II-region distribution. There is a practical problem to be overcome in making such maps of external galaxies, since their distance and consequent small angular size requires the use of aperture-synthesis interferometers in order to obtain the necessary angular resolution. Since the interferometers have a small filling factor, the molecular-line emissions are weak, and long observing sessions are needed in order to obtain the necessary signal-to-noise ratio.

We can obtain much stronger signals from the ISM of our own Milky Way, and the fine details and small-scale motions are more easily observed, but we are immersed in its structure and the overall patterns are not traced easily. Nevertheless, the spiral structure of our own Milky Way can be deduced, under the circular approximation, by taking the observed line profiles such as those used to construct Figure 10.4, and transforming them through the use of the fundamental Equation (10.2) into a face-on view of the Milky Way. Figure 10.5 is the celebrated result obtained by Oort *et al.* (1958), who combined the single-dish hydrogen observations taken in the Netherlands and in Australia to delineate the spiral structure of our Galaxy. This representation is not easily improved upon, since there are large-scale departures from circular velocity.

There is a clear relationship between the pitch of spiral arms in galaxies and the richness of their ISM (Sa galaxies have large central bulges of stars and are tightly wound, with little star formation visible and relatively little gas; Sc galaxies have small central bulges, loosely coiled arms, much star formation and a relatively large proportion of gas). If such spiral patterns were purely kinematic, permanent structures, the differential galactic rotation would wind them up tightly over the lifespan of the Universe. It is natural, therefore, to look for a dynamical cause for spiral structure, as a large-scale radial oscillation in the form of a density wave.

The density-wave theory was presented in a rigorous way by Lin and Shu (1964), who showed that spiral waves could exist in a realistic model galaxy, within a linear approximation. Most of the analytical work can be found worked out in some detail by Binney and Tremaine (1987); see also Binney and Merrifield (1998). In its essence, the theory develops a dispersion relation for gravitational-wave motion in a thin disc consisting of both stars (obeying the Boltzmann equation) and gas (a fluid, subject to hydrodynamics), under the influence of its self-gravity. One starts with a mass distribution that, for circular orbits, would result in a rotation curve $\Omega(r)$. When perturbed, a stable spiral pattern may develop, characterized by an azimuthal number m (the number of spiral arms), the pattern speed Ω_p and a radial wave number k (which, with m, specifies the pitch of the spiral pattern). In the disc, with surface density μ, a perturbed particle will execute epicyclic motion at frequency κ. The particles in the disc will have a velocity dispersion σ, and, if the disc is to be stable, this dispersion must exceed

$$\sigma > 3.358 G \mu \kappa, \tag{10.9}$$

a condition known as the Toomre criterion (the two-dimensional analogue of the Jeans instability). For a given galactic rotation curve, the radial wave number vanishes at the

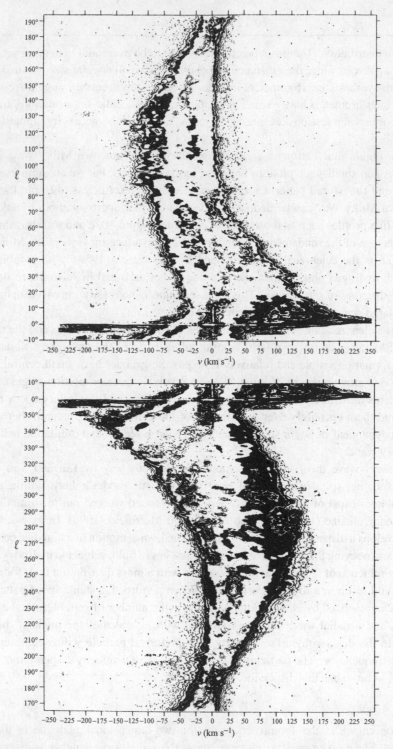

Figure 10.4. The spectrum of H I emission along the plane of the Galaxy. The velocity–longitude contours represent the neutral-hydrogen intensities along the Galactic equator, $b = 0°$ (Burton 1988).

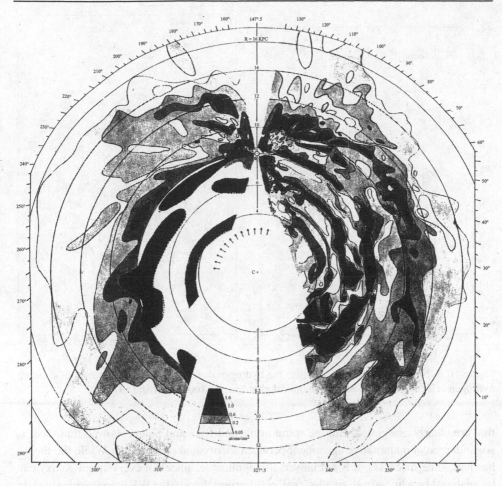

Figure 10.5. The Oort, Kerr and Westerhout map of the spiral arms of the Galaxy, constructed from a circular model (see Figure 10.1), showing the maximum densities projected onto the Galactic plane (Oort *et al.* 1958).

Lindblad resonances defined by

$$\Omega_p = \Omega \pm \kappa/m, \qquad (10.10)$$

where the nomenclature is derived from Lindblad's perceptive insight that interesting dynamical phenomena might arise when the difference between the rotation speed and the pattern matches an integral fraction of the epicyclic frequency.

The 21-cm hydrogen observations give a reasonable fit to the theory in the case of the regular spiral galaxy M81. The data of Rots and Shane (1975) are shown in Figure 10.6, where the H I density map is shown, and the observed radial velocities are superimposed on an optical photograph at the same scale. The lines of constant velocity exhibit variations

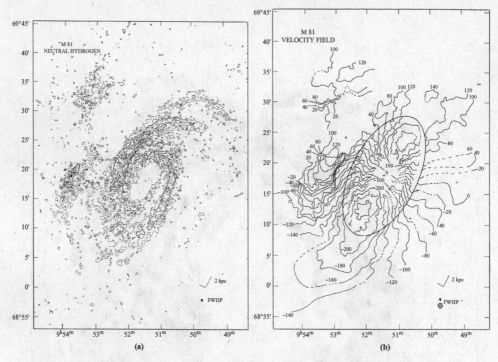

Figure 10.6. Neutral hydrogen in M81: (a) the hydrogen distribution and (b) contours of constant hydrogen velocity, superimposed on an optical photograph (Rots and Shane 1975).

that are clearly associated with the spiral arms, and Rots and Shane showed that there is some degree of quantitative agreement between theory and observation. While the theory has had some successes, it is not universally applicable, since not every spiral galaxy can be analysed by a linearized, small-signal, equilibrium theory (the WKB approximation). In particular, not even M81, despite its regular appearance, can be treated in isolation.

A comparable display of the CO observations, compiled by Dame *et al.* (1987), is presented in Figure 10.7. This set of data was taken with two identical 1.3-m telescopes in the northern and southern hemispheres, so providing complete sky coverage. Because of the shorter wavelength of the CO 1–0 line, the angular resolutions of the H and the CO surveys are comparable. The contours show brightness temperature, as in the hydrogen plot, and the same cautions concerning optical depth apply.

Several patterns show up clearly in both diagrams. There is a clear envelope, reaching its most extreme velocities at Galactic longitudes $\ell = \pm 20°$, that relates to the rotation of the Galaxy. Close to the Galactic centre ($\ell = 0$) there are markedly high velocities, far greater than the normal Galactic rotational velocity, that indicate that there may be anomalies associated with the central region. Along the entire Galactic circle, low-velocity material can be seen, but its distribution in CO is patchy, since much of the CO occurs in giant molecular clouds. We now address these various patterns, firstly under the assumption

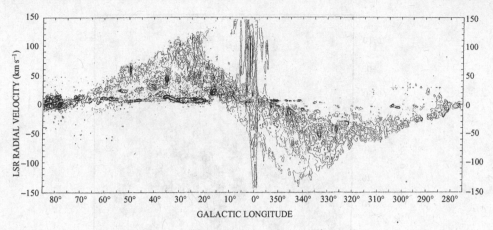

Figure 10.7. The spectrum of CO emission along the plane of the Galaxy, integrated over Galactic latitudes −3.25° to +3.25° (Dame *et al.* 1987).

that the material is in circular orbits, and then turn to the various departures from circular motion.

10.3 Non-circular motions

The spiral galaxy M51, often chosen as a beautiful example of spiral structure, is by no means a simple example of spiral density waves when one examines the optical photographs. There is a separate galactic system, NGC 4195, at the northern end of the left-hand spiral arm; this smaller galaxy is actually beyond the disc of M51, but might nevertheless have significant gravitational effects. Toomre and Toomre (1972), followed by Toomre (1981), carried out numerical simulations of two discs of stars passing by one another with only gravitational interactions; they found orbital parameters and disc inclinations that gave results strikingly like the form of M51. Since that time, many other interacting systems have been studied by Toomre and collaborators, and the importance of tidal interaction, and the temporary spiral perturbation that is a consequence, is now well established.

Even the regular spiral M81 does not appear to be free of the influence of neighbours, as shown by Figure 10.8. This map of the compact M81 group, with its neighbours M82 and NGC 3077, shows that over the entire 150-kpc region there is extensive hydrogen, with prominent features linking the three galaxies. This map typifies the problem of carrying out studies of nearby galaxies and groups of galaxies: high angular resolution is needed in order to study the dynamics of the individual galaxies, but the diffuse material is spread over such a large angular scale that much of it may require interferometric coverage at small spatial frequencies. In this case, the VLA was used in its most compact configuration (the D-array), supplemented by single-dish observations to detect the largest-scale components.

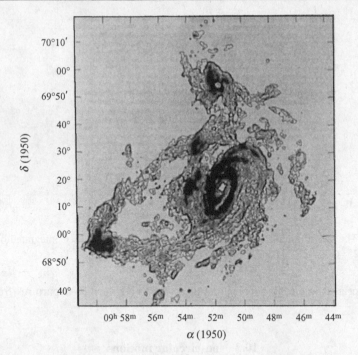

Figure 10.8. The M81 group and its associated H I complex. M81 dominates the group, but there are substantial hydrogen concentrations associated with M82 (top centre) and NGC 3077 (lower left) (Yun *et al.* 1994).

Large spiral galaxies frequently have no companions: however, the great Andromeda Galaxy, M31, is accompanied by two smaller elliptical galaxies, while the Milky Way system has two satellites, the Large and Small Magellanic Clouds (LMC and SMC). Collisions certainly can excite spiral density waves, and the Lin–Shu theory might not be appropriate because of the large amplitude of the perturbation. The field is still an active one, and requires the study of external galaxies in H I and molecular lines with aperture-synthesis arrays. Interactions and collisions may be essential in determining the observed structures of galaxies, as emphasized by Barnes and Hernquist (1992). Nevertheless, there are spiral galaxies that appear to have no companions, and there is a strong morphological connection in general between the presence of a gas disc and the form of the spiral structure. The final synthesis does not yet seem to have been accomplished.

Barred spirals (types SBa, SBb and SBc in the Hubble–Sandage scheme) form a particular morphological class, and nearly half of all spiral galaxies may possess a bar according to the study by Kormendy and Norman (1979). Numerical experiments have shown that a bar-like stellar structure forms naturally in a large stellar system, and the bar can be described as a stable density wave with $m = 2$. Binney and Tremaine have shown that there is a stable regime for bar formation if there is no inner Lindblad resonance in the system, allowing a bar to exist without deformation up to a corotation radius. The spiral arms of SB systems spring from the ends of the bar, and are always trailing; the numerical study by Sanders and Huntley (1976) showed that trailing spiral arms are formed naturally when a rigid bar of stars rotates within a gaseous disc.

The question of whether our own Milky Way system might have a stellar bar at the centre arises. The Milky Way maps of molecular-line emission, such as Figure 9.7, show a concentration about 3° long at the centre containing about 10% of the total molecular gas content of the whole Galaxy, but it is not obvious from total-brightness maps whether the structure is spheroidal or a bar. Star counts are of little use in resolving the shape because of the severe absorption. Velocities are, however, available from CO and other molecular measurements, and from OH maser stars that belong to the older population (mostly Mira-type variable stars). Within a bar there must be large radial components of velocity; these are indeed observed, showing that the Milky Way has a bar, with its long axis extending to 2.4 kpc and making an angle of about 20° to the Sun–centre line, the nearer end being in the northern direction. (See a review of the dynamics of the bar by Blitz *et al.* (1993).) The large components of orbital velocities of stars and gas along the bar may contribute to feeding the black hole at the centre. The Milky Way is now regarded as a typical barred spiral.

The dynamics of the central regions of the Milky Way are dominated by the gravitational field of the bar, which contains a mass of $(1–3) \times 10^{10} M_\odot$. (The effect of the black hole, which has a much smaller mass (Section 10.4 below), is important only close to the centre.) Several features of the velocity maps presented in Figures 10.4 and 10.7 have been interpreted as discrete entities; for example an anomalous feature that crosses the Galactic centre with a systematic velocity of -55 km s^{-1}, extending from longitude 338°, has been interpreted as a tangent point to an arm with radius 3 kpc: this was referred to as the '3-kpc expanding arm', but it should now be regarded as part of the bar. At larger distances from the centre, the 10–20-km-s^{-1} asymmetry of velocities between positive and negative Galactic longitudes is probably also an expression of the effect of the quadrupole gravitational component of the bar. Within 2° of the centre there is a more clearly defined feature, the Central Molecular Zone (CMZ), outlined in the CO-line intensity map of Figure 10.9.

The CMZ is a ring of radius 180 pc containing a high concentration of molecular species; it is comparable to the giant molecular clouds found elsewhere in the Galaxy, but with higher density and temperature (typically 70 K). The high-velocity gas in the CMZ shows up as a parallelogram in the plot of the CO spectrum against Galactic longitude in Figure 10.10, with velocities extending from -135 km s^{-1} to $+165$ km s^{-1}. The ring is tilted at about 6° to the Galactic plane; positions in Figure 10.10 are aligned with the plane of the ring.

Within the CMZ and approaching the Galactic centre itself, the dynamics are increasingly determined by the gravitation and energetics of the black hole, to which we turn in Section 10.4.

Returning to the large-scale features of the Galaxy, we draw attention to hydrogen clouds with velocities well outside those expected from the simple rotational model. Survey maps of H I by Hartmann and Burton (1995) show the distribution over the sky of neutral hydrogen in a series of velocity ranges, illustrating both the regularities of the hydrogen

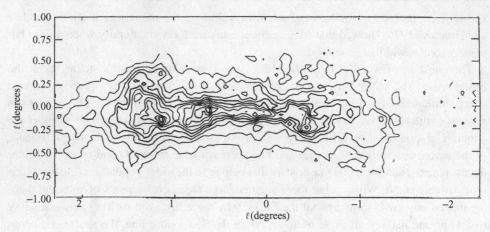

Figure 10.9. The ^{12}CO, $J = 1$–0 emission from the Central Molecular Zone of the Milky Way. (Observations by Uchida *et al.*; from the review by Morris and Serabyn (1996).)

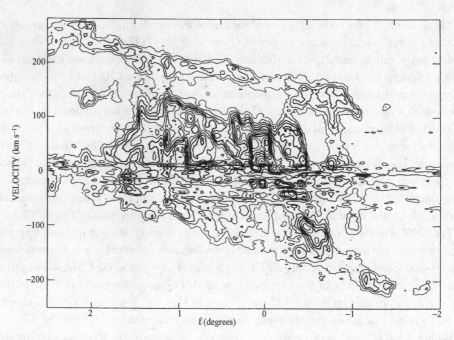

Figure 10.10. Velocities observed in ^{12}CO in the region of the Galactic centre, showing the parallelogram of velocities due to the Central Molecular Zone (CMZ). Contour units are 1.4, 2.8, 4.4, 6.2, 8.2 and 10.4 K. The positions labelled in Galactic longitude ℓ are along a line inclined by $6°$ to the Galactic plane, following the tilt of the plane containing the CMZ. (Observations by Uchida *et al.*; from the review by Morris and Serabyn (1996).)

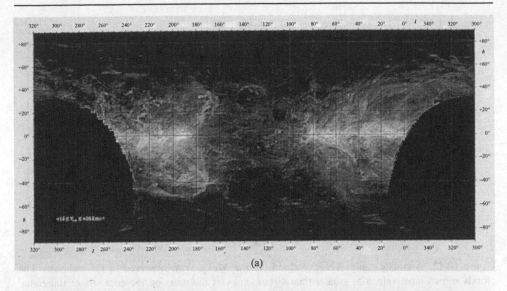

(a)

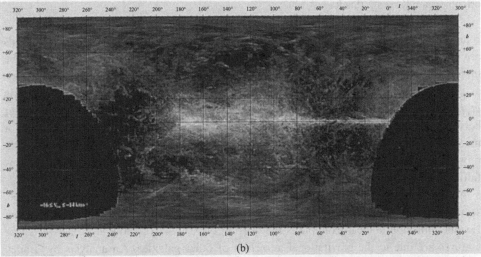

(b)

Figure 10.11. The hydrogen distribution at intermediate velocity ranges (Hartmann and Burton 1995).

motions and some of the significant local anomalies. Figure 10.11 presents two of these maps, showing hydrogen with approaching and receding velocities $|v| \approx 36\,\mathrm{km\,s^{-1}}$. These show anomalous 'intermediate-velocity' clouds at high Galactic latitudes. Their cirrus-like structure is a common feature at all velocities.

At much greater radial velocities, far outside permitted rotational velocities, there are strikingly anomalous 'high-velocity clouds', discovered by Muller *et al.* (1966). These are the clouds already shown in the total-hydrogen map of Figure 9.4. In Figure 10.12 Stark *et al.* (1992) removed all hydrogen with velocity within $40\cos b\,\mathrm{km\,s^{-1}}$ of the average

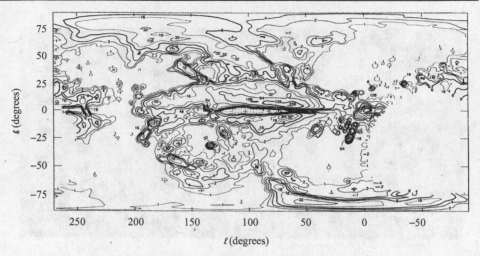

Figure 10.12. Anomalous high-velocity clouds in the Milky Way, observed in H I emission. The clouds shown have velocities greater than $40 \cos \ell \, \mathrm{km \, s^{-1}}$; the units on the contours of integrated brightness are $\mathrm{K \, km \, s^{-1}}$ (Stark *et al.* 1992).

rotation pattern, leaving only the high-velocity clouds. Most of these discrete H I clouds are moving towards us, some with velocities exceeding $-400 \, \mathrm{km \, s^{-1}}$; they cannot be understood as a result of normal Galactic rotation. There is one prominent feature, known as the Magellanic Stream, that stretches in a systematic way across the sky. Mathewson *et al.* (1979) noticed this feature, and proposed that it was a fossil remnant of an encounter between the LMC and the Milky Way system. This proposal may well be correct, but for many of the high-velocity clouds there is as yet no proven explanation. It is possible that most of them are Galactic 'rain', ejected by great explosions in the Galactic disc, and now falling back to the plane.

Another common feature of spiral galaxies is a *warp* in the otherwise flat galactic plane, seen at large distances from the centre. This is observed in our Galaxy and others through H I line emission; a clear example is the warp in NGC 3741 (Gentile *et al.* 2007). These features are due to the gravitational attraction of other galaxies. An analysis by Binney (1992) shows that, like the Magellanic Stream, the warp in our Galaxy may have been due to the gravitational attraction of the LMC in a close encounter with the Milky Way.

10.4 The Galactic centre

The central region of the Milky Way is highly disturbed, as the 10-GHz continuum map made by the 35-m Nobeyama telescope shows (Figure 10.13). The strong central source, Sgr A*, has long been known as a variable non-thermal radio source, probably at the dynamical centre of the Milky Way, and has now been established to be a black hole. The nuclei of many galaxies are believed to contain a black hole, which is the focus of some of the most dramatic radio phenomena in the Universe. The centre of the Milky Way system,

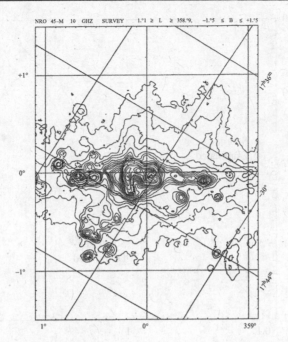

Figure 10.13. Continuum radiation at wavelength 10 GHz in a region of four square degrees in the direction of the Galactic centre (Handa *et al*. 1987).

while not showing the violent events exhibited by the quasars and active galactic nuclei, does contain a black hole, whose existence is clearly demonstrated by the dynamics of stars and molecular clouds in its vicinity. Although our Milky Way black hole is a weak radio source compared with the quasars which we discuss in Chapter 13, it is surrounded by an astonishing variety of phenomena that are revealed by radio observations both in the continuum and in molecular-line radiation.

The black hole at the Galactic centre is hidden by massive optical obscuration, but infrared observations with the 10-m Keck telescope at 2.2 μm have nevertheless produced the most decisive evidence for its existence. The infrared coordinate system needed calibration from the radio astrometry on OH stars near the galactic centre; with these, orbits of the infrared stars could be established. An extensive swarm of such stars has now been observed, some in compact, highly elliptical orbits, one with an orbital period of only 6.5 years (Schödel *et al.* 2003; Ghez *et al.* 2005). Several of these pass so close to the compact source Sgr A* that its mass can be clearly established at $(3.7 \pm 0.2) \times 10^6$ solar masses. This is undoubtedly the black hole whose existence was surmised by Lynden-Bell and Rees (1971). The location coincides with that of the radio source Sgr A*.

The VLBI measurements of the angular size of the compact radio source Sgr A* had been limited by interstellar scattering, as Figure 10.14 shows. At wavelengths longer than 3.5 mm, the apparent angular size varies as λ^2, a clear indicator of scattering in an ionized medium, but the measurements at 1.4 and 1.25 mm resolve the source (Doeleman *et al.* 2008). At

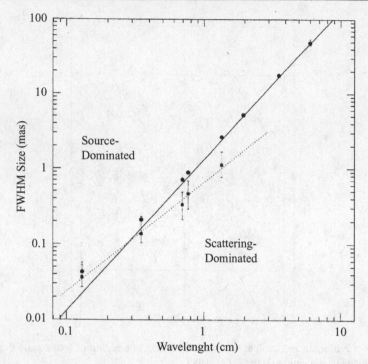

Figure 10.14. The wavelength dependence of the angular size of Sgr A* (Doeleman *et al.* 2008, arXiv:astro-ph/08092442). The sizes measured at wavelengths 3.5 mm and 1.25 mm lie above the extrapolated scattering line, showing that the source itself is resolved at these short wavelengths. The square points and the dotted line represent the deduced source diameter in the absence of scattering.

the shortest wavelength the linear size inferred from these observations corresponds to about four Schwarzschild radii for a black hole of 3.5 million solar masses. At a radius of $3r_S$, orbits are unstable, and thus the combination of the infrared astrometry and the millimetre-wave VLBI observations establishes beyond any doubt that Sgr A* is a black hole.

The central 3-arcminute map shown in Figure 10.15, covering the central 10 pc × 10 pc region, illustrates the interplay between high- and low-energy phenomena near the Galactic centre. Figure 10.15 shows the spectral indices derived from 6- and 20-cm observations, and highlights the steep spectrum of synchrotron radiation (with a negative spectral index) within an elliptical locus. Offset to the west, but embedded in the non-thermal structure, there is a curious three-armed spiral whose flat spectrum indicates thermal bremsstrahlung emission. This central 'spiral' (the three features are known as the northern arm, the eastern arm and the western arc) is believed to be infalling material that is also rotating about the nucleus. Surrounding this central complex, there is an elliptical ring of molecular gas, as shown in the HCN 1–0 line emission illustrated in Figure 10.16.

Although the Galactic black hole is regarded as dormant by comparison with the active galactic nuclei we discuss in Chapter 13, there is sufficient energy released by infalling

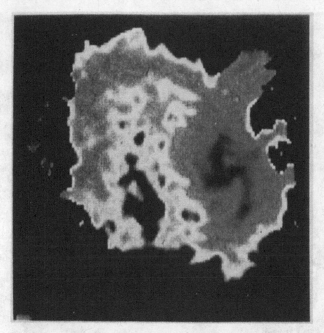

Figure 10.15. A spectral index map of the central 10 pc × 10 pc region of the Milky Way galaxy (map size 3 arcmin × 3 arcmin), derived from VLA maps at 20 cm and 6 cm. The grey and white regions are non-thermal; the areas with the steepest spectrum are white. The central spiral has a thermal spectrum, and in this representation it is dark (courtesy of C. H. Townes).

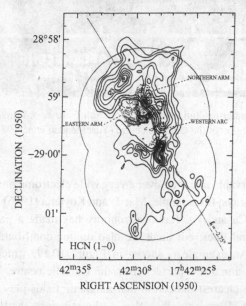

Figure 10.16. Superposition of HCN line emission (bold contours) on the 5-GHz radio continuum in the region of the Galactic centre. The HCN contours extend about 7 arcmin along the Galactic plane (Güsten *et al.* 1987).

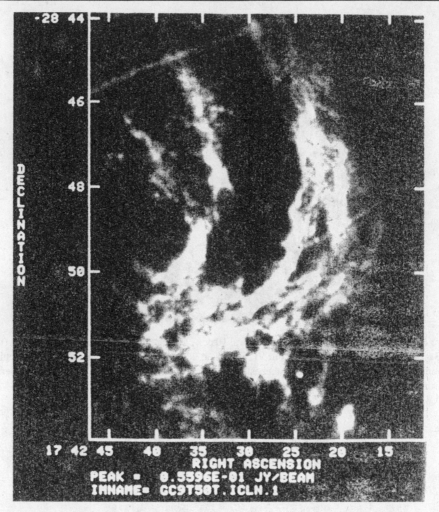

Figure 10.17. A high-angular-resolution map (beamwidth 5 arcsec × 9 arcsec) of features within 20 arcmin (60 pc) of the region of the Galactic centre (Yusef-Zadeh *et al.* 1984).

matter to generate observable radiation over a very wide electromagnetic spectrum, including gamma-rays and X-rays; for example, Maede and Koyama (1996) report X-ray spectral lines from Si, S, Ar, Ca and Fe. Radio astronomy has made a particular contribution in the molecular spectral lines, but there are also unique contributions from radio continuum observations. An example is shown in Figure 10.17, which shows remarkable filaments of radio emission a few parsecs from the Galactic centre. Their radio emission is polarized, giving the clearest indication that magnetic fields play an important part in the dynamics of the ISM, and possibly also in the structures and events closer to the centre.

10.5 The scale of the Galaxy

Conventional values of Θ and R, which set the scale of Galactic phenomena and influence the mass models, will be changing as measurements improve, so when one reviews the literature there should always be an awareness of the convention being used. From 1964 to 1986, conventional values of $250 \, \mathrm{km \, s^{-1}}$ and $10.0 \, \mathrm{kpc}$ were used; the IAU then recommended that $220 \, \mathrm{km \, s^{-1}}$ and $8.5 \, \mathrm{kpc}$ were preferable, but it was recognized that these values would certainly need revision. The source Sgr A* seems to mark the Galactic nucleus, and Sramek and Backer, using the VLA as an astrometry instrument, have been following the proper motion of Sgr A* for several years. Most recently, Reid *et al.* (1999) found using the VLBA that the proper motion is $5.90 \pm 0.4 \, \mathrm{mas \, yr^{-1}}$ in the plane of the Galaxy.

It appears, therefore, that we have a value for the angular velocity of the solar neighbourhood about the Galactic centre that is known to better than 10%. For a tangential velocity of $220 \, \mathrm{km \, s^{-1}}$, this corresponds to a distance of $8.0 \, \mathrm{kpc}$ to the centre of the Milky Way system. There are independent ways of measuring $R_\circ$, both optical and radio, and an overview of their status has been presented by Reid (1993). The principal radio measurements have relied on the dynamical properties of H_2O masers, which appear to be assemblages of compact, brilliant spots. These have measurable proper motion μ and, since their radial velocities v_z relative to the maser complex are known, one can derive a distance by taking the average $\langle v_z / \mu \rangle$. The complex of masers near the Galactic centre, in Sgr B2, gave a value of $7.1 \pm 1.5 \, \mathrm{kpc}$ (Reid *et al.* 1988), while measurements of the intense maser complex W49, which lies on the solar circle, yielded $8.1 \pm 1.1 \, \mathrm{kpc}$ (Gwinn *et al.* 1992). There are possible systematic errors in both of these values, since Sgr B2 is seen only in projection, and may be further out from the centre than we think. The W49 instance depends upon the assumption of circular motion at $R_\circ$, the solar radius; this, too, could be in error.

The best optical methods are now giving comparable values for $R_\circ$ (Genzel *et al.* 2000), and, while each method is subject to its own systematic errors, the consistency of all the results suggests that $R_\circ = 8.0 \, \mathrm{kpc}$ is close to the correct value, possibly to be revised to a slightly lower value if the radio results prove to be more accurate. For a general review of the state of knowledge of the principal constants of the Milky Way system, see Schechter (1996). The total hydrogen mass of the Galaxy is not easily determined. One cannot simply integrate the area under the line profiles, for in the plane the self-absorption effects can be substantial. The estimate of 5×10^9 solar masses by Fich and Tremaine (1991) is as good as any, and is probably valid to within 30%.

10.6 Atoms and molecules in other galaxies

Apart from the detailed information on the dynamics of nearby galaxies, as seen for example in Figure 10.6, H I observations show that neutral hydrogen often extends well beyond the visible content of a galaxy. Emonts *et al.* (2007), for example, show that in most of 21 nearby galaxies the total hydrogen mass exceeds $10^9 M_\odot$, extending to sizes of up to 190 kpc.

A similar large hydrogen content is found by Lah *et al.* (2007) in a survey of galaxies with large redshift ($z \approx 0.24$). The mean hydrogen content of 121 galaxies, selected from an optical survey of star-forming galaxies, was found to be $2 \times 10^9 M_\odot$. This shows that these distant galaxies are similar to our own Galaxy, and provides a useful measurement of the hydrogen density of the Universe at a look-back time of 3 Gyr. Studies of much earlier galaxies, to follow the course of galactic evolution from the age of reionization, require telescopes with much larger collecting area. An early study by Wilkinson showed that a collecting area of the order of a square kilometre would be needed (Wilkinson *et al.* 1997). This served as the motivation for the Square Kilometer Array project, which is now actively being pursued by an international consortium (Chapter 17).

Emission lines from molecular gas in high-redshift galaxies offer an interesting contrast. The study by Solomon and Vanden Bout (2005) described the observational results of current studies of relatively nearby systems. These had gas masses comparable to, or greater than, that of the Milky Way, in the range $(0.5–11) \times 10^{10}$ solar masses. They pointed out that, when observing at a constant frequency, the detectability of extragalactic molecular line radiation is almost independent of redshift. At large z, although a specific line is weakened by the redshift factor $1/[z(z + 1)]^2$, at a given frequency one will be observing lines of higher angular momentum. A combination of the higher probability of spontaneous radiation for those intrinsically higher-frequency transitions and an enhancement of the statistical weight roughly cancels out the redshift diminution.

A dramatic example is given by the galaxy associated with the quasar J1148+52, at a redshift of 6.4. The map of Walter *et al.* (2004) of the CO 3–2 transition showed that the total mass of gas in the system is 2×10^{10} solar masses; of that roughly 2×10^6 solar masses is in the form of CO. The galaxy TN J0924, at redshift 5.203, has approximately three times the gas content. The formation of heavy elements clearly had to occur at an early stage of evolution in the Universe, during the era of Population III stars (Chapter 8) during the 'dark ages' prior to the era of reionization.

Singly ionized carbon, C II, has a forbidden fine-structure transition, $^2P_{3/2}–^2P_{1/2}$, at 157 μm (1900.54 GHz), which is inaccessible from the ground, but, at a sufficiently large redshift, it should be observable in other galaxies. Together with the CO molecule, it provides the major cooling mechanism for the diffuse ISM. Neutral carbon, C I, has two fine-structure transitions, at 809.34 GHz ($^3P_2–^3P_1$) and 492.16 GHz ($^3P_1–^3P_0$); the latter is certainly accessible to ground radio observations by ALMA. All of these carbon lines, in redshifted systems, should eventually allow the detailed study of heat balance in early galaxies.

Further reading

Atlas of Galactic Neutral Hydrogen, Hartmann D. and Burton W. B., Cambridge, Cambridge University Press, 1997.
Dynamics of Galaxies and their Molecular Cloud Distributions, IAU Symposium No. 146, ed. Combes F. and Casoli F., Dordrecht, Kluwer, 1991.

Galactic Astronomy, Binney J. and Merrifield M., Princeton, NJ, Princeton University Press, 1998.

Galactic Dynamics, Binney J. and Tremaine S., Princeton, NJ, Princeton University Press, 1987.

Galaxies: Structure and Evolution, Tayler R. J., Cambridge, Cambridge University Press, 2nd edn, 1993.

HI in the Galaxy, Dickey J. M. and Lockman F. J., *Ann. Rev. Astron. Astrophys.*, **28**, 215, 1990.

The Centre of the Galaxy, IAU Symposium No. 136, ed. Morris M., Dordrecht, Kluwer, 1989.

The Dynamics, Structure and History of Galaxies, ASP Conference Proceedings No. 273, ed. Da Costa G. S. and Jerjen H., Washington, Astronomical Society of the Pacific, 2002.

Unsolved Problems of the Milky Way, IAU Symposium No. 169, ed. Blitz L. and Teuben P., Dordrecht, Kluwer, 1996.

11

Stars

The strongest discrete radio source in the sky, and the first to be discovered, is the Sun. At metre wavelengths, however, where the discovery was made, it was not the photospheric surface but the very extensive solar atmosphere, the corona, whose radio emissions were being observed. The visible discs of stars are generally best studied at optical and infrared wavelengths, but stars at a late stage of evolution may develop very large atmospheres, which are the seat of some of the most spectacular of radio emissions. These stars have left the main sequence, having exhausted hydrogen burning in their cores, cooling and expanding to form red giants.

This chapter is concerned with the surfaces of the Sun and the planets in our solar system, and with thermal radiation from stellar atmospheres. The widest significance for astronomy comes from the circumstellar masers, which occur both in the outflowing stellar winds and in infalling accretion.

11.1 Surface brightness

At millimetre wavelengths the radio Sun is closely similar to the visible Sun; it approximates to a disc at a temperature not much higher than 6000 K, which is the temperature of the photosphere. At metre wavelengths it is larger, brighter and more variable, with occasional flaring outbursts; here we are observing the solar corona rather than the photosphere. In either case, however, stars like the Sun would not be observable at the distance of the nearest stars, where the inverse square law would reduce the observed flux density by a factor of 10^{10} or more. Nevertheless, there are types of star with observable radio emission, and they include some of the most remarkable objects in our Galaxy. Some nearby stars are large enough for their thermal emission to be observable, either because they present a much larger disc or because they are surrounded by hot gas or dust. Others radiate very much more intense non-thermal radiation; these include X-ray binaries in which enormous energies are produced in the gravitational field of a black hole.

Most stars lie on the main sequence of spectral types[1] O, B, A, F, G, K, M. The position of an individual star on this sequence is determined primarily by its mass; O stars may have

[1] For which the traditional mnemonic is Oh Be A Fine Girl Kiss Me.

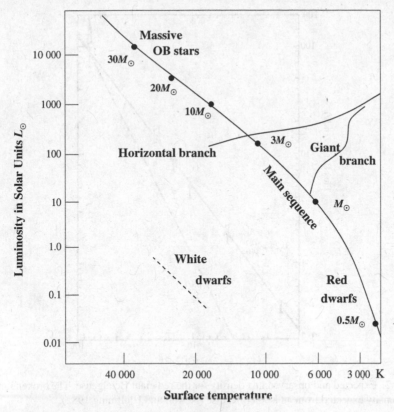

Figure 11.1. An outline Hertzsprung–Russell diagram, showing the locations of the main types of star in the luminosity–temperature plot.

a mass 50 times that of the Sun, $M_\odot$, while an M star may have a mass much less than a solar mass. The closest known star to the sun, Proxima Centauri, has a mass of $0.1 M_\odot$; if the mass of a star is less than $0.07 M_\odot$, nuclear reactions can no longer be sustained, and the body is no longer a star (Figure 11.1). The mass of a main-sequence star also determines its temperature and its radius, and consequentially its luminosity, with corrections that depend on the heavy-element abundance. The most massive stars evolve rapidly, and leave the main sequence; they are necessarily young compared with most low-mass stars, and are referred to as early-type stars. They eventually become the more luminous but cooler red supergiants; these may explode as supernovae, leaving behind a remnant neutron star or a black hole. The less luminous stars in the middle of the sequence, such as the Sun, evolve along well-defined tracks and eventually become white dwarfs. The lowest-mass stars are cool and inconspicuous; a star at this end of the sequence is only of interest in radio astronomy if it is a member of a binary system or if it is accreting matter from an interstellar cloud in which it is embedded.

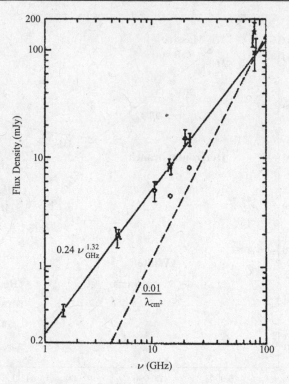

Figure 11.2. Expected and observed flux density for the red giant Betelgeuse. The broken line shows the flux density expected from an isothermal disc (Newell and Hjellming 1982).

Observational radio astronomy contributes little to the study of the visible surfaces of the stars, which provide the rich optical spectra that are the principal basis of our understanding of their composition and evolution. The fundamental problem is the low flux density to be expected from normal stellar surfaces at normal temperatures. For a star with surface temperature T, subtending a solid angle Ω, the flux density S to be measured is

$$S = 2(kT/\lambda^2)\Omega. \tag{11.1}$$

This practical limit is easily remembered as the minimum temperature $T_{\min}$ detectable with a radio telescope whose minimum detectable flux in janskys is $S_{\min}$; for a star with circular disc diameter θ arcsec, observed at wavelength λ cm,

$$T_{\min}\theta^2 = 1800\lambda^2 S_{\min}. \tag{11.2}$$

Not many normal stars have visible surfaces that satisfy this criterion. Many stars, however, have ionized atmospheres that are transparent optically but may be powerful sources of radio waves; an example is the red giant Betelgeuse (α Orionis), whose surface temperature is 3600 K, but which has an extended atmosphere at a temperature of about 10 000 K. The angular diameter of the visible disc is 0.075 arcsec. Figure 11.2 shows the expected flux density from the optical disc, with the observed flux density over a range of wavelengths (Newell and Hjellming 1982). At 3 mm (90 GHz) the atmosphere is transparent, and only the disc is observed; at lower frequencies the atmosphere becomes optically thick at progressively larger and larger radii, such that at 1 GHz the star appears as a disc of diameter about four times the visible diameter. The combined effect at the lower frequencies of increasing size and the higher temperature is to reduce the slope of the spectrum, fortuitously giving a good power-law fit,

$$S_\nu = 0.25\nu_{\text{GHz}}^{1.32} \, \text{mJy}. \tag{11.3}$$

Much more extensive and energetic atmospheres are generated around the components of many binary star systems, which are the seat of novae and visible flares; many are also X-ray binaries. In many cases there are outflows with complex structure, and in some there are obvious interactions with a surrounding nebula or the ISM. The radio emission is often much more powerful than the thermal free–free emission of a hot corona; its intensity, spectrum and polarization then indicate gyrosynchrotron or synchrotron radiation.

The angular resolution of large-synthesis radio telescopes now allows detailed observations of these energetic stellar atmospheres and binary systems. In this chapter we start with the Sun and progress through stars with more extensive, but still quiescent, atmospheres to the transient and more spectacular explosive outflows of the various types of novae.

11.2 The Sun

The Sun is a star of spectral type G, near the middle of the main sequence. It has a mass of 2×10^{31} kg and a radius of 7×10^5 km. The temperature of its surface, the photosphere, is 5770 K, as determined bolometrically. Its visible spectrum is approximately that of a black body at that temperature, cut into throughout its length by absorption lines originating in the cooler chromosphere which lies immediately outside. The shortest radio-wavelength radiation originates in the upper regions of the chromosphere, where the temperature is rising towards that of the very much hotter corona; the corona is optically thin, and contributes little to the overall surface brightness. In contrast, metre-wavelength radio waves originate in the corona. Now the opacity of the surrounding solar corona is large, and the brightness temperature of the radiation is that of the corona, not that of the photosphere or chromosphere. At intermediate (centimetre) wavelengths both the chromosphere and the corona are involved. Although the opacity of the corona is small at centimetre wavelengths, it has an appreciable emission because of its high temperature. The effect at the centre of the Sun has been measured by Zirin *et al.* (1991), who measured the brightness temperature T_b at the centre of the Sun for frequencies ν between 1.4 and 18 GHz. They showed that T_b can be modelled as the sum of two components: a constant contribution from the upper chromosphere at 11 000 K and an added component from a corona at 10^6 K, proportional to $\nu^{-2.1}$, the frequency dependence of the optical depth of the corona (see Section 7.5). Absorption of the chromospheric component by the corona is negligible in this region of

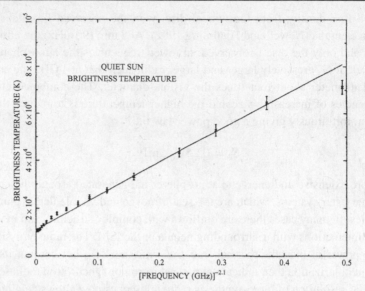

Figure 11.3. The brightness temperature at the centre of the Sun plotted against $\nu_{\mathrm{GHz}}^{-2.1}$, which follows the frequency dependence of the optical depth of the corona (Zirin *et al.* 1991).

small optical depth. The combination gives

$$T_{\mathrm{b}} = A\nu^{-2.1} + T_{\mathrm{chrom}}. \tag{11.4}$$

The measured temperatures are shown plotted against $\nu^{-2.1}$ in Figure 11.3. There is a good fit to the simple model in which the chromosphere contributes a constant 11 000 K and the thin corona, at 10^6 K, raises the brightness temperature to 70 000 K at 1.4 GHz.

At long wavelengths, for which the corona is optically thick, the Sun appears larger as well as brighter. At about 0.5–1 GHz the effect of the corona can be seen outside the photospheric disc as a *limb-brightening*, where the long line of sight through the hot corona produces a higher brightness temperature, while at metre wavelengths the disc is lost behind the bright corona. The corona is variable in shape and extent; there is no map of an ideal symmetrical quiet Sun, but Figure 11.4 (Sheridan and McLean 1985) shows the increase in diameter at lower frequencies. The brightness temperature at the centre of the disc is approximately $(8–10) \times 10^5$ K over this range of frequencies. At the lowest reliably observable frequencies, about 30 MHz, the radiation originates further out in the corona and the brightness temperature falls to about 5×10^5 K.

The high temperature of the solar corona is also observable in X-rays and in optical emission lines with a high excitation temperature. From these spectral lines, and from the radio and X-ray brightness temperatures, it is clear that the whole corona, extending out to several solar radii, is at a temperature of between 0.5×10^6 and 3×10^6 K.

Inside the Sun, the gradient of temperature from the hydrogen-burning core out to the photosphere is determined by radiation and convection. In contrast, the dramatic rise from the photosphere to the corona is due instead to mechanical and/or hydromagnetic

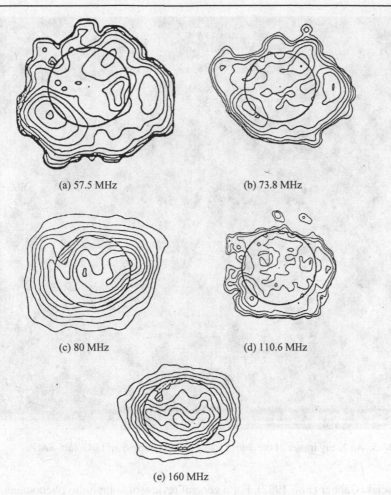

(a) 57.5 MHz   (b) 73.8 MHz

(c) 80 MHz   (d) 110.6 MHz

(e) 160 MHz

Figure 11.4. Solar radio brightness temperature at low radio frequencies (Sheridan and McLean 1985).

oscillations generated by instabilities below the Sun's surface, which propagate outwards and dissipate their energy above the chromosphere. Although the total energy involved is small compared with that of the normal solar radiation, the coronal heating has a profound effect both on the solar radio emission and on the behaviour of the outer corona, which is so hot that it streams outwards as a solar wind, eventually leaving the Sun with a velocity of some hundreds of km s^{-1}. The transport of energy into the corona is controlled by the surface magnetic field, which is very non-uniform; the result is that the corona and the solar wind are very patchy, as seen in the X-ray image of Figure 11.5. Discrete streams of the solar wind may be detected at and beyond the distance of the Earth from the Sun; these are encountered by spacecraft, but are best mapped by observing the scintillation of small-diameter radio sources (the quasars) seen through the solar wind at low radio

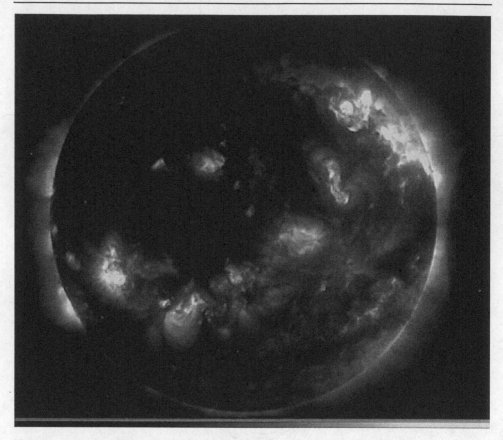

Figure 11.5. An X-ray image of the Sun, 11 July 1991 (courtesy of L. Golub, SAO).

frequencies (Gapper *et al.* 1982). For a general review of solar radio phenomena, see Dulk (1985).

Solar flares, which are caused by rapid changes in the large magnetic fields over sunspots, inject bursts of particles with high energy into the plasma of the solar corona; these particles then travel outwards along magnetic-field lines at an appreciable fraction of the velocity of light. Among a variety of radio bursts that they excite are those designated Type III, which are plasma oscillations in the range of low-frequency radio astronomy, in which the total radio luminosity of the Sun can be enhanced many thousands of times. Figure 11.6 shows the track of metre-wavelength outbursts from the solar corona, originating in flares near the limb of the Sun. The radio flare radiation is non-thermal and largely narrow-band, sweeping downwards in frequency as the burst travels outwards through the corona. The velocity can be deduced if there is a model of the density (and hence the plasma frequency) versus height in the corona. Figure 11.7 (Wild *et al.* 1954) shows an early example in which the velocity was found to be $475 \, \text{km} \, \text{s}^{-1}$; later improved models give a velocity about double that shown in this figure.

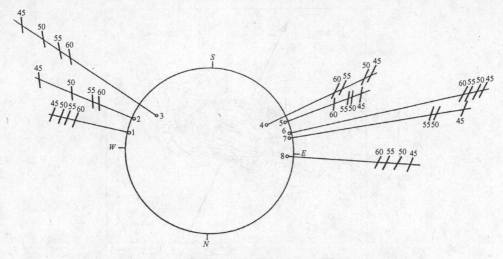

Figure 11.6. Tracks of metre-wavelength bursts originating near the limb of the Sun, showing the peak radio-emission frequency (MHz) along the track. The fall in radio frequency corresponds to decreasing plasma frequency as the energizing particles leave the Sun (Wild *et al.* 1959).

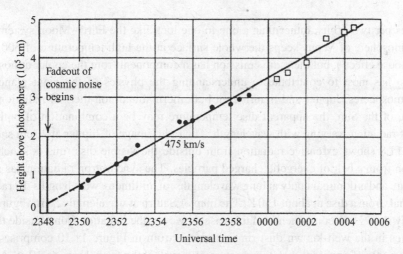

Figure 11.7. Heights in the solar corona for a Type III burst, deduced from the peak frequency as the burst travelled outwards. The velocity, here found to be 475 km s^{-1}, depends on the density model.

11.3 The planets

Thermal radio emission has been detected from all of the planets and even from a number of asteroids. An average surface brightness temperature can be measured from the flux density and the angular size, but this apparent disc temperature might not be very meaningful. The actual surface temperature may vary greatly according to rotational phase; the temperature variations on Mercury are particularly complex because it is in a tumbling mode (three

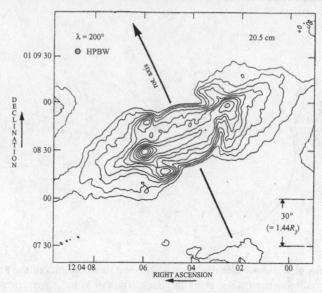

Figure 11.8. A contour map of Jupiter at wavelength 6 cm (de Pater 1990).

rotations per two orbits, rather than a one-to-one lock like the Earth–Moon system). The CO_2 atmosphere of Venus keeps the whole surface at the high temperature of 600 K (the 'greenhouse effect'), but the radio emission has a component from the cooler atmosphere.

Radio has most to contribute to understanding the physics of the giant planets and their atmospheres. Jupiter and Saturn emit both thermal and non-thermal radiation; as in the case of the Sun, the apparent disc temperature may be a combination of both these types of radiation, varying with wavelength. The radio image of Jupiter at 20 cm shown in Figure 11.8 shows extensive radiation from outside the visible disc; this is synchrotron radiation from a belt of energetic charged particles. The synchrotron radiation has a steep spectrum, and is dominant only at long wavelengths; at millimetre wavelengths the radiation is thermal, from a disc at about 150 K. The map of Saturn at wavelength 2 cm (Figure 11.9) similarly shows two components; in this case, however, the radiation from outside the disc originates in the well-known dust rings. The spectrum in Figure 11.10 comprises many measurements of apparent disc temperature at wavelengths from 1 mm to 10 m. At short wavelengths the variations are at least partly due to the varying aspect of the rings; at longer wavelengths the atmosphere has an appreciable optical depth, and the disc temperature rises accordingly (the solid curve shows the calculated effect of a model atmosphere at a higher temperature than the surface).

Several of the planets exhibit bursts of non-thermal low-frequency radio emission. Kilometre-wavelength radiation from the Earth, generated by plasma oscillations above the terrestrial ionosphere, is easily detected by spacecraft. Jupiter radiates decametre-wavelength bursts due to plasma oscillations excited by the dynamo action of the satellite Io, which generates currents that flow down to the ionosphere of the planet. The collision

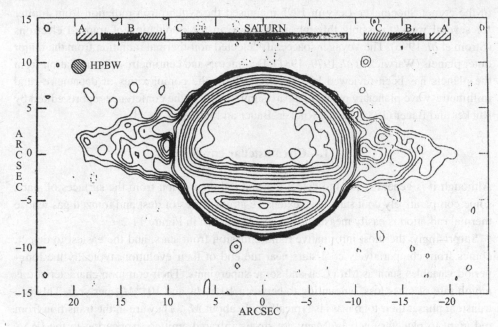

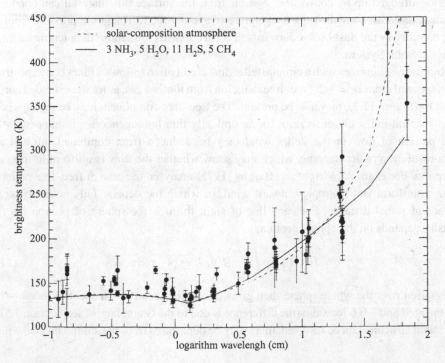

Figure 11.9. A contour map of Saturn at wavelength 2 cm. The ring inclination was 12.5°; the positions of rings A, B and C are shown above. The contours of brightness temperature range from 2 to 145 K (de Pater and Dickel 1991).

Figure 11.10. The radio spectrum of Saturn, expressed as an equivalent disc temperature. The full line is calculated by Briggs and Sackett (1989) for a model atmosphere (de Pater 1990).

of the comet Shoemaker–Levy in 1995 increased the synchrotron radiation from Jupiter for several weeks, possibly through an increase in the population of trapped electrons (Strom *et al.* 1996). The Voyager spacecraft detected non-thermal radiation from the major outer planets (Warwick *et al.* 1979, 1981). Decimetre- and centimetre-wave radiation from the planets has been reviewed by de Pater (1990); the complexities of decametre- and millimetre-wave planetary phenomena are summarized in the conference reports edited by Rucker and Bauer (1984) and by Rucker, Bauer and Peterson (1988).

11.4 Circumstellar envelopes

Although it is generally hard to detect thermal radio emission from the surfaces of stars, some comparatively cool stars have extensive atmospheres of dust and ionized gas whose thermal radiation is easily measurable, as we have seen in Figure 11.2.

Surprisingly, the most informative radio emission from stars, and the easiest to detect, comes from comparatively cool stars near the end of their evolution, typically the long-period variables such as Mira Ceti and some supergiants. Their common characteristic is a high rate of mass loss, amounting to between $10^{-8}M_\odot$ and $10^{-4}M_\odot$ per year. This is a transient phase; their total mass is typically only about $M_\odot$. They are in the transition from red giant to planetary nebula. Many are strong infrared emitters, appearing in the IRAS satellite survey.

The intense stellar wind blowing out from these cool stars contains material from the interior, dredged up by convection. Not far from the surface this material can condense, forming molecules and dust at a temperature of about 1000 K. Radiation pressure from the star then drives the dusty wind outwards to form a dense cloud more than ten times larger than the Solar System.

Radio emission from such a circumstellar dust cloud often follows a blackbody spectrum, with spectral index $\alpha = -2$. Free–free radiation from ionized gas, as for the extended corona of αOri (Figure 11.2), may also be present. The free–free component may be distinguished by a spectral index closer to zero; for an optically thin homogeneous plasma $\alpha = -0.1$. The pattern of flow in the stellar wind may be deduced from combined infrared and high-resolution radio mapping, which may show whether the flow is uniform in time and direction. For example, Wright and Barlow (1975) analysed the case of free–free radiation from a uniform and isotropic outward wind, in which the density falls inversely as the square of radial distance. For each line of sight through the sphere of plasma the flux density depends on the optical depth as

$$I(\nu, T) = \int_0^{\tau_{max}} B(\nu, T)\exp(-\tau)d\tau. \tag{11.5}$$

Integration over the whole sphere then gives a total luminosity with spectral index -0.66 for infrared and -0.6 for radio (the difference is due to the Gaunt factor: see Section 7.5). If the radial structure can be resolved, there may be departures from the simple inverse-square law of particle density. The radial distribution may then be related to the history of emission over a period of the order of 10^5 years; a discrete episode of emission, for example, would give a hollow shell. Structure of this kind is more easily seen in the brighter, non-thermal maser emission which occurs in many circumstellar envelopes.

11.5 Circumstellar masers

In many molecular clouds, and particularly in the mass outflow from red giant stars, we observe the astonishing and beautiful phenomenon of maser radio emission. Masers, the acronym for microwave amplification by the stimulated emission of radiation, can occur when the energy levels of a molecular species are populated in a non-Boltzmann distribution (see Chapter 7). Very small energy differences are involved; for radio emission the factor $h\nu/k$ is usually only of the order of 0.1 K. The molecular cloud contains material and radiation with several different temperatures, and a population inversion may be due either to collisions or to radiation. Several different molecules, each with several different masering transitions, may be observed in such stars; several of the strongest and most informative emitters are the OH, H_2O, CH_3OH (methanol) and SiO masers. Isotopic varieties, such as ^{28}SiO, ^{29}SiO and ^{30}SiO can be observed; these are obviously useful in determining relative abundances, but they may also be useful in determining temperatures and velocities when the more prolific lines are saturated.

Many other masers are observed in interstellar clouds (Chapter 9); the distinction between circumstellar masers, originating in outflow, and interstellar masers, where material is condensing to form stars, might not always be clear, since some stars may simultaneously accrete material in their equatorial regions and expel a wind from their poles. Condensing material may form a disc, whose dynamics may be investigated by analysing the Doppler shift of maser lines. In these cooler regions more complex molecules may be involved; Minier *et al.* (1999) found that methanol (CH_3OH) maser lines at 6.7 and 12.2 GHz traced Keplerian motion in discs in three such star-forming regions. The most powerful masers are observed in other galaxies; these are the megamasers, an example of which is described in Section 13.6. There is extensive literature on the subject of masers; see, for example, the tutorial by Cohen (1989), the textbook by Elitzur (1992) and two conference proceedings, IAU 206 (2002) and IAU 242 (2008).

11.6 The silicon oxide masers

The energy levels involved in the SiO masers are comparatively simple since the molecule is diatomic. The principal radio line, at 43 GHz, is between rotational levels $J = 1$ and 0 in a vibrationally excited state. Many other lines have been observed, involving rotational levels up to $J = 6$ and vibrational levels up to $v = 3$. The non-Boltzmann distribution, giving an excess population in the higher level, is due to collisions with neutral molecules (the maser 'pump'), in a region at a temperature of about 1000–2000 K with a total particle density of about 10^9–10^{10} cm^{-3}. SiO is itself a minor constituent in this dense cloud, which occurs within a few stellar radii of the surface. The maser cloud velocities in the maser emission extend only over less than 15 km s^{-1}.

This SiO maser region is more of an extended stellar atmosphere than a stellar wind. The maser emission is patchy, as can be seen by examining the several different components in a typical spectrum: for the star R Cas the separate components have been resolved spatially, and are clearly seen over an area not much larger than the disc of the star's surface (McIntosh *et al.* 1989). The intensities of the components vary greatly. The SiO molecule is non-paramagnetic, with only a small magnetic moment; nevertheless, the circular polarization due to Zeeman splitting can be observed in several stars, indicating fields of some tens of gauss. Such a field is strong enough to control the dynamics of the stellar atmosphere, since the energy density $B^2/(8\pi)$ in such a field is greater than the particle energy density nkT. Strong linear polarization is often observed; this is less easily explained than the circular polarization, but is related in some way to structure within the SiO clouds.

The SiO masers do not extend beyond about five stellar radii above the surface. At this distance SiO becomes incorporated into dust particles; these are then accelerated outwards by radiation pressure, carrying the gas with them and forming the stellar wind.

The proper motion of maser components is difficult to measure, since the emission is usually variable on a timescale of months only, and also because very high angular resolution is necessary. Boboltz *et al.* (1997), using the VLBA, showed that the SiO masers round the star R Aqr were moving inwards with a velocity of about 4 km s^{-1}; this is a star that is accreting rather than creating an outward stellar wind.

11.7 The water masers

The H$_2$O masers, like the SiO masers, are frequently found in Mira variables. They appear further from the star, at distances of 6–30 stellar radii. They are also found in active *star-formation regions* (SFRs), and particularly powerful examples, often called megamasers, are found in certain external galaxies such as NGC 5128. The most prominent line, at 22 GHz, is a transition between rotational states designated 6_{16} and 5_{23}. Excitation is by collisions with molecules; the temperature is of the order of 750 K, and the particle density is of the order of 10^8 cm^{-3}. The total energy output is often phenomenal: the brightest H$_2$O masers include, within a few kilohertz, more energy than the total bolometric output of the Sun.

The extension of receiver and interferometer techniques into the millimetre and sub-millimetre bands has allowed the observation of numerous transitions between rotational states both in SiO and in H$_2$O; for example the 650-GHz $J = 15$–14 transition in SiO and the $J = 5$–4 transition at the higher vibration level $v = 4$ are observable (Menten and Young 1995).

11.8 The hydroxyl masers

Farther out again from the star the outflow from a red giant is established typically as a symmetrical radial flow with a velocity of $10-20\,\mathrm{km\,s^{-1}}$. The OH masers are observed at distances of $10^{16}-10^{17}$ cm, more than ten times greater than for H_2O masers. They occur in a well-defined and comparatively thin shell, where the ambient ultraviolet light from the Galaxy is sufficiently strong to dissociate the dust. The density and temperature are of the order of $1000\,\mathrm{cm^{-3}}$ and 450 K. The OH molecule is paramagnetic, with a dipole moment 10^3 times greater than those of SiO and H_2O, so Zeeman splitting can be observed even though the magnetic field at the large radial distance is only of the order of some milligauss.

More than 1000 such OH stars have been observed. The strength of the OH masers is related to the rate of outflow in the stellar wind; the mass loss of 10^{-8} solar masses per year from a short-period Mira variable leads to a weaker maser emission than in a supergiant with an outflow of 10^{-4} solar masses per year. Since the maser emission occurs in the older population of stars, they are, like SiO masers, a useful dynamical probe for studying the structure of the Galaxy (see Chapter 10).

Four transitions are observed at wavelengths near 18 cm; these are all transitions within the ground state, which is split by lambda doubling and again by hyperfine splitting (Figure 9.3). The excitation is due to infrared radiation at a wavelength of about 35 μm, which selectively excites the upper level of the doublet. In stars, the strongest maser action is found for the 1612-MHz line, especially for the stars with the greatest rates of mass loss. The masers are generally found to be saturated, with an intensity corresponding to a maser amplification by a factor of the order of e^{20}.

The thin expanding shell structure of circumstellar OH masers is demonstrated by the structure of the dominant spectral line at 1612 MHz. Figure 11.11 shows examples of double and single lines; the double lines are due to Doppler shifts in the front and back of an expanding shell. The whole spectrum typically covers a range of Doppler shifts of $30\,\mathrm{km\,s^{-1}}$, while the individual components are typically only $0.1\,\mathrm{km\,s^{-1}}$ wide. A maser attains its full brightnesss only if its line of sight traverses a sufficiently long region with the same Doppler shift. If there is a radial gradient in the wind, this long line of sight occurs preferentially at the edges of the cloud; a ring of masers is then seen at the radial velocity of the star. If there is no gradient, the line may be strongest in the centre of the cloud, on the line of sight to the star itself; in this case, both the front and the back may be seen, at velocities separated by twice the expansion velocity. This is seen in the maps of OH masers around the star OH 127.8 (Figure 11.12); here the maps at different frequencies across the line profile effectively show a series of cross-sections of the expanding cloud, the lowest and highest frequencies originating from the back and the front, respectively.

Circular polarization is frequently observed in OH masers, indicating magnetic field strengths of order 1 milligauss. Extrapolated back to the star surface, this would be a field of some tens of gauss, in agreement with the fields deduced from the SiO and H_2O masers. Zeeman splitting is easily observed in OH masers, since the OH molecule is paramagnetic; a field of 1 milligauss may split a line by more than its intrinsic bandwidth. The H_2O and

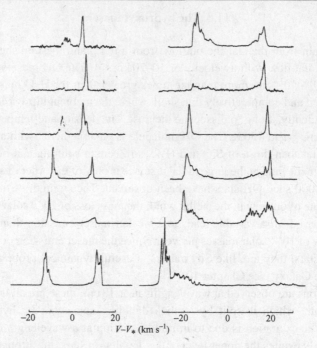

Figure 11.11. Double OH maser lines from expanding shells. The lines are split by the different Doppler effects on the front and back of an expanding shell (Cohen 1989).

CH_3OH molecules are non-paramagnetic, and linear rather than circular polarization is usually observable. Field strengths of some tens of milligauss are observed in the infalling molecular gas of SFRs; examples are the OH masers in Ceph A (Cohen *et al.* 1990) and methanol masers in W3 (Vlemmings *et al.* 2006).

The relation of the OH masers to the SiO and H_2O masers in a single star, VX Sgr, is shown in Figure 11.13 (Chapman and Cohen 1986). Besides demonstrating the general picture set out in the preceding sections, this map shows that some complications remain to be understood. In particular, the outflow might not be spherically symmetrical and the outward velocity gradient is not yet fully explained by invoking radiation pressure on dust.

11.9 Classical novae

The continuum radio emission from a quiet star like the Sun is dramatically outshone by the free–free radiation from the expanding stellar envelope resulting from a classical nova explosion. The 'nova' itself is not a new star; it is a white dwarf that is accreting material from a normal-star companion in a close binary system. The white dwarf itself is a star that has collapsed after completing its hydrogen burning, and is now mainly composed of helium. Accretion brings a new supply of hydrogen to its surface, forming a degenerate and unstable layer. A mass of about $10^{-4} M_\odot$ may accumulate before the layer ignites.

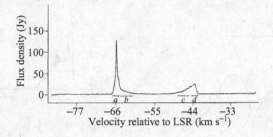

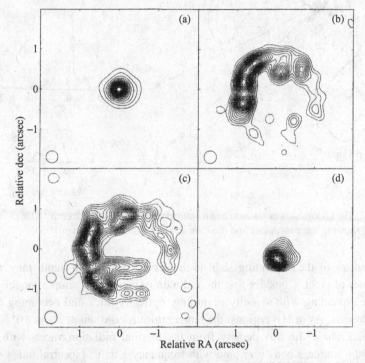

Figure 11.12. MERLIN maps of OH masers around the star OH 127.8. Four velocity ranges have been selected from the spectrum (above) and mapped separately (Booth *et al*. 1981).

An increase in temperature in the degenerate hydrogen brings no increase in pressure; the hydrogen layer therefore explodes rather than expanding smoothly, and the whole layer then blows outwards as an expanding shell with a velocity of some hundreds of or a few thousand km s^{-1}. The optical emission lasts a few days; the radio emission is from an outer shell that is ionized by intense ultraviolet light. This radio emission grows and decays over a typical period of some years. It is thermal radiation, which may be understood in terms of the theory of free–free radiation set out in Section 7.5, including both optically-thick and optically-thin regimes.

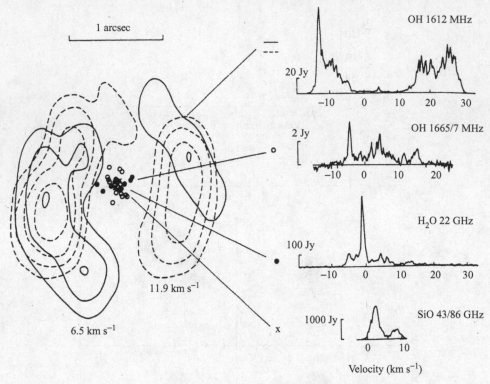

Figure 11.13. The OH masers in relation to SiO and H_2O masers in the circumstellar envelope of the star VX Sgr, showing their locations and their spectra (Chapman and Cohen 1986).

The geometry of the expanding shell is important: the optical depth may vary along different lines of sight. Consider first the radio emission from a simple model: a cube of ionized gas expanding with velocity v, initially optically thick and becoming thin as the density decreases. As in H II regions, the temperature is constant at about 10^4 K, so in the optically-thick phase the flux density from the thermal radiation varies with time t as the area of the source, i.e. as $v^2 t^2$, and with frequency v as v^2 (spectral index $\alpha = 2$). In the later optically thin phase the total radio emission depends only on the density, and it falls as t^{-3}; the spectrum is flat with index $\alpha = -0.1$. The transition occurs later for lower radio frequencies; all novae show a series of radio peaks that follow sequentially from high to low frequencies. More realistic models of shells may have tangential lines of sight that are optically thick, while those near the centre are thin. The transition from thick to thin therefore occurs over some time, and the spectral index may change more slowly than in the simple cube model. The details depend on the geometrical thickness of the shell and on the fall of temperature as the shell expands (Seaquist 1989).

Figure 11.14 shows the radio emission at frequencies between 3 and 90 GHz from the nova FH Serpentis 1970, with theoretical curves computed from such a simple model, fitting initial values of density, temperature and velocity to the expanding shell (Hjellming

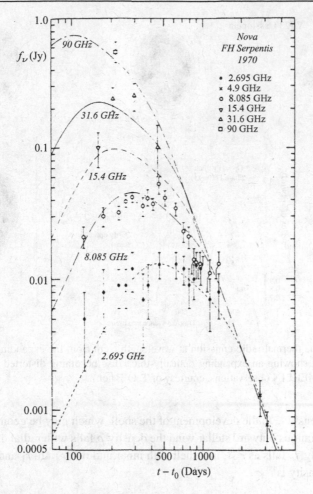

f_ν (Jy)

*Nova
FH Serpentis
1970*

- • 2.695 GHz
- × 4.9 GHz
- ○ 8.085 GHz
- ▽ 15.4 GHz
- △ 31.6 GHz
- ▫ 90 GHz

90 GHz

31.6 GHz

15.4 GHz

8.085 GHz

2.695 GHz

$t - t_0$ (Days)

Figure 11.14. The thermal radio emission from the nova FH Serpentis 1970 (Hjellming *et al.* 1979).

et al. 1979). For about 100 days the shell is optically thick at the lower radio frequencies, and appears as a disc at the temperature of the ionized gas; although the temperature is falling, the overall intensity is rising due to the expansion of the disc. As the optical depth τ falls, the brightness temperature along various lines of sight must be computed from

$$I_\nu(\tau) = (2kT/\lambda^2)(1 - e^{-\tau}). \tag{11.6}$$

In the optically-thin regime, which as expected occurs earlier for higher frequencies, the emission is calculated (see Chapter 4) from the volume emissivity,

$$j_\rho = 7.45 \times 10^{-39} N_e^2 T_e^{-0.34} \nu_{GHz}^{-0.11}. \tag{11.7}$$

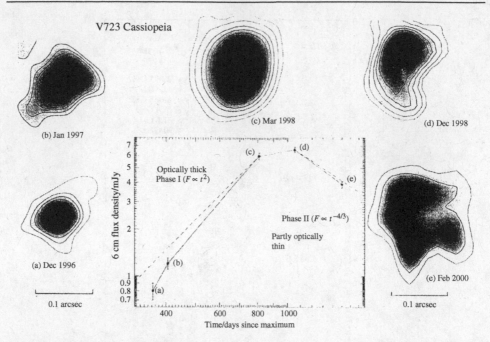

Figure 11.15. The thermal radio emission at wavelength 6 cm from the expanding shell ejected by Nova V723 Cas, showing an expanding optically-thick disc becoming distorted and partially thin after 800 days (MERLIN observations, courtesy of T. O'Brien).

The model is sensitive to the development of the shell, which may be geometrically thin or thick. In a continuous outward stellar wind the density ρ falls with radial distance r as r^{-2}, so the emissivity j_ρ falls as r^{-4}; in a shell with inner and outer radii r_1 and r_2 containing a mass M the density is

$$\rho = \left(\frac{M}{4\pi r^2}\right)(r_2 - r_1). \tag{11.8}$$

It is remarkable that such a simple model accounts so well for the emission over such a wide frequency range. The expansion of the thermally emitting shell has been observed for Nova V723 Cas at 5 GHz, as shown in Figure 11.15. The optically-thick phase, with an expanding nearly uniformly bright disc and flux density increasing as t^2, lasted for almost 1000 days; the optically-thin shell was less symmetrical, and after 4 years the fall in flux density was still not as steep as the expected t^{-3}.

For Nova Cygni 92 the angular diameter of the shell was resolved only 80 days after the outburst (Pavelin *et al.* 1993); the brightness temperature eventually reached 45 000 K, higher than expected from radiative excitation but still consistent with thermal radiation from shock-excited gas. The expanding shell is seen in a series of MERLIN 6-cm maps (Figure 11.16) published by Eyres *et al.* (1996). The shell is far from symmetrical; this

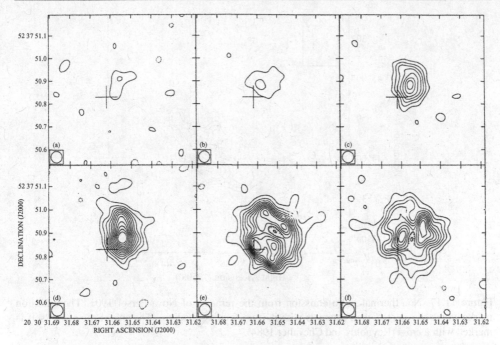

Figure 11.16. The expanding shell of thermal radio emission from Nova Cyg 92, mapped using MERLIN at 6 cm (Eyres *et al.* 1996).

asymmetry may reflect an anisotropic explosion, or it may occur when the expanding cloud collides with structure in the surrounding ISM.

In other types of nova the radio spectrum and high brightness temperatures indicate non-thermal radiation; this may be due to high-energy particles in the outflow itself. This may also occur for a classical nova; Figure 11.17 shows the remains of Nova Persei 1901, in which we now see synchrotron radiation to one side of the site of explosion, where the expanding cloud is colliding with an invisible nebula. The high-energy electrons responsible for this non-thermal emission are accelerated in shock fronts at this collision.

11.10 Recurrent novae

RS Ophiucus is the best observed example of a small class of novae that recur at intervals of several years. Its outburst in 2006 was the sixth since it was first observed in 1898. The increase in brightness at outburst is dramatic; optically it typically increases by six magnitudes, while the radio emission, which was undetectable before the 2006 outburst, rose rapidly to intensities that could be explained only in terms of non-thermal radiation. The lowest radio frequencies were the last to be observed; the GMRT was used to detect emission at 610 MHz on day 20 after the outburst and at 325 MHz on day 38 (Kantharia *et al.* 2007). These novae are associated with symbiotic binaries; as in the classical novae,

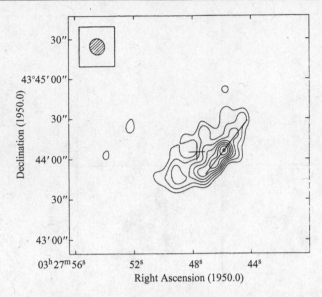

Figure 11.17. Non-thermal radio emission from the remains of Nova Persei 1901. The emission marks where the ejected shell interacts with the surrounding medium. The site of the nova itself is marked with a cross (Reynolds and Chevalier 1984).

these consist of a white dwarf accreting from a red giant. The difference is primarily that the outburst occurs after a smaller accretion of hydrogen; this may be related to the large difference in binary period, which is 230 days for RS Oph and 0.6 days for a typical classical nova.

The cloud of material resulting from the thermonuclear explosion in RS Oph is ejected with a velocity of $4000 \, \text{km s}^{-1}$. Without any interaction with other circumstellar material, this cloud might be expected to radiate free–free radio emission corresponding to a temperature of about 10^4 K, as for a classical nova but with a much lower intensity because of the lower mass. Instead, in the 1985 outburst, intense radio emission was observed, growing for the first 100 days as the cloud expanded. Furthermore, VLBI observations soon produced a map, which gave an angular diameter showing that the brightness temperature was at least 10^7 K. This high temperature, and the radio spectrum, must be interpreted as synchrotron emission resulting from shock excitation of the cloud at a collision with circumstellar gas, probably in the form of a continuous wind from the red giant. Soft X-rays were also observed from this outburst; the spectrum again indicated an origin in high-temperature gas derived from the collision.

The radio maps of several of these recurrent novae show ejected clouds, which are very different from the predictions of the isotropic model which we described for the classical novae. The MERLIN map of R Aqu (Figure 11.18) shows clouds aligned along an axis, and a thermally emitting core (Dougherty *et al.* 1995). This is our first example of many

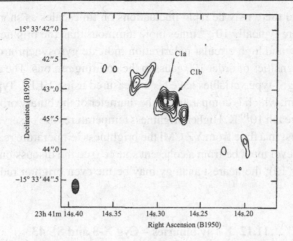

Figure 11.18. A MERLIN image at 5 GHz of the symbiotic star R Aquarii. The star is located in the double source C1, and ejected clouds are strung out over three arcseconds (Dougherty *et al.* 1995).

types of radio source in which the radiating material is in the form of twin jets; in this case, as in many others, they probably leave the star along the rotation axis. The well-collimated jets, rapid intensity fluctuations and discrete structure within the jets suggest that the clouds in the jets are travelling outwards at high velocity. We shall see in the next section that there are indeed outward velocities approaching the velocity of light, in the case of X-ray binaries; for these very energetic objects we observe synchrotron radio emission. There are also, however, high-velocity-jet sources that are thermal radio emitters. An example is the radio jet in the Herbig–Haro objects HH 80 and HH 81 (Marti *et al.* 1995). The actual source may be hidden behind a dark molecular cloud that is the seat of recent star formation. Structure in the jet, which is 5 parsecs long, was observed at intervals of several years using the VLA at wavelength 3.5 cm. Proper motions in the range 70–160 mas yr^{-1} were easily observed over this time, giving outward velocities of 600–1400 km s^{-1}.

11.11 Non-thermal radiation from binaries and flare stars

Binary star systems with a condensed star as one component include some of the most spectacular radio sources in the Galaxy. There are, however, many examples of binaries among the main-sequence stars that show strong radio emission. The well-known star Algol (α Persei) is quoted as a type for binaries consisting of late-type main-sequence stars, although it is actually a triple system. CC Cas is in the same class of radio stars; it consists of a pair of interacting massive O stars. Algol itself has a continuous radio emission of about 0.01 Jy, detected at 2.7 and at 8 GHz, which rises rapidly by a factor of 30 or more within hours at the time of a flare. The RS CVn variables, which include AR Lac and UX Ari, have more frequent flares, of the order of one per day; here the rise time may be only one hour, and there may be rapid fluctuations on timescales as short as five minutes. These flare stars are typically 10^{2-4} times more luminous than the brightest solar outbursts. The high intensity and high circular polarization indicate gyrosynchrotron radiation, with magnetic-field intensities of order 100 gauss in the emitting regions. The angular diameters of some of the Algol-type variables have been measured using VLBI. Typically the source diameter is 10^{12} cm, which is comparable to the diameter of the binary orbit. The brightness temperature may reach 10^{10} K. Higher brightness temperatures are observed for some of the red dwarf variables; in a flare from YZ CMi the brightness temperature reached 7×10^{12} K. Radiation at this level must be from a coherent source (see the discussion of self-Compton effects in Chapter 13); the nearest analogy may be the even brighter radio emission from pulsars.

11.12 X-ray binaries – Cyg X-3 and SS 433

Many of the X-ray binaries are detectable as radio sources, and some produce spectacular outbursts. Cygnus X-3 is one of the most remarkable: it is a low-mass X-ray binary (LMXB), which is also detectable as a gamma-ray source. The binary period is 4.8 hr. The neutron star's companion is a helium star, in the last stages of evolution before collapse. Cygnus X-3 has been observed during an outburst by a combination of interferometer radio telescopes (Spencer *et al.* 1986). The peak of the radio emission occurs in time sequence from high to low radio frequencies, as might be expected from almost any emission mechanism for disturbances travelling outwards from an origin on an accreting neutron star. Such travelling disturbances are indeed seen in the form of twin jets; they move outwards with velocities variously estimated as 0.3–0.6 times the velocity of light. Observations with milliarcsecond resolution show a velocity of $0.63c$ and precession in the jets with a period of 5 days (Miller-Jones *et al.* 2004). The effects of such high velocities, which are encountered in quasars as well as in these sources within our Galaxy, require a relativistic analysis, which we give in Section 11.13 below.

Synchrotron radiation from an expanding spherical shell, or from a sector of such a shell, may be analysed in a similar way to thermal radiation from the classical novae (Section 11.9 above). If a cloud initially containing a power-law distribution $KE^{-\gamma}\,dE$ of electron energies expands adiabatically to radius r, then $r^3 K E^{-\gamma}\,dE = \text{constant}$, and at time t

$$K(t) = K_0 \left(\frac{r}{r_0}\right)^{-(\gamma+2)}. \tag{11.9}$$

At the same time, however, the magnetic field will decrease during the expansion, and the geometry of the expansion has to be taken into account. A model by Marti *et al.* (1992) shows that the 1972 outburst of Cyg X-3, which was observed at eight different radio frequencies, can be accounted for by such a model (Figure 11.19). Synchrotron radiation alone is not, however, sufficient to account for the whole of these observed flux densities;

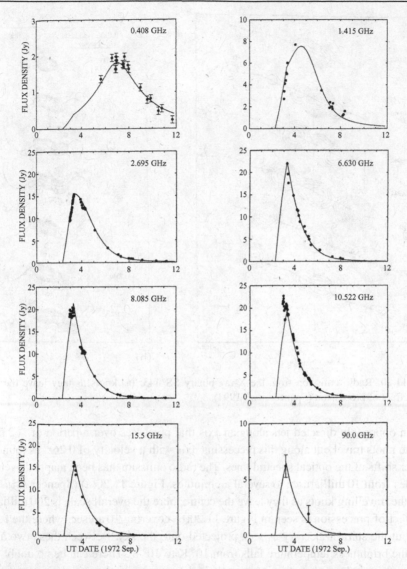

Figure 11.19. The 1972 outburst of Cyg X-3, observed at eight radio frequencies and modelled by Marti *et al.* (1992).

in the initial stages of expansion, when the cloud is optically thick, the spectral index is lower than expected for synchrotron self-absorption, indicating free–free absorption from a larger mass of thermal electrons.

Finally, perhaps the most spectacular of these relativistic twin-jet galactic sources is the binary jet SS 433. This is observed as an optical as well as an X-ray and radio source; the optical emission includes spectral lines, showing that the jets contain hadronic material. There

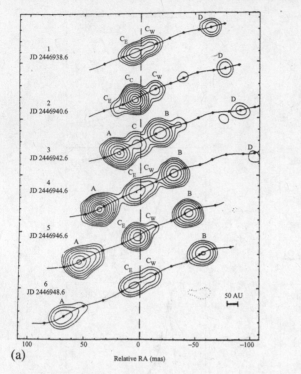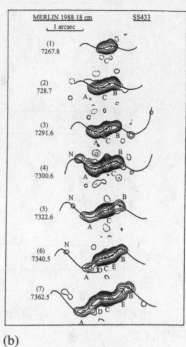

Figure 11.20. Radio emission from the X-ray binary SS 443: (a) knots as they leave the centre; (b) the effect of precession (Spencer *et al.* 1993).

are twin oppositely directed jets along an axis that precesses over a period of 162.5 days. Discrete knots travel out along this precessing axis with a velocity of 0.26*c*, as found from Doppler shifts of the optical spectral lines. The radio emission has been mapped over angular scales from 10 milliarcsec to several arcminutes. Figure 11.20(a) (Spencer *et al.* 1993) shows the travelling knots as they leave the centre; here the overall scale is 200 milliarcsec. The effect of precession is seen in Figure 11.20(b) covering 10 arcsec, where the full line traces out the spiral trajectory as seen projected onto the sky. Between these two angular scales the brightness temperature falls from 10^8 K to 10^5 K. There can be no doubt that in this case the whole of the emission process is determined by the initial ejection from the neutron star, and that it is entirely synchrotron radiation.

In all these twin-jet sources one may speculate that the ejection process is not unrelated to that responsible for the similar jets observed on a much larger scale in quasars. The mechanisms will be discussed in that connection in Chapter 13.

11.13 Superluminal motion

The effects of superluminal velocities in astrophysical jets were first analysed in relation to the twin jets of extragalactic quasars (Chapter 13). These jets remain narrowly collimated over distances of up to a megaparsec, and condensations within the jets leave the central source with velocities close to the velocity of light. The relativistic velocities may be observed as apparent velocities greater than c, depending on the aspect of the jet; these are the *superluminal* velocities. It is remarkable that this cosmic phenomenon is also seen, in miniature, in the X-ray binaries.

The first galactic superluminal source, GRS 1915+105, was discovered by Mirabel and Rodriguez (1994). Figure 11.21 shows maps made over a period of only 13 days. Four separate condensations are seen to leave the source and move outwards; the three to the left are moving with a proper motion of 23.6 ± 0.5 mas per day, which corresponds to a velocity of $1.5c$, while the weaker one on the left is apparently moving with velocity $0.6c$.

The theoretical possibility of such superluminal velocities was pointed out by Rees (1966). He considered a spherical shell expanding isotropically with relativistic velocity v, as in Figure 11.22. The locus of radiating points as seen by the observer is a spheroid. Radiation from the front of the shell is seen by a distant observer to be Doppler shifted by a factor $\gamma(1 + v/c)$, where the relativistic factor $\gamma = (1 - v^2/c^2)^{-1/2}$, while the apparent velocity at the observed edge of the source is γv, which may be greater than c.

Consider now a discrete radiating cloud moving out from the central source with velocity $v = \beta c$ along a line of sight at angle θ to the line of sight (Figure 11.23). In a time interval τ between two observations the sideways motion is $v\tau \sin\theta$. In this time the source moves a distance $v\tau \cos\theta$ towards the observer, such that the observed time interval is reduced to $\tau - \beta\cos\theta$. The apparent velocity is then c times a 'superluminal' factor F, where

$$F = \beta \sin\theta(1 - \beta\cos\theta)^{-1}. \tag{11.10}$$

The maximum effect on the apparent velocity is at angle θ_m given by

$$\sin\theta = \gamma^{-1}, \qquad \cos\theta_m = \beta^{-1}. \tag{11.11}$$

The apparent velocity is greater than $v = \beta c$ for angles in the range

$$\frac{2\beta}{1 + \beta^2} > \cos\theta > 0 \tag{11.12}$$

and the maximum superluminal factor F_{max} is

$$F_{max} = \gamma\beta. \tag{11.13}$$

Relativistic beaming increases the flux density; it also Doppler shifts the frequency of the radiation by a factor $\delta = \gamma^{-1}(1 - \beta\cos\theta)^{-1}$, so the observed flux density depends on the emitted spectrum. If the emitted spectrum is $S_0(\nu) \propto \nu^\alpha$, the observed flux density is

$$S(\nu) = S_0\left(\frac{\nu}{\delta}\right)\delta^{3-\alpha}. \tag{11.14}$$

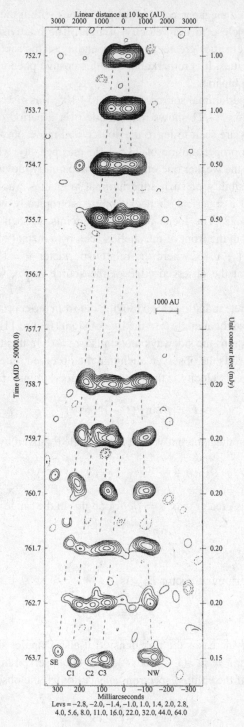

Figure 11.21. A series of 5-GHz MERLIN maps of the superluminal X-ray binary GRS 1915+105 (Fender *et al.* 1999, arXiv:astro-ph/9812150).

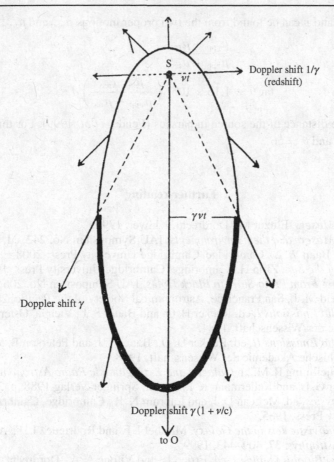

Figure 11.22. A relativistically expanding shell expanding isotropically with relativistic velocity v, seen by a distant observer (Rees 1966).

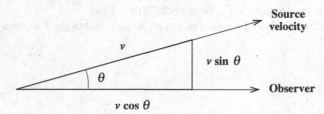

Figure 11.23. The geometry of superluminal motion.

This applies to an isolated discrete source: if there is a continuous stream, the apparent density of sources along the stream is increased by a factor δ^{-1} and the observed flux scales as $\delta^{2-\alpha}$.

If in the case of GRS 1915+105 we assume that the receding jet (on the right in Figure 11.21) is the same as that of the three approaching jets (on the left), a complete solution for θ and v can be found from the two proper motions μ_{app} and μ_{rec}, since

$$\beta = \frac{\mu_{app} - \mu_{rec}}{\mu_{app} + \mu_{rec}}, \tag{11.15}$$

$$\tan\theta = 1.16 \times 10^{-2} \left(\frac{\mu_{app}\mu_{rec}}{\mu_{app} - \mu_{rec}} \right) d, \tag{11.16}$$

where d is the distance of the source in parsecs (Fender *et al.* 1999). For this source this gives $\beta = 0.9$ and $\theta = 66°$.

Further reading

Astronomical Masers, Elitzur M., Dordrecht, Kluwer, 1992.

Astronomical Masers and their Environments, IAU Symposium No. 242, ed. Chapman J. M. and Baan W. A., Cambridge, Cambridge University Press, 2008.

Astrophysics of the Sun, Zirin H., Cambridge, Cambridge University Press, 1988.

Cosmic Masers: From Proto-Stars to Black Holes, IAU Symposium No. 206, ed. Migenes V. and Reid M. J., San Francisco, Astronomical Society of the Pacific, 2003.

Planetary Radio Emissions I, ed. Rucker H. O. and Bauer S. J., Vienna, Österreichische Akademie der Wissenschaft, 1984.

Planetary Radio Emissions II, ed. Rucker H. O., Bauer S. J. and Peterson B. M., Vienna, Österreichische Akademie der Wissenschaft, 1988.

Radio Stars, Hjellming R. M., in *Galactic and Extragalactic Radio Astronomy*, ed. Verschuur G. L. and Kellermann K. I., Berlin, Springer-Verlag 1988, p. 381.

Solar Radiophysics, ed. McLean D. J. and Labrum N. R., Cambridge, Cambridge University Press, 1985.

Sources of Relativistic Jets in the Galaxy, Mirabel I. F. and Rodriguez L. F., *Ann. Rev. Astron. Astrophys.* **37**, 409–443, 1999.

Stellar Jets and Bipolar Outflows, ed. Errico L. and Vitone A. A., Dordrecht, Kluwer, 1993.

Stellar Radioastronomy, Gudel M., *Ann. Rev. Astron. Astrophys.* **40**, 217–261, 2002.

The Physics of Stars, Phillips A. C., New York, Wiley, 1994.

The Stars: Their Structure and Evolution, Taylor R. J., Cambridge, Cambridge University Press, 1994.

12

Pulsars

In 1934 Baade and Zwicky suggested that the final stage of evolution of a massive star would be a catastrophic collapse, leading to a supernova explosion and leaving a small and very condensed remnant, a neutron star. This theory was immediately successful in explaining the observed supernovae, especially in relation to the Crab Nebula, which was identified as the remains of a supernova observed in the year 1054. Neutron stars seemed, however, to be hopelessly unobservable; they would be cold and only about the size of a small asteroid. More than 30 years later, neutron stars were discovered both in X-ray and in radio astronomy. As an X-ray source, a neutron star usually behaves as an intense thermal source of radiation; many are in binary systems, and the heating is then due to the accretion of matter from a binary companion. As a radio source, a neutron star is seen as a pulsar, when it behaves completely differently; it now radiates an intense beam of non-thermal radiation, rotating with the star, which is detected as a radio pulse as it sweeps across the observer. The intriguing and complex behaviour of the pulsars relates not only to condensed-matter physics but also to many aspects of stellar evolution, galactic structure and gravitational physics.

Early radio telescopes often use receivers with long integration times, which would not detect the periodic short pulses from pulsars. The discovery of pulsars (Hewish *et al.* 1968) was the result of a deliberate measurement of the rapid intensity fluctuations of radio sources due to scintillation in the solar corona. A large receiving antenna was used, at the wavelength of 3.7 m; the sensitivity and time resolution of this system were sufficient to detect individual pulses at intervals of 1.337 s from the pulsar which we now designate PSR B1919+21 (PSR stands for pulsating source of radio; the numbers refer to its position in right ascension and declination[1]). In most subsequent searches for pulsars the technique has been to look for a long and very regular sequence of pulses, allowing a long effective integration time to be used. Pulsars nevertheless require large radio telescopes for their discovery and detailed study, and the compilation of a catalogue of almost 2000 pulsars in the 40 years after the initial discovery has involved substantial technical development and some tens of thousands of hours of telescope time (Lyne *et al.* 1998).

---

[1] The positions of most astronomical objects are designated either in 1950 coordinates (B) or 2000 coordinates (J). The present convention is for all pulsars to have a J designation and to retain the B designation only for those published prior to about 1993.

This chapter provides an introduction to several aspects of pulsar research: the condensed matter of the neutron star itself, electromagnetic emission from the stellar atmosphere, the origin and evolution of pulsars as a constituent of the galactic population of stars, and the use of binary systems in tests of relativity theory.

12.1 Neutron-star structure

After the exhaustion of nuclear fuel has led to the collapse of a normal star from its equilibrium state, there are two stable condensed states short of a total collapse to a black hole; these are represented by the white dwarf stars and the neutron stars. In both, a total collapse under gravitation is prevented by the pressure of degenerate matter: electrons in a white dwarf, neutrons in a neutron star. Each type can exist only over a limited range of mass, and their properties are determined solely by their individual masses. Temperature has no effect on density or size; even though a white dwarf is still hot enough to be visible, its diameter does not change as it cools. The only sources of energy in a neutron star are the remaining heat from its formation, a large internal magnetic field and rotational energy; it is the rotational energy that provides the power for the lighthouse radiation of radio pulsars.

The degenerate electron gas within a white dwarf obeys Fermi–Dirac statistics, in which only two electrons (one for each spin) can occupy any element of phase space (with six dimensions: three spatial and three momentum coordinates). At the very high pressure and comparatively low temperature of the interior, all cells of phase space are filled, up to a limiting momentum and energy. In a normal gas, obeying Maxwell–Boltzmann statistics, the occupation of phase space depends on their velocities, which are determined by temperature via probability theory. A degenerate electron gas behaves as though there are only static forces between the electrons: the equation of state (the relation between density and pressure) contains no reference to temperature.

In a degenerate neutron gas the forces between neutrons are again static, but they addition-ally involve more distant interactions than those between closest neighbours. Laboratory measurements of forces between pairs of particles are insufficient to determine a realistic equation of state, except to a very rough approximation, which assigns an effective diameter to a single neutron. In this approximation a neutron star may be regarded as a single giant nucleus, in which neutrons are packed together, as they are in any massive nucleus such as iron. A more sophisticated approach reveals that the diameter of a neutron star decreases with increasing mass, as it does for a white dwarf. The relation between mass and diameter of a neutron star is not as well known as it is for a white dwarf; it depends on the poorly determined 'hardness' of the equation of state, that is, on the effective inter-neutron force at the highest densities. The radius is usually taken to be in the range 10–12 km. The moment of inertia is less dependent on the mass, because of the relationship between mass and radius; it is usually taken to be 3×10^{44} g cm^2, which coincidentally is nearly the same as the moment of inertia of the Earth.

White dwarf stars can exist only over the mass range from about $0.02 M_\odot$ to an upper limit, known as the Chandrasekhar limit, of $1.4 M_\odot$. Within this range the diameter decreases

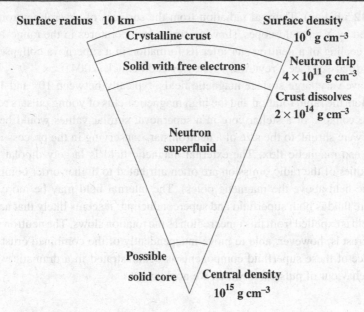

Figure 12.1. A section through a typical model neutron star, of mass $1.4M_\odot$.

with increasing mass; at the Chandrasekhar limit the star tends to collapse, and the electrons and protons are squeezed together to form neutrons. Theoretical models for neutron stars show that they also exist only over a small range of mass; the precise range depends on the uncertain equation of state, but the models allow a range of 0.2–2.0 solar masses $M_\odot$ (Lattimer and Prakash 2001). A larger mass would lead to further collapse into a black hole. Several neutron-star masses have been measured accurately from the characteristics of binary systems: remarkably, and so far without explanation, most are close to $1.4M_\odot$.

Figure 12.1 shows the structure of a typical model neutron star with mass $1.4M_\odot$. The outer part forms a solid crust, whose density at the surface is of the order of 10^6 g cm^{-3}, similar to that of a white dwarf. The density increases rapidly within the outer crust, which is a rigid and strong crystal lattice, primarily of iron nuclei. At higher densities, electrons penetrate the nuclei and combine with protons to form nuclei with unusually large numbers of neutrons. This continues until, at a density of about 4×10^{11} g cm^{-3}, the most massive of these nuclei become unstable and free neutrons appear. These neutrons form a superfluid (Migdal 1959), which, as we will see from the rotational behaviour of the pulsar, can move independently of the solid crust. Finally, the core itself contains only neutrons with a small equilibrium proportion of electrons and protons; it is superconducting and superfluid. Although the temperature of a neutron star has no effect on its structure, the surfaces of young neutron stars may be at temperatures greater than 10^6 K. Their thermal energy may be a relic of the energy released in the original supernova collapse; it may, however, be augmented by an internal release of rotational energy (in the glitches, described in Section 12.5 below). Thermal radiation from the surfaces of several neutron stars has been detected by X-ray telescopes, showing surface temperatures in the range 10^5–10^6 K. The rate of cooling of a neutron star after its formation in a supernova collapse has been extensively discussed (see a review by Yakovlev and Pethick (2004)).

Pulsars have very large surface magnetic fields, typically between 10^8 and 10^{13} gauss (10^{4-9} T). Both the rapid rotation and the high magnetic fields of young pulsars reflect their origin in the collapse of a stellar core in a supernova; similar values would be seen if a normal star were shrunk to the size of a neutron star, conserving in the process its angular momentum and magnetic flux. The external magnetic field is largely dipolar (although the complexities of the radio emission are often attributed to higher-order components of the magnetic field above the magnetic poles). The internal field may be more complex, since the core fluid is both superfluid and superconducting; it seems likely that any internal magnetic field is expelled from this inner region as the rotation slows. The neutron superfluid within the crust is, however, able to move independently of the combined crust and core. The existence of these superfluid components is demonstrated in a dramatic way by the rotational behaviour of pulsars, to which we now turn.

12.2 Rotational slowdown

The rotation periods of radio pulsars range from 1.4 milliseconds to 8.5 seconds. The rotation of millisecond pulsars can be monitored with extreme accuracy, although a precise ephemeris must be used to account for the orbital motion of the Earth (Edwards *et al.* 2006). All pulsar periods are observed to be increasing with time; the rotation is slowing, and the rotational energy is the source of the energy observed in the radio pulses. Rotational energy is primarily lost either directly by magnetic-dipole radiation, or by an outflow of energetic particles accelerated by electromagnetic induction; the energy lost in the form of radio emission is very much less than either of these. For the pure magnetic dipole the rate of loss of energy is related to the angular velocity ω by

$$\frac{\mathrm{d}}{\mathrm{d}t}\left(\frac{1}{2}\,I\omega^2\right) = I\omega\dot{\omega} = \frac{2}{3}M_\perp^2\omega^4 c^{-3}, \tag{12.1}$$

where $M_\perp$ is the component of the magnetic dipole moment orthogonal to the spin axis and I is the moment of inertia. This relation probably applies reasonably well even if part of the energy outflow is carried by energetic particles, and it is therefore customary to use Equation (12.1) to assign a value of dipole moment to each pulsar from measurements of the rotation rate $\nu = \omega/(2\pi)$ and its derivative $\dot{\nu}$ (or period P and $\dot{P}$), using a standard value for the moment of inertia of $I = 3 \times 10^{44}$ g cm^2. The dipole moment is usually quoted as a value B_0 of the polar field at the surface for an orthogonal dipole, giving the widely used relation

$$B_0 = 3.3 \times 10^{19}(-\dot{\nu}\nu^{-3})^{1/2} \text{ gauss}. \tag{12.2}$$

The slowdown may be expressed as a power law,

$$\dot{\nu} = -k\nu^n,$$ (12.3)

where k is a constant and n is referred to as the *braking index*; if the slowdown follows Equation (12.1) the index is $n = 3$. If we assume that a pulsar has an angular velocity that is initially very high, and a constant dipolar magnetic field, its age τ may be found by integrating Equation (12.3) to give

$$\tau = -(n-1)^{-1}\nu\dot{\nu}^{-1} = (n-1)^{-1}P\dot{P}^{-1}.$$ (12.4)

For a braking index $n = 3$, this calculated age becomes $\frac{1}{2}(\nu/\dot{\nu}) = \frac{1}{2}(P/\dot{P})$; this is often quoted as τ_c, the *characteristic age* of the pulsar. Given an initial rotation rate ν_i, the age becomes

$$\tau = -(n-1)^{-1}\dot{\nu}[1 - (\nu/\nu_i)^{n-1}].$$ (12.5)

A direct measurement of n is obtainable only if the second differential $\ddot{\nu}$ can be measured. Differentiation of Equation (12.3) gives

$$n = \frac{\nu\ddot{\nu}}{\dot{\nu}^2}.$$ (12.6)

A useful value of $\ddot{\nu}$ can be found only for young pulsars; even for these it requires a long run of timing measurements to allow for any irregular rotational behaviour, and few results have been obtained. Characteristic ages should be interpreted as actual ages only with some caution, since measured braking indexes are often found to be very different from $n = 3$. The difficulty is well illustrated by the behaviour of the two young pulsars B0531+21 and B0833-45, the Crab and Vela pulsars.

12.3 Rotational behaviour of the Crab and Vela pulsars

The slowdown in rotation rate of the Crab pulsar is shown in Figure 12.2. Over 25 years, during which time the rotation rate has fallen from 30.2 Hz to 29.9 Hz, the pulsar has completed more than 2×10^{10} rotations; despite the rotational irregularities described below, the observations are claimed to give an almost complete account of all rotations over this time. The characteristic age, which is found from the slope of the plot of rotation rate ν in Figure 12.2(a), can be measured to a precision of 1% in a single day; it is 1250 yr, which may be compared with the actual age of 950 yr since the supernova explosion was observed (AD 1054). The difference is a warning that present-day slowdown rates are only indications of actual age, not infallible measures of it.

The expanded plot of Figure 12.2(b), in which the initial slope is removed and the vertical scale expanded by 5000, shows the steps known as *glitches* which we discuss later. Measured values of $\ddot{\nu}$ both between the glitches and averaged overall give a braking index $n \sim 2.5$ (using Equation (12.6)); timing measurements now extend for 38 years, and indicate that the index is falling and is now approaching 2.0.

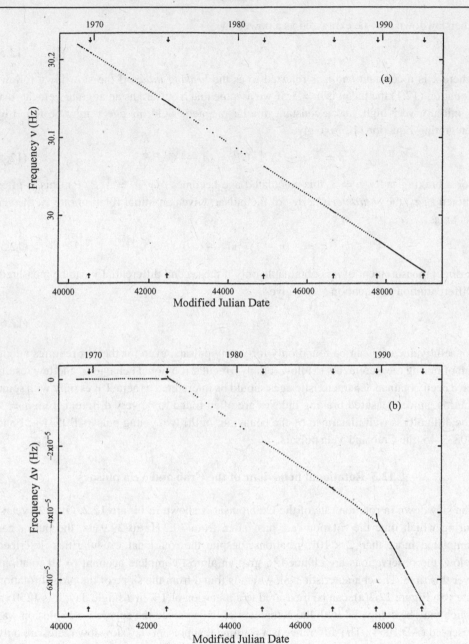

Figure 12.2. The slowdown in rotation rate ν of the Crab pulsar: (a) as recorded over 25 years; (b) with the vertical scale expanded, after subtraction of the average slope $\dot{\nu}$ (Jodrell Bank Observatory).

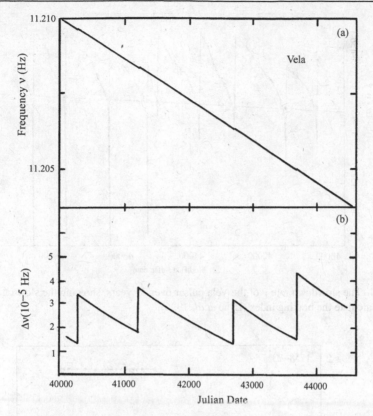

Figure 12.3. The rotation rate of the Vela pulsar: (a) as recorded; (b) with expanded vertical scale after subtraction of the average slope (data from Cordes *et al.* (1988)).

The rotation rate of the Vela pulsar (Figure 12.3(a)) also shows a clearly defined value of slowdown rate, but the expanded curve of Figure 12.3(b) is dominated by much larger glitches, with steps in rotation rate (of the order of 1 in 10^6) and in slowdown rate (of the order of 1 in 10^2). Here the only consistent value of $\ddot{\nu}$ is found from a plot of $\dot{\nu}$ over a long period, as shown in Figure 12.4; the value of braking index obtained from the slope of the line in this plot is $n = 1.4 \pm 0.2$.

These and other measured values of the braking index are significantly less than the expected value $n = 3$. The explanation may be that in these young pulsars there is a secular change in the slowdown law (Equation (12.3)) on a timescale comparable to the characteristic age. This might be due to an increase in the dipolar magnetic moment or a decrease in the effective moment of inertia (see Section 12.5 below). In these cases Equation (12.4) shows that the actual age of the pulsar may be greater than the characteristic age; for example, the age of the Vela pulsar may be over 20 000 yr, more than double the characteristic age.

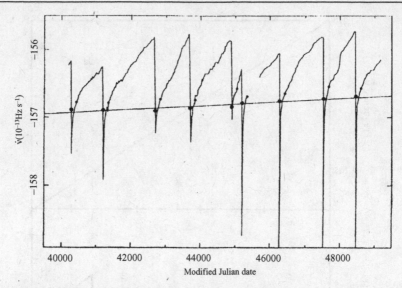

Figure 12.4. The slowdown rate $\dot{\nu}$ of the Vela pulsar over 25 years, showing the slow change which is used to calculate the braking index (Lyne *et al.* 1996).

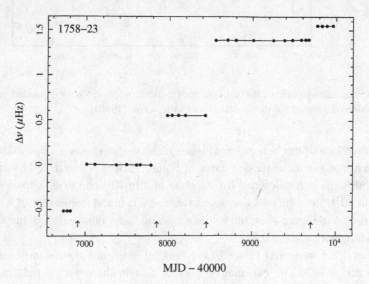

Figure 12.5. Step increases in rotation rate of PSR B1758-23, after subtraction of the mean values of the slowdown rate and its first derivative (Shemar and Lyne 1996).

12.4 Glitches in rotation rate

Glitches also occur in older pulsars, but spaced at longer intervals. The main effect is a step increase of rotation rate, as seen in the series of glitches in PSR B1758-23 (Figure 12.5).

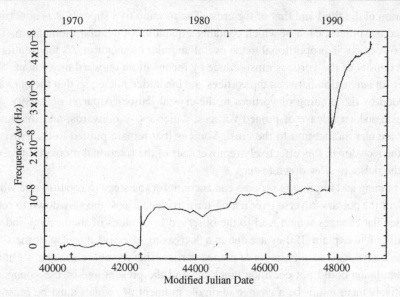

Figure 12.6. Steps and transients in the rotation rate of the Crab pulsar. Constant values of rotation rate and slowdown rate have been removed from the data (Jodrell Bank Observatory).

For most pulsars (including the Vela pulsar but not the Crab pulsar) the overall effect of the step increases in rotation rate at the glitches is that the averaged rate of slowdown is about 2% less than the rate between glitches (Lyne *et al.* 2000).

Most pulsars are observed only occasionally, so glitches are usually noticed as a change in period at an observation some time after the event. More frequent monitoring often reveals transient effects lasting some hours or days after the glitch. Transients at Crab pulsar glitches are seen in Figure 12.6; these add to the step increases in rotation rate. Flanagan (1990) has observed three distinct components in a glitch of the Vela pulsar, decaying with time constants of 32, 3.2 and 0.4 days (see also Dodson *et al.* (2007)).

The most frequent glitches occur for J0537-6910, a pulsar in the Large Magellanic Cloud with period 16 ms (Marshall *et al.* 1998), for which 20 glitches were recorded in $6\frac{1}{2}$ years by the Rossi X-ray timing satellite. A record of the several hundred glitches observed in radio pulsars is maintained on a communal website at the CSIRO, Australia.

12.5 Superfluid rotation

The explanation of glitches has involved some remarkable physics. The steps in rotation rate show that at least 2% of the moment of inertia is attributable to a separate component of the neutron star: the long time constant of the exponential recoveries shows that this component is a superfluid. There are two possible superfluid regions that might be responsible for this behaviour, located respectively in the core and in the crust. Both may be involved, but in any case there must be an explanation of the variable coupling between the angular momentum of the fluid and that of the crust. The rotation of a superfluid is abnormal; it is expressed as vortices, each of which contains a quantum of angular momentum. The area density of vortices is proportional to the overall angular momentum. As the rotation slows, the area density of the vortices must reduce by means of an outward movement. There is, however, an interaction between the vortices and the lattice nuclei, so that the outward flow is impeded by the pinning of vortices to the crystal lattice (Alpar *et al.* 1989). A glitch indicates a sudden release of pinned vortices, which move outwards, and the outer ones transfer angular momentum to the crust. Vortices that remain pinned to the crust take no part in the slowdown; this effectively removes part of the rotational moment of inertia and allows the pulsar to slow down faster.

This pinning and unpinning process can account for the steps in rotation rate which, for the bulk of the pulsars, reverse over a long-term average 2% of the slowdown in rotation.

The secular changes which lead to the observed low values of the braking index n are more difficult to explain. If they are due to a decreasing effective moment of inertia I, this indicates a continued accumulation of pinned vortices that are not released at a glitch; such an accumulation could not continue much beyond the present age of the young pulsars. Alternatively there might be a change in dipole moment M, which must be *increasing* at a rate comparable to the characteristic slowdown lifetime. The magnetic field within the neutron-fluid core of the star is quantized; it forms flux tubes that can interact with the rotational vortices. An interaction between the dipolar magnetic field and the rotational vortices can lead to an accumulation of stress that is released at a glitch, leaving a changed magnetic field configuration. The expansion of the rotational vortex network can carry the magnetic flux from the core into the crust, and increase the dipole moment. It may also stress the surface of the crust, so that the glitch may involve crust cracking and a readjustment of the surface distribution of the magnetic field (Ruderman 1991).

The distribution of intervals between glitches, and of their magnitudes, appears to follow those of avalanche dynamics in *self-organized criticality systems* (Melatos *et al.* 2008). This describes the collective behaviour of spatially connected regions subjected to an increasing stress, as in the familiar 'sandpile' effect.

12.6 Radio and optical emission from pulsars

The strong magnetic field completely dominates all physical processes outside the neutron star. The force of the induced electrostatic field acting on an electron at the surface of a rapidly rotating pulsar like the Crab pulsar exceeds gravitation by a very large factor, which would be as much as 10^{12} if there were no conducting atmosphere. The magnetic field remains approximately dipolar out to a radial distance $r_c = c/\omega$, which is the distance at which a corotating extension of the pulsar, with angular velocity ω, would have a speed equal to the velocity of light. This radial distance defines the *velocity-of-light cylinder*. Within this cylinder is a corotating *magnetosphere* of high-energy plasma (Goldreich and Julian 1969), in which the strong magnetic field allows charged particles to move along but not across the field lines. Field lines originating near the poles cross the velocity-of-light

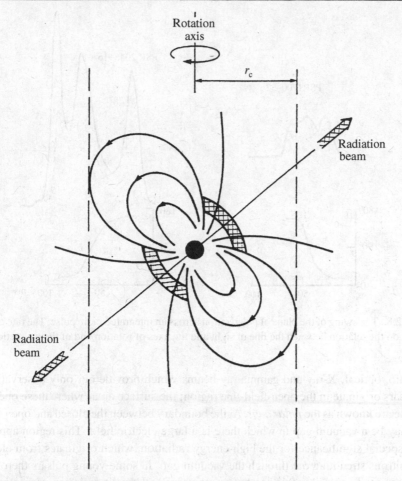

Figure 12.7. The magnetosphere, showing magnetic-field lines crossing the velocity-of-light cylinder. Within a distance $r_c = c/\omega$ of the rotation axis there is a charge-separated, corotating magnetosphere. The magnetic field lines which touch the velocity-of-light cylinder at radius r_c define the edges of the polar caps. Radio-emitting regions in the polar caps are shown cross-hatched.

cylinder (Figure 12.7), allowing energetic particles to escape; these particles are then able to energize a surrounding nebula such as the Crab Nebula (see Chapter 9).

In the closed equatorial region, the high conductivity allows the induced electric field to be cancelled out by a static field, so that

$$\mathbf{E} + c^{-1}(\omega \times r) \times \mathbf{B} = 0. \tag{12.7}$$

This corresponds to a charge density in the plasma, where the difference in numbers of positive and negative charges is

$$n_- - n_+ = \omega \cdot \mathbf{B}(2\pi ec)^{-1}. \tag{12.8}$$

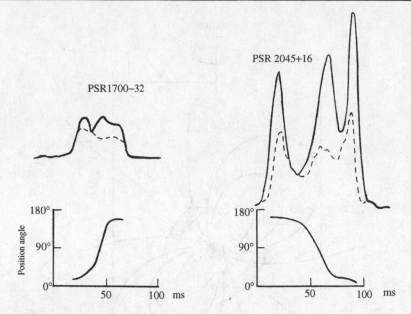

Figure 12.8. The swing of the plane of polarization across an integrated radio pulse. The rate of swing depends on the relation between the line of sight and the axes of rotation and of the magnetic field.

The radio, optical, X-ray and gamma-ray beams which provide our only observations of the pulsars originate in the open-field-line region; the surface areas where these open lines originate are known as the *polar caps*. At the boundary between the closed and open regions there may be a vacuum gap in which there is a large electric field. This region appears to have a special significance for the high-energy radiation, which originates from electrons and positrons streaming out through the vacuum gap. In some young pulsars there is also radio emission from within this gap, but more generally the radio emission originates closer to the surface; its source is distributed over the polar cap, as shown in Figure 12.7.

The distribution of plasma density within the polar cap is unknown, but it is probably lowest at the edge and at the magnetic pole; this transverse distribution has a profound effect on the outward propagation of radio waves, determining some prominent features of the overall radiation pattern (see Section 12.7 below).

The strong linear polarization of both the radio and the optical emission provides valuable clues to the geometry of the emitting regions. In a typical radio pulse (Figure 12.8) the plane of polarization swings monotonically through an S-shape; this is interpreted as the successive observation of narrowly beamed radiation from sources along a cut across the polar cap (Radhakrishnan and Cooke 1969). The plane of polarization is determined by the alignment of the magnetic field at the point of origin (or, more precisely, at a higher point where the ray detaches from a magnetic-field line: see Section 12.7 below), so that the sweep of polarization can be related to the angle between the magnetic and rotation axes and their relation to the observer. Lyne and Manchester (1988) showed in this way that the

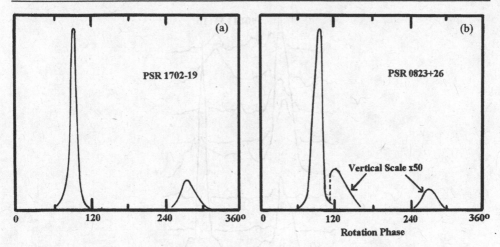

Figure 12.9. (a) A double pulse: emission from both poles. (b) An extended pulse from a pulsar seen nearly pole-on.

angles between the axes are widely distributed; they found no evidence, however, that the inclination angle changes during the lifetime of an individual pulsar. For those pulsars for which the axes are nearly perpendicular a pulse may be observed from both magnetic poles (Figure 12.9(a)), whereas for those for which the rotation and magnetic axes are nearly aligned the observer must be located close to the rotation axis; in this case the radio pulse may extend over more than half of the pulse period (Figure 12.9(b)).

The radio pulses vary erratically in shape and amplitude from pulse to pulse; however, the integrated profile obtained by adding some hundreds of pulses is usually reproducible and characteristic of an individual pulsar (Figure 12.10). Generally, these integrated profiles contain several distinct components, known as *subpulses*; these appear to be associated with different regions of the polar cap, each of which excites radio emission in one narrowly defined direction. If the excitation of each region varies randomly and independently of the others, the sum will vary from pulse to pulse, but adding many pulses will produce an integrated pulse profile that depends only on the average emission from each region. A selection of integrated profiles is shown in Figure 12.11.

In many pulsars the variations of excitation in the different regions are not independent. For example, sudden changes in the observed integrated profile seem to correspond to a switch between different patterns of mean excitation; this behaviour is known as *moding*. Again, in many pulsars, the variation of intensity is organized into a steady drift across the profile over a time of several pulse periods; this *pulse drifting* is regarded as a lateral movement of an area of excitation across the polar cap (Figure 12.12). In some pulsars the track of this movement appears to be closed, so that the same pattern of excitation can recur after an interval considerably longer than the time for a subpulse to cross the width of the pulse profile. This has been interpreted by Deshpande and Rankin (Deshpande 2000) as a pattern of excitation rotating round the polar cap in the form of a carousel, as indicated in

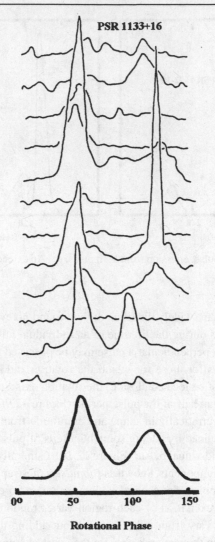

PSR 1133+16

Rotational Phase

Figure 12.10. Individual and integrated pulses from the pulsar PSR 1133+16.

the diagram of Figure 12.13. A systematic survey of pulsars bright enough for drifting to be detected (Weltevrede *et al.* 2006) showed that well-organized drifting is more likely to be observed in older pulsars; the authors suggest nevertheless that drifting of some kind may occur in all pulsars.

Figure 12.12 also shows the phenomenon of *pulse nulling*, in which a series of pulses is completely missing. This may involve a single pulse, or a sequence of pulses lasting various periods of time; some pulsars spend more time in an off state than on. As an extreme, some produce single pulses with very long gaps between; these are known as RRATs (rotating radio transients) (McLaughlin *et al.* 2006).

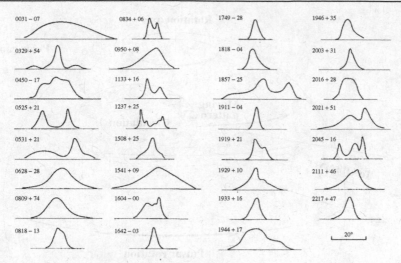

Figure 12.11. A selection of integrated radio profiles.

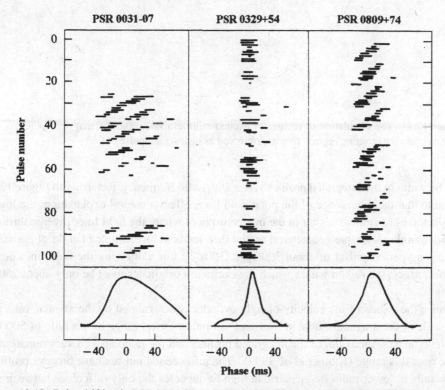

Figure 12.12. Pulse drifting in PSR 0031-07 and PSR 0809+74 contrasted to PSR 0329+54 (centre). Successive pulses contain subpulses that occur at times that drift in relation to the average pulse profile. These pulsars also exhibit the effect of pulse nulling, when sequences of pulses are missing (Taylor and Huguenin 1971).

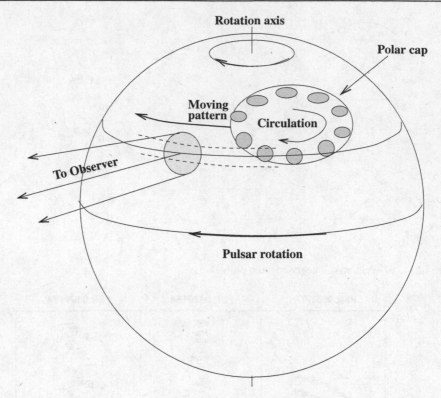

Figure 12.13. The circulation of regions of excited radio emission round a magnetic pole. As the regions cross the emitting regions they are observed as drifting subpulses.

The width of the integrated profiles varies with radio frequency. Referring to Figure 12.7, we note that the divergence of the polar field lines offers a natural explanation; the lower frequencies are emitted higher in the magnetosphere where the field lines diverge through a wider angle. A simple geometrical model then leads to an estimated height of emission for a long-period pulsar of about $300(\nu_{\mathrm{MHz}}/300)^{-1/2}$ km. Following the field lines down to the surface, the region within which the excitation originates must be only about 250 m across.

Since the radius of the velocity-of-light cylinder is determined by the angular velocity, the scale of the magnetosphere in slow and fast pulsars must range over a ratio of 5000 to 1. Nevertheless, the width of the integrated profiles and the radio spectra vary remarkably little over this range (Kramer *et al.* 1999). The millisecond pulsars have broader profiles, especially at lower radio frequencies; at high frequencies the only difference between the radio characteristics of the two classes is in luminosity, which is lower by a factor of ten for the millisecond pulsars. At other end of the scale, the 8.5-s pulsar (Young *et al.* 1999) has an exceptionally narrow beamwidth of only 1°: there is no explanation for this, and, if this is common, it suggests that there may be a substantial population of very slow pulsars, most of which cannot be detected at all because their narrow beams are never directed towards us.

Among the many other complexities of pulsar emission, we note especially the so-called *giant pulses* occasionally observed from some young pulsars, and particularly from the Crab pulsar. These are the shortest and most intense pulses observed from any celestial source. Hankins *et al.* (2003) found that individual giant pulses were as short as 2 ns, with an intensity of 1000 Jy. The source of such a short pulse must be only a few metres in size, generating an energy of the order of 2×10^{14} erg cm^{-3}, which is comparable to the total energy density in that part of the magnetosphere.

The detailed characteristics of radio pulses are reviewed and discussed by Seiradakis and Wielebinski (2004).

12.7 The radiation mechanism and refraction

The intensity of the radio emission shows at once that it must be coherent and not thermal in any sense; the brightness temperature in some cases exceeds 10^{30} K. The optical and other high-energy emission from the Crab and Vela pulsars, in contrast, can be accounted for as incoherent curvature or synchrotron radiation from individual high-energy particles streaming out along field lines in the vacuum gap at the edge of the polar cap. Furthermore, the radio pulses show a very high degree of polarization, which on occasion may approach 100%; this cannot be explained in terms of incoherent synchrotron or curvature radiation.

The radio emission is therefore *coherent*, and the association of a particular frequency with a definite radial distance shows that this distance is determined by a resonance in the plasma of the magnetosphere. Melrose (1992) argues that a two-stage process is involved, in which the coherence derives from bunching in an unstable stream of particles, and the radiation is a resonant coupling at a critical density to a propagating mode directed along a field line. Two propagating modes can ultimately escape from the magnetosphere; they have different refractive indices and different polarizations, so the polarization of the emergent wave is determined at the location of escape from the field lines.

The original acceleration of the particles takes place near the surface of the polar cap, in a cascade process. In this cascade, as suggested by Sturrock (1971), electrons or positrons are accelerated to a high energy and radiate gamma-rays via curvature radiation; these gamma-rays then create electron and positron pairs as they encounter the strong magnetic field, and the new particles are accelerated to continue the cascade. This process is obviously ineffective at the magnetic pole itself, where the field lines are straight, so the plasma density is expected to be low at the centre of the polar cap: this has been a puzzle since many pulsars have a peak in their radiation at the centre of the integrated profile. The explanation may lie in refraction, which takes place in the lateral gradient of plasma density, as shown in the cartoon of Figure 12.14. Following Petrova (2000), rays are focussed towards a peak at the centre and an outer ring; these are observed as *core* and *conal* components, as in the profiles of Figure 12.11.

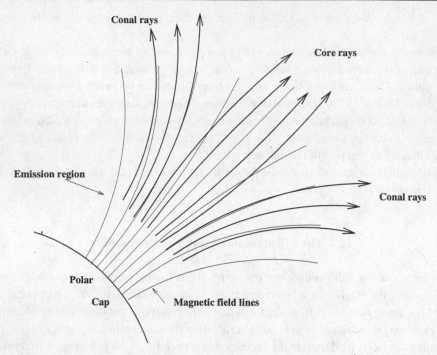

Figure 12.14. A sketch to show the effect of refraction in the magnetosphere. Rays initially aligned with the magnetic-field lines deviate in the directions of lower plasma density, towards the centre and edge of the polar cap.

Pulsars are most often and most easily observed at radio frequencies in the range 100 MHz to 2 GHz. Their radio emission has a steep spectrum, with spectral index $\alpha \sim -2 \pm 1$, but for most pulsars observations at lower frequencies are less informative because of the distortions due to propagation effects in the ISM (Section 12.9). Some spectra have been measured over a wide range, from 40 MHz to 80 GHz (see a compilation by Malov *et al.* (1994)). There is often a turnover at frequencies below about 100 MHz (Figure 12.15), but these spectral characteristics cannot be directly related to the energy spectrum of the emitting particles, as can be done for incoherent sources of synchrotron radiation (Chapter 4).

12.8 The population and evolution of pulsars

The first 500 pulsars to be discovered appeared to constitute a homogeneous population, which could be accounted for by an origin in supernovae followed by a monotonic decline in rotation rate. The Crab and Vela pulsars, with periods less than 100 ms, were the youngest, while the oldest, with periods of about 1 s, had characteristic ages of up to 10^8 yr. Beyond a period of a few seconds, which was reached typically within about 10^7 yr, the radio pulses ceased or became too weak to be detectable. Comprehensive surveys gave an estimated population of about 10^5 active pulsars in the whole of the Galaxy. Their distribution over

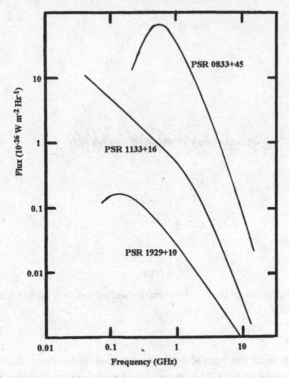

Figure 12.15. Radio spectra of three typical pulsars. The spectra are generally convex and often have a low-frequency cut-off.

the sky showed that they were concentrated in the Galactic plane (Figure 12.16), and were therefore associated with Stellar Population I. Their comparatively short lifetime suggested a birthrate consistent with an origin in Type II supernovae.

The discovery of a pulsar with the astonishingly short period of 1.6 ms (Backer *et al.* 1982) showed that the limitations of the early search techniques might be concealing a large population of undiscovered short-period pulsars. Searches using techniques sensitive to shorter periods soon revealed a large number of *millisecond pulsars*. More recent techniques have extended the surveys to include periods as short as 1 ms (Johnston *et al.* 1992), but, as we shall see, the bias against the discovery of short-period pulsars is still important. There is also a natural bias against the discovery of pulsars with a narrow radiation beam, since the observer is more likely to be outside the solid angle swept by a narrow beam.

The population known at present is displayed in Figure 12.17, which is a plot of the period P and its rate of increase $\dot{P}$ for pulsars catalogued in mid 2008. The original population of so-called normal pulsars extends from the top left towards the lower right of this diagram. Pulsars following the simple slowdown law of Equation (12.3) would follow a straight-line track towards the lower right; this evolution must eventually carry them over a boundary and out of the region where they are energetic enough to radiate. This boundary is known

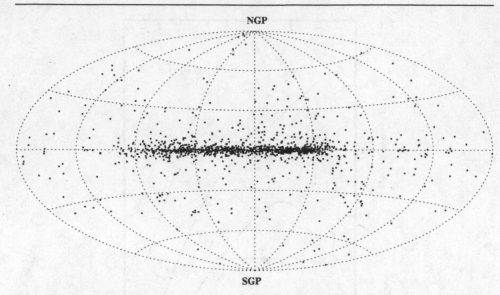

NGP

SGP

Figure 12.16. The distribution of the first 1000 catalogued pulsars over the sky (courtesy of G. Hobbs, Jodrell Bank Observatory).

as the *death line*; it need not be a sharp boundary, and it has been termed a *death valley* by Chen and Ruderman (1993). The lower left of the diagram contains the more recently discovered millisecond pulsars, arbitrarily defined as those with periods less than 10 ms. The majority of the millisecond pulsars were found to be members of binary systems; in contrast almost all of the main population are solitary. The binary systems are shown circled in Figure 12.17. Many of the known millisecond pulsars, including PSR J1748−2446ad with the shortest known period of 1.39 ms (Hessels *et al.* 2006), are in the globular clusters (see Section 12.15 below). Figure 12.17 also shows the X-ray sources known as magnetars and AXPs.

Lorimer *et al.* (2006) have used a well-defined sample of over 1000 pulsars to deduce the distribution of luminosity among the pulsars and their distribution throughout the Galaxy. This analysis requires distances to be obtained from the observed dispersion measures via a model of electron distribution in the Galaxy, which leads to some uncertainty, especially near the centre of the Galaxy. Their results are shown in Figure 12.18. This figure shows the deduced populations as functions of Galactocentric radius R and height z above the Galactic plane, and the distributions in luminosity L and pulse period P. The figure also shows the observed number distributions in the sample. The model curves are smooth analytic functions fitted to the data; for example, the distribution in height z is shown as an exponential with scale height 330 pc. The total population of active pulsars above a lower limit of luminosity (0.1 mJy kpc^2 at 1.4 GHz) is found to be $155\,000 \pm 6000$.

The birthrate of pulsars can now be found by observing the rate of flow of observed pulsars across the $P/\dot{P}$ diagram (Figure 12.17), multiplied by the ratio of total to observed population. Allowing also for a beaming factor, which effectively hides about a quarter of

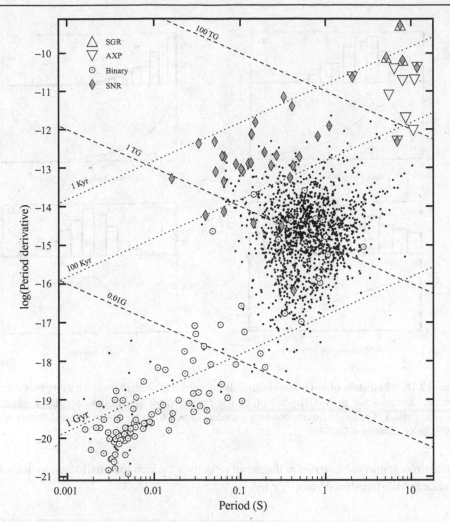

Figure 12.17. The distribution of pulsars in period P and its derivative $\dot{P}$. Binary systems are shown circled; AXP denotes anomalous X-ray pulsars, and SGR soft gamma-ray repeaters, two related categories of neutron stars (courtesy of C. Espinoza, Jodrell Bank Observatory).

the active pulsars, Lorimer *et al.* find the birthrate to be 1.4 ± 0.2 per century, somewhat less than the (as yet uncertain) supernova rate in our Galaxy.

12.9 Searches and surveys; the constraints

Observational selection, which hid the millisecond pulsars from the early observers, must always be borne in mind when the luminosity and spatial distributions of pulsars in the Galaxy are to be described. Besides the limitations in instrumental sensitivity related to antenna size, receiver noise, bandwidth and integration time, there are effects of propagation in the ionized ISM to be considered. The most obvious is the frequency dispersion in pulse

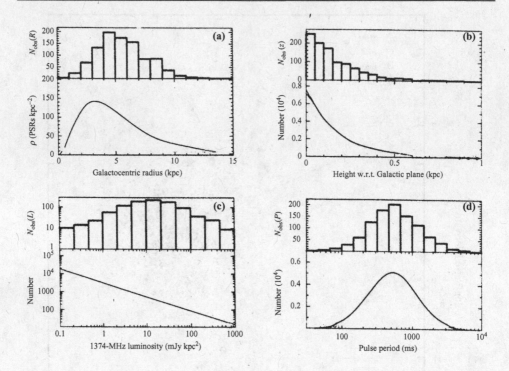

Figure 12.18. The distribution of pulsars in (a) Galactocentric radius R; (b) height z above the Galactic plane; (c) luminosity L at 1.4 GHz; and (d) period P. The upper panels show the distributions of observed pulsars. The lower curves represent a model derived from the observations (Lorimer *et al.* (2006), arXiv:astro-ph/0607640).

arrival time. Radio pulses travel at the group velocity v_g, which for small electron densities is related to the free-space velocity c by

$$v_g \approx c \left(1 - \frac{n_e e^2}{2\pi m \nu^2} \right), \tag{12.9}$$

where n_e is the electron number density and e and m are the electronic charge and mass. The second term gives a frequency-dependent delay τ for propagation over a distance L:

$$\tau = 1.345 \times 10^{-3} \nu^{-2} \int_0^L n_e \, dl \text{ s}. \tag{12.10}$$

The integral $\int_0^L n_e \, dl$, which measures the total electron content between the pulsar and the observer, is known as the *dispersion measure*, DM, usually quoted with units cm^{-3} pc. For a radio frequency ν_{MHz} the delay is

$$\tau = \frac{DM}{2.410 \times 10^{-4} \nu_{MHz}^2} \text{ s}. \tag{12.11}$$

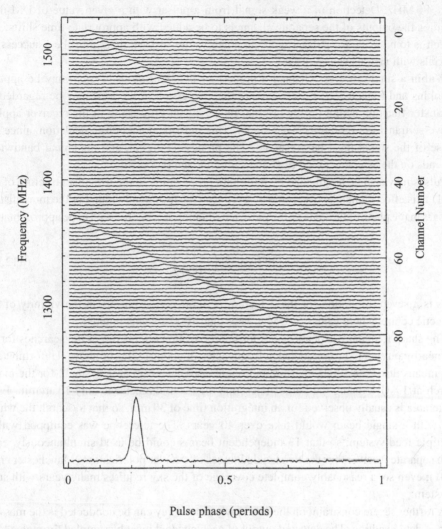

Pulse phase (periods)

Figure 12.19. A single pulse from PSR B1641−45 recorded in 96 adjacent frequency channels, each of width 3 MHz, centred on 1380 MHz. The sum of all channels, with appropriate delays for dispersion, is shown below.

Within the bandwidth B of a receiver the delay will vary and a pulse will be spread by

$$\Delta\tau = 8.3 \times 10^3 \mathrm{DM} \nu_{\mathrm{MHz}}^{-3} B_{\mathrm{MHz}} \text{ s.} \tag{12.12}$$

This spread limits the time resolution of the receiver by lengthening the pulse. The effect is overcome by restricting the bandwidth of the receiver, or by splitting the receiver band into a number of sub-bands, each with its own detector. Figure 12.19 shows the arrival of a single pulse from B1641-45 (DM $= 485$ cm^{-3} pc) in 96 adjacent receiver channels centred on 1380 MHz. Detection of a weak signal from a pulsar with a given value of DM then requires the outputs of the separate channels to be added with appropriate time shifts; if a search is to be made for pulsars with unknown values of DM, there must be a succession of trials with various values of DM.

Within a single receiver channel a process of *coherent de-dispersion* may be applied (Hankins and Rickett 1975). Recognizing that dispersion in the ISM may be regarded as a transfer function acting on the original pulse, signal processing in the receiver applies a reverse transform on the received signal. This must be done before detection, since the phase of the signal must be preserved. The process is digital, and the signal bandwidth depends on the available speed of signal sampling and processing.

Pulses may also be lengthened by multipath propagation in the irregular structure of the ISM. This effect cannot be overcome by a reduced receiver bandwidth. For the more distant pulsars in the plane of the Galaxy the pulse lengthening τ_{scat} is observed to be approximately

$$\tau_{\text{scat}} \approx 0.3 \nu_{\text{GHz}}^{-4} \left(\frac{\text{DM}}{1000} \right)^{4.4} \text{ s.} \tag{12.13}$$

This is a severe limitation for the detection of short-period pulsars in the vicinity of the Galactic centre, where values of DM greater than 1000 are encountered.

The sharp frequency dependence of these propagation effects forces the searches for all but nearby pulsars to be conducted at frequencies of about 1 GHz or above. Unfortunately, this means that for a single-aperture radio telescope the beam area is small; for the major search at 1.4 GHz using the Parkes telescope the beam diameter is only 13 arcmin. Each beam area is usually observed for an integration time of 30 min, so that to search the whole sky with a single beam would take over 40 years. The telescope was equipped with a multiple feed system, so that 13 independent beams could be used simultaneously, each with separate receivers for the two handednesses of circular polarization (Manchester *et al.* 2001); even so, a reasonably complete coverage of the sky requires many years with such a system.

A further severe constraint on the rate at which a survey can be conducted is the massive task of data handling. The detected output of each channel must be sampled through a long integration time at a rate several times faster than the shortest-period pulsar to be discovered. The set of samples at different radio frequencies must then be added with delays appropriate to some value of dispersion measure; the sum must then be Fourier transformed, and the resulting periodogram searched for significant signals. The process must be repeated for a series of values of dispersion measure. The duration of each process increases rapidly with integration time, so a balance has to be struck among survey time, sampling interval, range of dispersion measure and computation time. The observational constraints on any large survey inevitably impinge mainly on the sensitivity to pulsars with the shortest periods and the largest dispersion measures; millisecond pulsars close to the Galactic centre may never be detected as such, even though they may be identified as candidates by their spectra and polarization.

12.10 Trigonometric distance and proper motion

The trigonometric parallax (giving distance) and the proper motion of nearby pulsars can be measured by interferometry. The most accurate measurements are made by comparing the positions of the pulsar and an adjacent quasar, in which case only differential measurements need be made. Assuming that the position of the quasar is related to a fundamental frame, the pulsar's position may be found in this way to sub-milliarcsecond accuracy. Continued observations over a year or more provide measurements of distances through parallax, while proper motion is also measurable for many pulsars (Brisken *et al.* 2003). Geometrically determined distances, and proper motions, may also be available from accurate timing observations continued over a year or more. Velocities of solitary pulsars determined from their proper motion and distance are remarkably high; these pulsars received a kick at birth amounting to some hundreds of $km\,s^{-1}$.

Distances are obviously available for those pulsars which have optical identifications, such as those in supernova remnants. Some distance information is available for pulsars close to the Galactic plane that are observed through clouds of neutral hydrogen (see Section 9.1). These H I clouds absorb at wavelengths near 21 cm; the actual absorption wavelength depends on their location, giving an indication of the distance of an individual pulsar.

These distances are useful as calibrators of the distances provided by the dispersion measure (DM). Given a model of the distribution of electron density in the ISM, the DM found from Equation (12.10) immediately gives the distance. For nearby pulsars an average electron density $n_e = 0.025\,cm^{-3}$ pc may be used, with an accuracy of around 20%; for example the Crab pulsar with $DM = 57$ is known to be at a distance of about 2 kpc. At distances greater than 1 kpc it is, however, necessary to take account of the structure of the Galaxy; a model distribution (Section 9.4) proposed by Taylor and Cordes (1993) and improved by Cordes and Lazio (2002) is generally used for such distance determinations.

12.11 X-ray pulsars

The explanation of the new class of binary and millisecond pulsars, and their linkage to the main population, came from X-ray astronomy. The strongest X-ray sources are binary systems in which a neutron star is accreting matter from a companion star; in the process the neutron star is spun up and can eventually become a millisecond pulsar. These are thermal sources of X-rays; the accreting gas is concentrated by the magnetic field, forming hot spots above the magnetic poles. The X-ray flux then varies as the star rotates, bringing the hot spots in and out of view. (A different class of X-ray pulsars, including the Crab pulsar, consists of radio pulsars whose spectrum of pulsed radiation extends continuously from radio up to optical, X-ray and even gamma-ray energies.)

The identification of an X-ray source with a condensed star was first demonstrated by the discovery of pulsations at intervals of 4.8 s from the source Cen X-3; furthermore, a periodic Doppler shift in the pulse interval showed that the source was in a binary system (Giaconni *et al.* 1971). More than 100 such X-ray sources are now known, distributed throughout the Galaxy. All are neutron stars in binary systems; their companions have masses between $M_\odot$ and $20M_\odot$. Most of the companions are young stars with masses between $5M_\odot$ and $20M_\odot$, which are evolving through the red giant phase towards a supernova collapse, and whose remnant core would probably become another neutron star. Others, the low-mass X-ray binaries (LMXBs), have companions with masses about $M_\odot$, which are evolving towards a collapse to a white dwarf. Many LMXBs have been found in globular clusters, including 19 in 47 Tuc alone (Bogdanov *et al.* 2006), which were found at the positions of the known millisecond radio pulsars in the cluster. The radiation has a soft X-ray thermal spectrum, indicating a surface temperature of $1–3 \times 10^6$ K. In some X-ray pulsars there are observable absorption lines, attributed to cyclotron resonance, indicating magnetic fields comparable to those deduced from slowdown rates (Özel *et al.* 2008).

Thermal radiation has also been observed from isolated neutron stars, giving a surface temperature that must derive from internal energy, either the remaining energy of the stellar collapse or from sources such as glitches. Temperatures of several times 10^5 K are found; remarkably this favours ultraviolet wavelengths, and several observations have been made by the Hubble Space Telescope (Mignani *et al.* 2004).

Some X-ray pulsars have no radio counterparts. A search at the centres of some supernova remnants has found so-called *compact central objects*, with periodicities of the order of some hundreds of milliseconds and small slowdown rates, indicating neutron stars with low magnetic-dipole fields (Gotthelf *et al.* 2005). It seems likely that accretion onto these stars is rapid, suppressing any radio emission.

12.12 Binary radio pulsars

As shown in Figure 12.17, most binary radio pulsars are among the millisecond pulsars. Their initial detection is sometimes hampered by the periodic Doppler shift of their pulse periods, but once detected their orbital characteristics become measurable with remarkable accuracy. Figure 12.20 shows this for two examples with orbits of small and high eccentricity. All of those with small eccentricities have a low-mass white dwarf companion; these binaries are the likely outcome of evolution from LMXBs. Those with high eccentricity usually have a neutron-star companion; these appear to have evolved from the higher-mass systems, the HMXBs. Solitary millisecond pulsars may have evolved from either class; in the case of the HMXBs the binary may have been disrupted at the time of the supernova explosion, while for the LMXBs the white dwarf may have been evaporated by intense radiation from the pulsar itself. The latter scenario is supported by observations of several pulsars with very-low-mass companions, in which an occultation occurs over a large part of the binary orbit; this is attributed to a cloud of ionized gas streaming away from the white dwarf.

The increased rotation rate of the millisecond pulsars is driven by transfer of angular momentum from the orbit during the mass transfer. There seems to be a lower limit of about 1.4 ms on the period, although this may still be an effect of observational selection. If

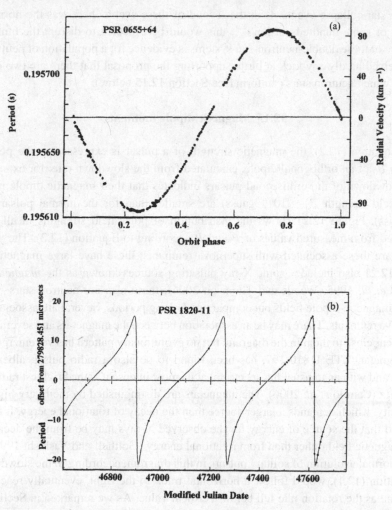

Figure 12.20. Doppler shifts of pulse periods in binary pulsars: (a) for an orbit with small eccentricity (PSR 0655+64) (Damashek *et al.* 1982); (b) for an orbit with large eccentricity (Lyne and McKenna 1989).

this limit becomes established, it may be related either to the influence of the neutron-star magnetic fields on the accretion process, or to the stability of the neutron star itself. A 'hard' rather than 'soft' equation of state leads to a larger diameter, with a larger disruptive centrifugal force. If searches confirm the present limit on rotation rate, this will support the hard equation of state.

The orbits of the binary pulsars include both some with very high eccentricity ($e \sim 0.8$) and some with very low eccentricity ($e \sim 0.00002$). The low-eccentricity systems, with white dwarf companions, are evidence of rotational speed-up by accretion. The high-eccentricity systems, the neutron–neutron-star binaries, must have been formed in two separate star-collapse events. If in the second of these events there was the normal kick velocity of some hundreds of $km\,s^{-1}$, this would be expected to disrupt the binary. The existence of these double-neutron-star systems is evidence for a population of neutron stars that received hardly any kick at birth, supporting the proposal that there are two different ways in which neutron stars can form (see Section 12.15 below).

12.13 Magnetic dipole moments

As in Equation (12.2), the magnetic strength of a pulsar is expressed as the polar field strength B_0 of an orthogonal dipole, calculated from the slowdown rate; for example, the small slowdown of the millisecond pulsars indicates that their magnetic dipole moments (polar field strength $B_0 \sim 10^{8-9}$ gauss) are smaller than for the normal pulsars ($B_0 \sim 10^{12}$ gauss). Figure 12.21 shows the distribution of polar field strengths in all pulsars, calculated from measured values of P and $\dot{P}$ according to Equation (12.2). The youngest pulsars are those associated with supernova remnants; these have large magnetic fields. Figure 12.21 also includes some X-ray pulsating sources known as the *magnetars* (van Paradijs *et al.* 1995; Woods and Thomson 2006); these also are neutron stars, with the highest magnetic-dipole fields but comparatively long periods. Several are associated with supernova remnants. There may be an association between the magnetars and several pulsars that appear close to them in the diagram, but no evolutionary pattern has yet emerged. Only one magnetar, XTE J1810-197, has been found to be also a radio pulsar, albeit highly variable and with an unusual flat spectrum: at its most intense it is the brightest radio pulsar at 20 GHz (Camilo *et al.* 2006). The magnetars are distinguished by their very high X-ray luminosity, which demands a larger source than the decay of rotational energy. It has been proposed that the source of energy for the observed X-rays may be from the decay of the large magnetic field rather than from rotational energy (Gotthelf and Vashisht 1997).

The normal evolution of solitary pulsars in this diagram, according to the slowdown law of Equation (12.1), would follow a horizontal track to the right, eventually reaching the death line as the rotation rate fell below a critical value. As we remarked in Section 12.5, the slowdown regime of the youngest pulsars may be evolving, but there is no indication that this is happening in the main population. There must be some decay of luminosity as pulsars traverse this diagram; there would otherwise be a large concentration towards the death line. The millisecond pulsars, which are much older, have much lower fields; there may be some process related to the binary spin-up that reduces the field from the order of 10^{12} gauss to about 3×10^8 gauss.

12.14 Velocities

For many years the only direct evidence of the origin of the normal population of pulsars in supernovae was the identification of the Crab and Vela pulsars. These were also the youngest known pulsars; this distinction is now shared by the 16-millisecond pulsar J0537-6910 which is clearly associated with a supernova remnant in the Large Magellanic Cloud (Marshall *et al.* 1998). There are also several more associations of young pulsars with

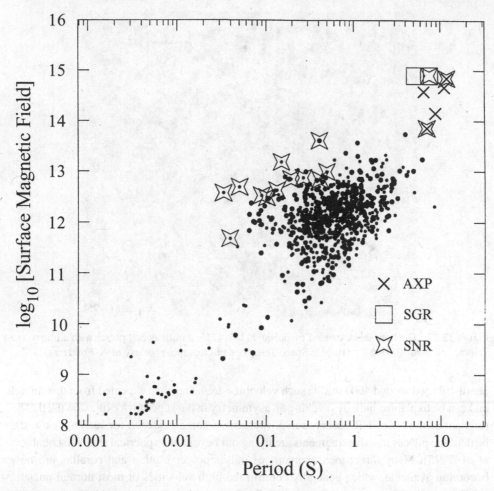

Figure 12.21. The polar magnetic-field strengths of pulsars and other neutron stars: AXP are anomalous X-ray sources and SGR are soft gamma-ray repeaters and SNR are associated supernova remnants (courtesy of M. Kramer, Jodrell Bank Observatory).

previously identified supernova remnants (Lorimer *et al.* 1998). In some of these associations, however, the pulsar is displaced from the centre of the nebula, suggesting that the pulsar left the scene of the explosion with a very large velocity; in one example the transverse velocity is 1700 km s^{-1}. Some remarkable examples of collisions between high-velocity pulsars and a local nebula have been seen optically (Cordes *et al.* 1993); Figure 12.22 shows the shock associated with pulsar J0437-4715. This pulsar has a transverse velocity of around 100 km s^{-1} in the direction of the arrow; the shock forms 10 arcsec in front.

The high velocities might be accounted for by the disruption of a binary system resulting from the explosion of one of its components; in the process of disruption a large fraction of the orbital velocity might be left with the pulsar. An assembly by Hobbs *et al.* (2005) of 233 measured proper motions finds, however, that the velocities of young normal pulsars

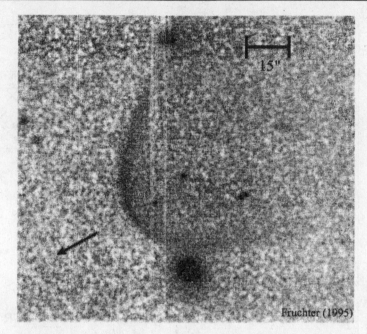

Fruchter (1995)

Figure 12.22. The bow shock created by pulsar J0437-4715, a millisecond pulsar with a transverse velocity of about $100\,\mathrm{km\,s^{-1}}$ (Hubble Space Telescope photograph, courtesy of A. Fruchter).

are distributed around $400\,\mathrm{km\,s^{-1}}$; such velocities are larger than expected from this model and can be explained only by invoking an asymmetry in the supernova explosion itself. The remnant of the Vela SNR observed in X-rays shows some evidence for such a kick at the birth of the pulsar: massive fragments are flying out beyond the spherical shell (Aschenbach *et al.* 1995). Many direct measurements of pulsar proper motion and parallax are now becoming available, which generally confirm the high velocities of most normal pulsars; a programme of astrometry with the VLBA has yielded the remarkably high velocity of $1083 \pm 100\,\mathrm{km\,s^{-1}}$ for PSR 1508+55, with a parallactic distance of $2.37 \pm 0.23\,\mathrm{kpc}$ (Chatterjee *et al.* 2005).

The velocities of the millisecond pulsars are generally an order of magnitude lower than those of the normal population (Toscano *et al.* 1999). This is probably related to their age: faster-moving pulsars would have left the gravitational field of the Galaxy within a fraction of their lifetime, leaving only those with velocities less than about $100\,\mathrm{km\,s^{-1}}$. The orbits of binary millisecond pulsars, to which we now turn, do, however, contain further evidence on the velocities imparted by the birth kicks.

12.15 Binary orbits and interactions

Among the binary pulsars we find examples of the highest and the lowest ellipticities of any astronomical orbiting systems. There are also binaries with two neutron stars; most of these have low eccentricities and one partner with an unusually low mass in the range $(1.2–1.3)M_\odot$. The younger partner must have formed from a supernova collapse while in its binary orbit, and cannot therefore have had a birth kick greater than about $50\,\text{km}\,\text{s}^{-1}$. This is a strong indication that a different type of supernova can occur in a binary system but not in a single star. This second type of supernova is believed to be the collapse of the degenerate O–Ne–Mg core of a helium star with low mass $((1.6–3.5)M_\odot)$, which can occur only in a binary system, in which the envelope of the helium star is removed by accretion onto the older neutron-star partner (Podsiadlowski *et al.* 2004). The difference in kick velocity, compared with the iron-core collapse, is presumably inherent in the different dynamics of the collapse. For a discussion of the two different formation mechanisms, see van den Heuvel (2007).

Two recently discovered binaries with white dwarf companions provide remarkable illustrations of the interactions occurring between the components of a close binary in this class. PSR 1957+20 consists of a binary with period 1.61 ms in an orbit with period only 9 h. Its companion is a white dwarf with mass $0.02M_\odot$. This is close to the lower limit for a white dwarf mass; furthermore, it is surrounded by an ionized cloud that occults the pulsar for one-tenth of the orbit: the pulsar is seen in the act of evaporating its companion. In another example, PSR 0655+64, the orbit is remarkably circular, with eccentricity less than 2×10^{-5}. This again is the result of interaction; tidal forces dissipate the energy of orbital revolution and reduce the eccentricity. A useful review of binary orbits and their interpretation is given by Bhattacharya and van den Heuvel (1991).

A substantial population of binary and millisecond pulsars is found in the *globular clusters*; for example in 47 Tuc no fewer than 20 were discovered in the Parkes survey (Camilo *et al.* 2000), and this total is growing rapidly through surveys at the large Arecibo and GBT telescopes. Several millisecond pulsars in 47 Tuc have negative values of $\dot{P}$; this is interpreted as an acceleration of the pulsar in the gravitational field of the cluster, which may provide a useful estimate of the mass within the cluster. Most of the globular-cluster pulsars are binaries, but there is a large variety of types of binary, indicating that interactions between the comparatively densely packed members of the cluster may have interfered with the evolutionary sequence. An extreme example in the cluster NGC 6342 is an apparently normal pulsar, with period 1 s and $B_0 = 10^{12}$ gauss, in a 6-h orbit with a $0.1M_\odot$ white dwarf companion. There is no obvious sequence of events that would lead to a normal pulsar within a globular cluster, which contains no young stars, and no explanation of its partnership with a white dwarf.

Outside the globular clusters there are some interesting cases of binary partnerships, some of which represent systems that were very nearly disrupted by the supernova explosion in which the neutron star was created. An extreme example is the binary PSR 1259−63, with $P = 47$ ms, in a very elliptical orbit ($e = 0.976 \pm 0.25$) round a young tenth-magnitude Be star (Johnston *et al.* 1993).

The pulsar PSR 1257+12 is the first discovered example of a star with a planetary system (Wolszczan and Frail 1992); it has three planetary bodies in orbit, two with masses close to four Earth masses $(4M_\oplus)$ and a third of mass about $0.015M_\oplus$. Their orbital periods are

98.2, 66.6 and 25 days, respectively (Wolszczan 1994). These were the first planets to be discovered outside the solar system.

12.16 Tests of general relativity

Millisecond pulsars make remarkably good clocks; they are generally free of glitches and their rotation can be monitored over long periods with microsecond accuracy. Several are now known to be in binary association with another neutron star, and in all such binaries the orbit is moderately or highly elliptical. The orbital periods are short, so the pulsar is moving in a large gravitational potential gradient at a high velocity. The first such system to be investigated was the 59-ms pulsar PSR B1913+16, which is in a binary system with an orbital period of only $7\frac{3}{4}$ hours. Despite its comparatively long period this pulsar has the characteristic stability of millisecond pulsars, and its rotation and the binary orbit could be traced with unprecedented accuracy. The whole orbit would fit within the diameter of the Sun; it has an ellipticity $e = 0.617$, and the orbital velocity varies from $1.33 \times 10^{-3}c$ to $0.316 \times 10^{-3}c$. There are large observable effects of special and general relativity, and the timing observations of this pulsar have provided unique tests of relativity theory (Taylor and Weisberg 1989). The most obvious post-Newtonian effect is the precession of the orbit, at the rate of $4.2°$ per year. The crucial factor which determines the importance of gravitational effects is the ratio between gravitational potential and rest-mass energy; for B1913+16 this factor $(GM/(c^2r))$ reaches 10^{-6}, while for the planet Mercury, whose orbit round the Sun had previously provided the classic test of precession, the ratio is 100 times smaller.

For most binaries, without observable relativistic effects, the inclination of the orbit to the line of sight is unknown, and the timing observations only yield the *mass function*

$$m_c^3 \sin^3 i (m_p + m_c)^{-2}. \tag{12.14}$$

This is derived from the propagation time across the orbit, which is of the order of v/c times the orbital period. The accuracy of timing of this millisecond binary allows terms of order $(v/c)^2$ to be determined; these relate to the varying gravitational potential round the orbit and to the transverse Doppler effect. There is also a measurable term of order $(v/c)^3$, related to propagation across the orbit as the line of sight crosses the varying gravitational potential of the companion. A combination of these factors allows a precise determination of the orbit of B1913+16 and of the two masses; both masses are known to an accuracy better than 1% and both are close to $1.4M_\odot$.

The orbital precession of this binary pulsar now provides the classical test of general relativity. Even more fundamentally, the orbital period has been observed to be decreasing, showing that the orbit is shrinking at exactly the rate expected from the loss of orbital energy due to gravitational quadrupole radiation, as predicted by relativistic theory (Figure 12.23). This is the first demonstration of the reality of quadrupole gravitational radiation. The direct detection of gravitational radiation on Earth is an observational challenge that has still to be met; in contrast, the relativistic binary has already given the expected value to an accuracy better than 1%. The set of measurements even provides a test for the theory of general

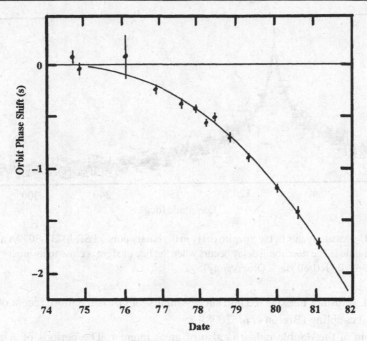

Figure 12.23. The change in orbital period of PSR 1913+16 due to gravitational radiation (Taylor and Weisberg 1989).

relativity itself, since the number of measurable parameters of the system exceeds that of the unknowns (Kramer 2005). Einstein's general theory has passed a series of penetrating tests with complete success.

Ironically the coupling between the binary orbit and the pulsar's spin is such that the pulsar axis is precessing at a rate of $1.2° \, \mathrm{yr}^{-1}$, and within a few decades the radiation beam is expected to pass outside our line of sight. There are already observable changes in the pulse profile attributed to this effect (Kramer 1998). Fortunately, there are other neutron–neutron-star binaries that will allow these basic tests of gravitational theory to continue. Outstanding among these is a double-pulsar system in which both neutron stars are pulsars.

The double-pulsar system PSR J0737-3039A/B has higher mean orbital velocities and accelerations than the Hulse–Taylor binary PSR 1913+16. In consequence the rate of relativistic precession is $\sim 17° \, \mathrm{yr}^{-1}$, four times that of PSR 1916+16. By good fortune the line of sight is close to the plane of the orbit, so that each pulsar in turn is seen through the strong gravitational field of the other; the resulting effect on the pulse arrival time is known as the Shapiro delay. Figure 12.24 shows the measured delay for the A pulsar as a function of orbital longitude; the maximum delay is $\sim 100 \, \mu s$, when the line of sight is closest to pulsar B. (Incidentally, this is a direct demonstration of the curvature of space-time.) In this double-pulsar system, there are four independently measured confirmations of general relativity (GR), each of which confirms the validity of GR at the 0.05% level (Kramer

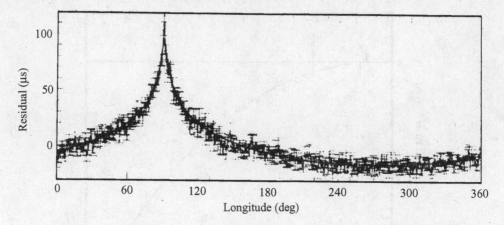

Figure 12.24. Measurements of the Shapiro delay in the binary pulsar PSR J0737-3039A as a function of orbital longitude. The maximum delay occurs when the line of sight is close to its binary companion PSR J0737-3039B (Jodrell Bank Observatory).

et al. 2006). A further unexpected and successful test of GR is the more recent observation of spin-orbit coupling (Breton *et al.* 2008).

The origin of the double pulsar is also of great interest. The periods of A and B are, respectively, 22 ms and 2.7 s. B is much younger than A, and it appears that the system must have been a neutron-star–white dwarf binary in which the neutron star was spun up by accretion from the progenitor of the white dwarf, while the white dwarf eventually became the slower pulsar B through core collapse (see Dewi and van den Heuvel (2004) and Section 12.12 above).

Further reading

40 Years of Pulsars: Millisecond Pulsars, Magnetars and More, AIP Conference Proceedings No. 983, New York, American Institute of Physics, 2008.
Pulsar Astronomy, Lyne A. G. and Smith F. Graham, Cambridge, Cambridge University Press, 3rd edn, 2006.
Pulsars, Manchester R. N. and Taylor J. H., New York, Freeman, 1977.
Pulsars as Physics Laboratories, ed. Blandford R. D., Hewish A., Lyne A. G. and Mestel L., Oxford, Oxford University Press, 1992.

13

Radio galaxies and quasars

Cygnus A (Figure 13.1) is one of the strongest radio sources in the sky. It was discovered by Hey in his early survey of the radio sky (Hey *et al.* 1946a, 1946b); its trace can even be discerned on Reber's 1944 map (see Appendix 3 for a brief history). It is, however, an inconspicuous object optically, and it was not identified until its position was known to an accuracy of 1 arcmin (Smith 1951; Baade and Minkowski 1954). Its optical counterpart was found to be an eighteenth-magnitude galaxy with a recession velocity of $17\,000\,\mathrm{km\,s^{-1}}$, that is, with redshift $z = 0.06$, indicating a distance of almost 1000 million lightyears. The source was shown to be double, with an overall size of more than a minute of arc, through the pioneering interferometry observations of Jennison and das Gupta (1953). Furthermore, the large angular size and double-lobed shape of Cygnus A were such distinctive features that similar radio sources might well be recognizable at very much greater distances; this was confirmed in 1960 when the radio source 3C 295 was identified by Minkowski with another similar galaxy, this time with a redshift of 0.45. Another galaxy, NGC 5128, had already been identified as a radio source known as Centaurus A (Bolton *et al.* 1949); this again showed the characteristic double-lobed shape, but with an angular diameter of $4°$ it was obviously much closer.

The radio galaxies were greeted both as a key to cosmology, where they offered the first possibilities of observing well beyond a redshift $z = 1$, and for their extraordinary astrophysics. These very luminous objects were apparently deriving their energy not from nuclear sources within stars but from gravitational potential energy on a galactic scale. The gravitational potential was due to a large concentration of mass, now accepted to be a black hole, at the centre of the galaxy. The radio emission itself was soon understood to be synchrotron radiation, but the processes of particle acceleration and the collimation into a double-lobed structure were for many years completely mysterious.

A second, even greater, revelation came in 1963, when Schmidt discovered the large redshifts of three star-like objects that had already been identified with three bright radio sources. The first of these 'quasi-stellar radio objects', soon known as QSOs or quasars, was 3C 273, whose redshift z was found to be 0.158. There followed 3C48, with $z = 0.37$, and 3C 147, with $z = 0.57$. By rapid stages the astonishing record of $z = 2.012$ was reached for 3C 9. Radio astronomy had indeed opened a new window on the Universe.

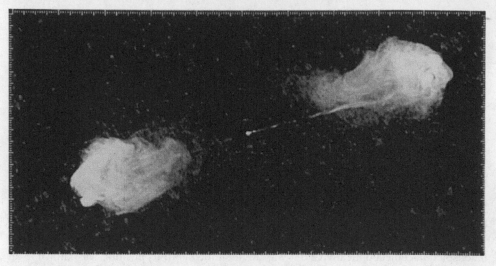

Figure 13.1. The radio galaxy Cygnus A (Perley *et al.* 1984).

These first quasars were optically bright; 3C 273 has magnitude 13, and can be seen in a small amateur telescope. It soon turned out that many others could be found in optical sky surveys, without the help of radio astronomy. Many thousands of quasars are now known, with redshifts ranging up to $z = 6$ and beyond; most of these have no detected radio counterparts.

All of the 320 extragalactic radio sources discovered in the 3CR survey (Bennett 1962) have been identified with optical counterparts with measured redshifts; most are radio galaxies or quasars. Deeper surveys of the whole sky and of selected areas have been made (see Chapter 16), and a separate population of radio sources has emerged. These are often referred to as *star-burst galaxies*, deriving their energy from recently formed stars and from supernovae. The most energetic are the youngest, in which the population of rapidly evolving massive stars is greatest. This is a wide category, defined here simply as those galaxies which do not have a 'monster' at the centre, i.e. a concentrated gravitational engine created by a black hole.

In this chapter we are concerned mainly with the physical processes acting in these two classes of radio source. Both classes are important in cosmology since they can be detected at very large distances. The cosmological significance of distinguishing large numbers of young galaxies at large redshifts, i.e. at early stages of evolution of the Universe, is discussed in Chapter 16.

13.1 Radio emission from normal galaxies

The radio emission of a normal galaxy, including our own Galaxy, consists mainly of synchrotron radiation from electrons with cosmic-ray energies, diffusing throughout the galaxy and radiating as they are accelerated in a magnetic field. The total radio power

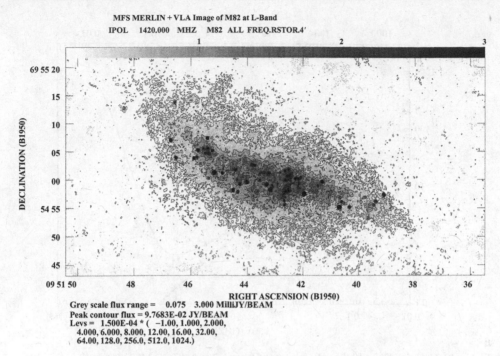

Figure 13.2. The starburst galaxy M82. The discrete sources are supernova remains (Muxlow *et al.* 1994).

has a remarkably close relation to two entirely different observable parameters, both of which are in turn related to the youth of the galaxy. The link is through the process of star formation, which is most frequent in young galaxies; the most energetic are the star-burst galaxies. The two observable parameters, apart from the radio emission, are the rate of occurrence of supernovae and the emissivity in the far infrared (FIR). This relationship has been demonstrated in most detail for the galaxy M82, whose radio emission is shown in Figure 13.2.

The spectrum of the star-burst galaxy M82 in Figure 13.3 extends from low radio frequencies to the FIR. Three separate components are identified by the broken lines: synchrotron, free–free and thermal. The free–free radiation is significant only at millimetre wavelengths (near 100 GHz); at lower frequencies the spectrum is dominated by synchrotron radiation, while in the FIR it is dominated by thermal radiation from dust. It might be expected that the free–free component would be closely related to the population of young stars; the surprise is that this is true also for the other two components.

The relation between the radio and the FIR emissivities of normal galaxies, which was first pointed out by Helou *et al.* (1985), is shown in the plot of Figure 13.4. This remarkably tight correlation is due to the common origin in star-formation activity. The rate of star formation is directly proportional to the rate of ultraviolet emission (which heats the interstellar dust) and to the supernova rate (which provides the energetic electrons

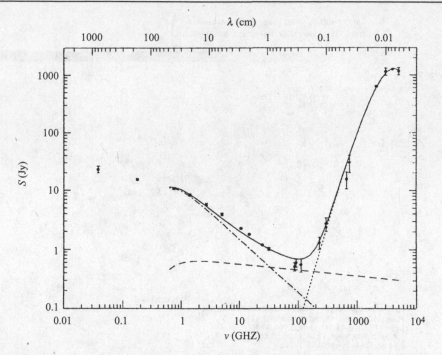

Figure 13.3. The radio and infrared spectrum of the star-burst galaxy M82. The broken lines show the three separate components of synchrotron, free–free and thermal radiation (Condon 1992).

for the synchrotron radiation). In both cases it is the total energy input that is significant, while factors such as the strength of the interstellar magnetic field have little effect. All that is required is that a simple proportion of the energy supplied by the young massive stars should be converted without significant loss into the overall radio and FIR emission.

Ultraviolet light is another indicator of young stars in galaxies, and might seem to provide a more direct way of distinguishing the star-burst galaxies; it turns out, however, that ultraviolet absorption within the galaxy is a severe limitation. In contrast there is no self-absorption in the high-frequency radio emission, as may be seen in the maps of radio emission from M82 (Figure 13.2), where individual supernova remnants are clearly distinguishable throughout the galaxy. Details of the relations among the ultraviolet, radio and FIR emission from normal galaxies are given in a review by Condon (1992).

13.2 Spectra and dimensions

The normal galaxies include our Galaxy, the Milky Way, which was the first radio source to be discovered. Similar radio emission can be detected from most visible spiral galaxies, usually at a comparatively low level, about 10^{37} erg s^{-1}. The young star-burst galaxies typically radiate 10^{40} erg s^{-1}. Radio galaxies and quasars, powered by the 'monsters', extend this range up to more than 10^{45} erg s^{-1}.

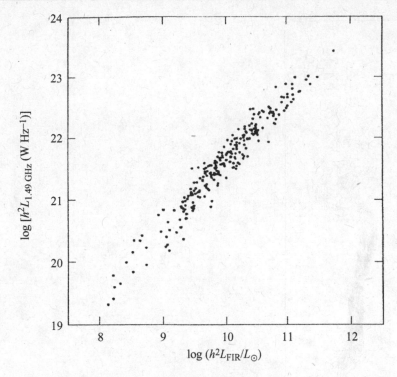

Figure 13.4. The relation between the radio and the FIR emissivities of normal galaxies (Condon *et al.* 1991).

It was for some time a puzzle that two apparently very different classes of object, the quasars and the radio galaxies, could be among the most powerful radio emitters, both standing out from normal galaxies by a factor of 10^8. The quasars are extremely compact, with core dimensions less than 1 pc; some are as small as 0.01 pc, comparable to the size of the solar system. Radio galaxies, in contrast, are up to 3 Mpc across, much larger than the visible galaxy with which they are associated. The link between quasars and radio galaxies is, however, obvious in many radio galaxies such as 3C 449 and NGC 1265 in Figure 13.5, where thin jets are emerging from a quasar-like nucleus, feeding energy into the twin lobes of the main radio source. The different physical regimes of the central source, the jets and the extended lobes are linked by the common energy source in the nucleus; this is now believed to be the release of gravitational energy of stellar material falling into a black hole.

The radio emission both from radio galaxies and from quasars is synchrotron radiation. In the extended lobes of radio galaxies the emission follows a classical regime, with typical electron energies of order 1 GeV accelerated in a field of order 10^{-5} gauss. The total energy in the electrons and the magnetic field is up to 10^{60} erg. The spectral indices α of the radio emission are distributed in a fairly narrow range around -0.8. Recalling (from Section 8.6) the relation $\alpha = (p - 1)/2$ between the index α, where $S(\nu) \propto \nu^{\alpha}$, and the electron energy

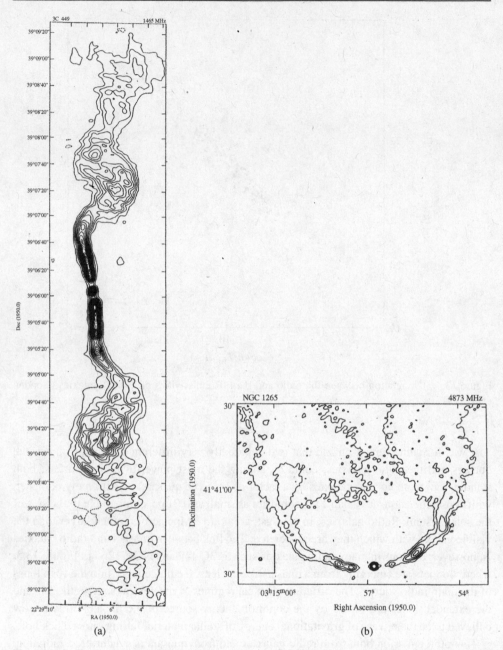

(a) (b)

Figure 13.5. Two examples of FR I radio galaxies: (a) 3C 449 (Perley *et al.* 1979) and (b) NGC 1265 (O'Dea and Owen 1986). In order to display the complex structures seen in Figures 13.1 and 13.5, a high degree of sophistication is needed, since both a high angular resolution and a high dynamic range (about 10^4 : 1) are needed. These maps were made by the VLA at wavelengths 21 cm and 6 cm, respectively; in order to achieve the necessary u, v coverage several different array sizes of the VLA had to be used. The signal processing used the methods outlined in Chapter 6; in particular, self-calibration was an important factor in the processing.

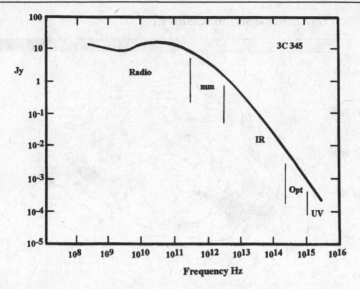

Figure 13.6. The continuum spectrum of 3C 345, extending from radio to ultraviolet (after Kellermann and Owen (1988)).

spectrum[1] $N(E) \propto E^{-p}$, we find an index $p = 2.6$, similar to that of cosmic-ray electrons in our Galaxy.

In the compact quasars the radio source can be optically thick, giving a nearly flat spectrum, with a spectral index approaching zero. There is often a transition between optically-thick and -thin regimes in quasars, such that the spectrum is curved, with a change of slope $\Delta\alpha \simeq 0.5$ between the flat spectrum at low frequencies and the steep spectrum at millimetre or even infrared wavelengths. Synchrotron radiation from quasars can extend through the visible into the ultraviolet and X-rays, as seen in the continuum spectrum from 3C 345 (Figure 13.6). For these quasars most of the radiated power is in the infrared.

13.3 Structures

The most luminous quasars completely outshine their host galaxies, in both their radio and their optical emission; in the brightest and most distant quasars it is often very difficult even to detect the surrounding galaxy. Where the host galaxy is sufficiently 'radio-loud', and can be distinguished from the quasar at its centre, it usually appears to be an elliptical or a very disturbed system, while, for quasars of lower luminosity, there are some indications that the host galaxies may be spirals. At the low end of the luminosity scale for galaxies driven by monsters, there are identifications with Seyfert galaxies, which are prominent optical objects. Although they may be four or five orders of magnitude down in radio

---

[1] Note the opposite sign convention in use for cosmic-ray particles: see Chapter 7.

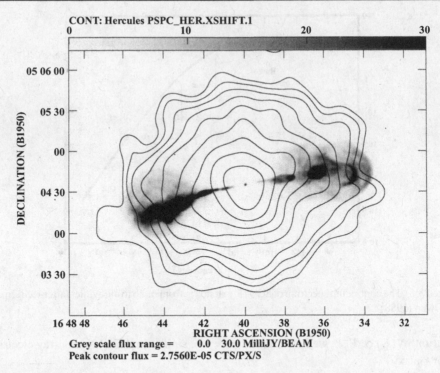

Figure 13.7. A contour map of the X-ray emission from the Hercules A cluster superposed on a grey-scale image of the radio galaxy at wavelength 18 cm. The hot gas emitting the X-rays extends through the cluster of galaxies, whose diameter is greater than 1 Mpc (Gizani and Leahy 1999).

luminosity, the Seyferts are related to the quasar phenomenon, and their active nuclei are generally at the centres of spirals. Powerful radio galaxies are hardly ever ordinary spirals, although even apparently 'normal' spiral galaxies can exhibit nuclei that emit excess radio power: M51 is a prominent example, and in Section 10.5 we have seen that even our own Milky Way galaxy appears to have an anomalous source, Sgr A*, at its centre. The hosts of the most luminous radio galaxies are sometimes ellipticals, but many of these appear to have anomalous forms, with prominent dust lanes. The relation of ellipticals to spirals is still unclear: there are suggestions that both have evolved from irregular galaxies, which appear to be common in the early Universe. It may be that spirals should be regarded as the typical population, while ellipticals are the products of collisions and mergers between spiral galaxies.

Outside the nucleus, the structure of a typical radio galaxy is determined by the interaction of energetic jets with an ionized gas that fills and surrounds the host galaxy, often extending well outside the visible galaxy. The existence of this hot diffuse gas is often known from the X-ray emission of a cluster of galaxies. Figure 13.7 shows the radio galaxy 3C 348, or Hercules A, superposed on a map of the X-ray emission from the surrounding cluster of galaxies. The jets themselves, especially the very narrow jets seen in radio galaxies such as

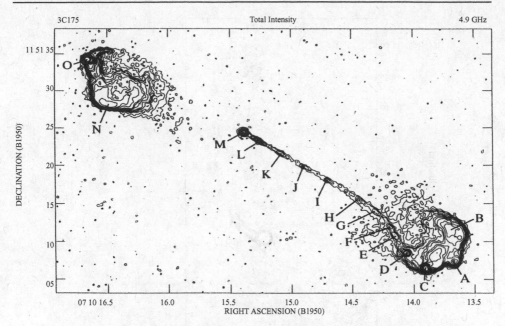

Figure 13.8. The FR II radio galaxy 3C 175 mapped at 5 GHz with resolution 0.4 arcsec. M is the central feature; other labels indicate hotspots in the jet and lobes (Bridle *et al.* 1994).

Cyg A (Figure 13.1), must initially follow trajectories related to the structure of the nucleus, but ultimately they must be confined by a medium through which they drive a tunnel. In this process ambient gas is entrained in the jets, but the main source of the radiation continues to be the stream of energetic particles ejected from the nucleus. The radio lobes develop when the jets are stopped by the ram pressure of the diffuse gas; at this point 'hotspots' of radiation develop, from which the more diffuse lobes spread out and are left behind as the hotspots advance.

There are two broad patterns of jets and lobes, divided fairly neatly by overall luminosity. In the FR II sources (the classification was introduced by Fanaroff and Riley (1974)), the jets are supersonic and give up their energy mainly in the hotspots. Jets in the weaker FR I sources become subsonic closer to the nucleus, creating a more diffuse structure with no obvious concentration at the extremes. The radio galaxies in Figures 13.1 and 13.5 show this distinction clearly. In many FR I radio galaxies the effect of the surrounding gas may be to distort the jet stream, as in NGC 1265 (Figure 13.5(b)) and 3C 120 (shown in Figure 13.9 later), where the jets are curved by a relative motion between the nucleus and the gas, and perhaps by the influence of the confining plasma as well.

The main lobe structure of radio galaxies is usually fairly symmetrical. Closer to the nucleus, however, at distances requiring the resolution of VLBI (see Section 6.7), there is often a marked asymmetry, as in 3C 175 (Figure 13.8). There may in some sources be a genuine asymmetry, with only one jet active at a particular time; in most, however, the

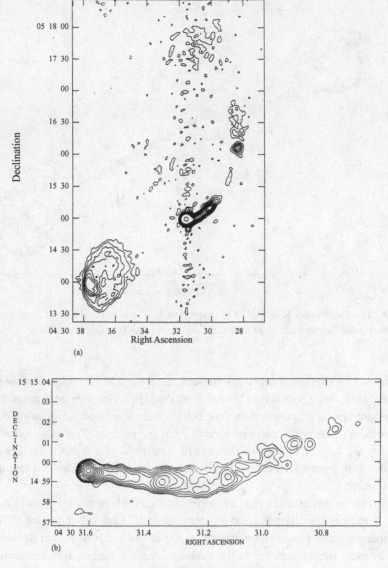

Figure 13.9. Maps of 3C 120: (a) the nucleus and outer lobes (Walker *et al.* 1987); (b) detail of the nucleus and jet (Muxlow and Wilkinson 1991).

effect is believed to be due to relativistic beaming (see Section 13.9 and Chapter 11). The jets leave the nucleus with relativistic velocities, boosting the intensity of radiation in the direction of the beam and reducing the intensity of the oppositely directed beam, which may even be unobservable. An example is 3C 120 (Figure 13.9); the main feature is a one-sided jet, while further from the nucleus there are lobes on both sides (Walker *et al.* 1987).

A well-observed example of a radio galaxy which is asymmetric on all scales is 3C 273, shown mapped at 408 MHz in Figure 13.10. Not unusually, the core is asymmetric, with 'superluminal' motion (see Section 13.9) indicating relativistic velocities. The single radio lobe, however, has been mapped with an excellent dynamic range, and the brightness ratio between the jet and any opposite counterpart is greater than 5500 : 1 (Davis *et al.* 1985). This appears to be a genuine case of a quasar nucleus with a one-sided jet; there is, however, an alternative explanation in terms of a curved source seen end-on, with one jet seen behind the other. Details of the jet, mapped at 1660 MHz, are shown in Figure 13.10(b) superposed on an optical photograph taken by the HST.

Finally, and to indicate that much remains to be revealed by observations with higher angular resolution and greater dynamic range, we draw attention to a distinct class of radio sources, the *compact sources*. These appear to have the same structure as radio galaxies, complete with core, jets and hotspots, but on a kiloparsec scale, in contrast to the 100-kiloparsec scale of the typical FR II radio galaxies. These are described in Section 13.12 of this chapter.

13.4 A simple model of active galactic nuclei

It might seem an impossible task to construct a single model that would bring together such an apparently disparate collection of radio sources, ranging from the most powerful and distant quasars and radio galaxies to the Seyfert galaxies and the compact sources: even within each class of object there is a rich variety. A single model of the central driving energy source has, however, been proposed, which appears to accommodate the whole diverse menagerie. The model will doubtless prove to be an over-simplification as observations reach closer to the active nucleus, but it is probably a good approximation for many, if not most, of the phenomena, and is currently a focus of thinking on that subject (Barthel (1989); Antonucci (1993); see especially the review by Urry and Padovani (1995)).

In the model, an active galactic nucleus (AGN) has a central massive object, usually a black hole, or at least a mass concentration with total mass 10^6–10^9 solar masses within 0.01 pc or even less, that is, about the size of the Solar System. The energy of the AGN derives from the gravitational potential energy of the surrounding material, which is released as it falls into the black hole. A considerable fraction, perhaps 10%, of the rest-mass energy of this material is released in the process. The material may be gas from the ISM, but it may also include stars that are disrupted as they pass within the Roche limit of the black hole. The angular momentum of the infalling material concentrates it into an accretion disc with steep gradients of angular velocity. Further collapse occurs through frictional dissipation of the differentially rotating disc and by turbulent dissipation.

So far, apart from the scale, the model resembles the accretion disc of an X-ray binary within our Galaxy. We now add two essential extra components, seen in Figure 13.11. Surrounding the thin accretion disc, there is a thicker torus of accreting material, which is cooler and sufficiently opaque to make the thin disc invisible from the side. The central regions can be observed only from polar directions, not from edge-on. The second added

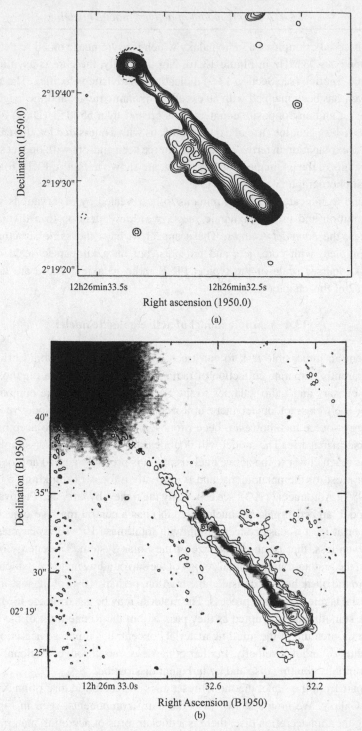

Figure 13.10. Maps of 3C 273 – a one-sided radio galaxy: (a) at 408 MHz (Davis *et al.* 1985) and (b) at 1660 MHz, combining data from MERLIN and the VLA, superposed on a grey-scale image from the Hubble Space Telescope and shown at double the scale (Bahcall *et al.* 1995), arXiv:astro-ph/9509028.

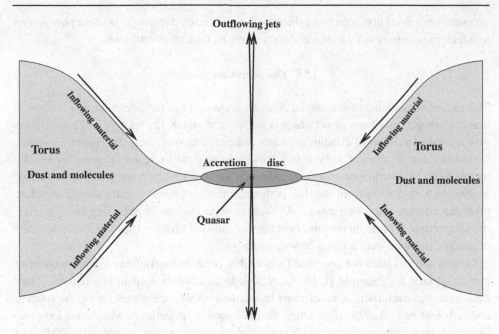

Figure 13.11. The unified model of an active galactic nucleus (AGN), centred on a black hole. The existence of an accretion disc, twin outward jets and a torus is well established, but the details of the inflow and the formation of the jets are still obscure.

component is the pair of jets of energetic material ejected from the nucleus along the polar directions; these are astonishingly narrowly collimated, and may travel very large distances compared with the size of the torus. These jets feed energetic particles and magnetic-field energy into the lobes of radio galaxies, forming the double structure first seen in Cygnus A.

The appearance of such a complex object, at any waveband, obviously depends on the relative importance of the various components; for example the jets and lobes of a radio galaxy may be more conspicuous than the central nucleus, whereas the reverse happens in a quasar. Many of the differences simply arise from intrinsic differences in strength of emission from the core, the disc, the jets and the lobes. The most important distinctions, however, arise from the differences between the polar and edge-on aspects from which such a complex object may be viewed. Seen edge-on, the torus may obscure the core and disc; seen pole-on, the brightness of the approaching jet, in which material is moving with high velocity, may be greatly enhanced by relativistic beaming. The dividing angle between pole-on and edge-on is about 45° from the pole.

It might be remarked that such a complex model, with so many adjustable parameters, looks artificial. It does, however, account for such a complex assembly of energetic objects that it is at least useful for discussing the whole range of AGN and quasars. Many survey papers have been devoted to relating observations at all wavebands to such a model; our account is only brief in comparison, since we concentrate on those aspects of the phenomena to which radio astronomy has made a particularly significant contribution.

13.5 The accretion disc

The easiest component of our model AGN to understand is the accretion disc. Provided that the orientation allows direct observation, optical and X-ray spectra and polarization give a detailed account of conditions in this compact and very energetic region. However, the whole disc is at most only a few parsecs across, so optical astronomy is usually unable to provide sufficient angular resolution (note that 1 pc corresponds to 1 milliarcsec at 200 Mpc). Radiation from the disc is thermal; the continuum spectra should therefore provide a measure of the temperature. As might be expected from an accreting disc, however, the temperature is far from uniform over the disc, and the observed spectra are composites of thermal radiation over a range of temperatures.

Optical spectral lines are observed from AGNs, predominantly from a small number of atomic species, for example H, He, C, N, O, Mg and Ne, as seen in H II regions. Their outstanding characteristic is an extreme broadening; this is accounted for by the rotation and turbulent motion in the disc. Often the lines show a spectrum in which they have two distinct components, one with width corresponding to velocities of order $10\,000\,\mathrm{km\,s^{-1}}$ (originating in a broad-line region (BLR)), the other with width corresponding to several hundred $\mathrm{km\,s^{-1}}$ (from a narrow-line region (NLR)). It has been proposed that the BLR is the accretion disc itself, while the NLR is a more extended region, extending high above the disc, which can be observed even if the dusty torus obscures the accretion disc. The line profile for both the BLR and the NLR may be asymmetric, extending towards the blue and indicating an outflow towards the observer. Continuum radiation from the BLR extends up to the keV X-ray region; above 3 keV there is a featureless continuum while at lower X-ray energies absorption can be seen, giving steps in the continuum. The BLR radiation is usually taken as the distinguishing characteristic of quasars and Seyferts.

When the model AGN is seen from the side, the torus obscures the direct radiation from the BLR, while the NLR is still observable. Part of the continuum radiation from the BLR can, however, be observed as scattered radiation from above the obscuring torus. This accounts for a low level of optical continuum that is highly polarized (about 15%–20%); the polarization is usually wavelength-independent, as expected from scattering by electrons. Some of the BLR X-rays are also scattered; this again implies that the scattering region is ionized, and suggests that it may be identified with the NLR itself. The NLR is the salient feature in Seyfert galaxies, where it may be resolved observationally into a region several hundred parsecs across.

13.6 The torus

The existence of the torus is more of a deduction from the observations outlined in the preceding paragraphs than the result of direct observation. There is, however, an excess of

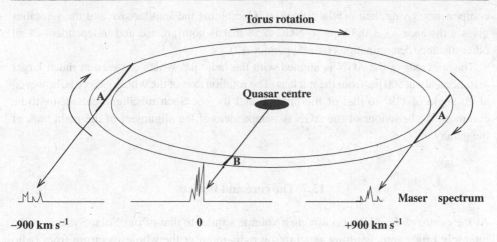

Torus rotation

Quasar centre

A

A

B

Maser spectrum

−900 km s^{-1} 0   +900 km s^{-1}

Figure 13.12. Maser emission from the inner regions of a torus in the AGN of the galaxy NGC 4258. The central maser, at the systematic velocity of the galaxy, is energized by the quasar core; the outer lines have sufficient path length to self-energize.

infrared radiation from AGNs, which indicates the presence of a dust cloud, and several AGNs contain huge H_2O masers, analogous to those observed in molecular clouds around some stars in the Milky Way (Chapter 12). The torus is therefore depicted as a cool molecular cloud.

Direct observation of the torus in a typical AGN requires an angular resolution of the order of one milliarcsecond. Fortunately, this is available from VLBI; furthermore, the narrowness of the water-maser lines may allow the dynamics of the torus to be revealed from the distribution of spectral components across the central regions of the AGN. This has been achieved by Miyoshi *et al.* (1995) in the nucleus of the spiral galaxy NGC 4258 (M106), where the water-maser sources are seen to be strung along a line, and their velocities vary progressively along the line, towards us at one end and away at the other (Figure 13.12). They are convincingly represented as located in a flattened inner part of the torus, which is rotating at velocities of up to 900 km s^{-1} round a massive central object. This has a mass of 40 million solar masses, and is a most convincing candidate for a black hole.

The water masers of NGC 4258 illustrate well the origin of maser emission, which may be either the selectively amplified emission from a continuum source lying behind the maser, or self-generated within the maser itself. Figure 13.12 shows the torus as a ring round a compact central source, which cannot itself be seen directly. The diagrammatic spectrum, shown below, has components at velocities around ±900 km s^{-1}, which derive from the edges of the rotating torus, where the line of sight is a maximum length. There is also a cluster of central components, the amplified radiation from the central source, which shows a gradient of velocities across the central source. This allows the acceleration to be determined, and hence the central mass, as noted above. Further VLBI observations (Herrnstein *et al.* 1999) have provided measurements of the lateral movement of the central components, giving their orbital velocities. Combining the angular size and the velocities gives a distance 7.2 ± 0.3 Mpc to NGC 4258 that is both precise and independent of all other distance determinations (see Appendix A2).

The spin axis of the AGN is aligned with the radio jets which are seen at much larger distances, about 500 pc, from the nucleus. The rotation axis of the whole galaxy is, however, at an angle of $119°$ to that of the nucleus and its jets. Such misalignment seems to be common; the behaviour of the AGN is independent of the alignment of the main bulk of the galaxy.

13.7 The core and the jets

At the centre of the AGN, occupying a volume similar to that of our Solar System, is the intensely bright core, emitting synchrotron radiation over the whole spectrum from radio to gamma-rays. Optically identified quasars are those whose cores are not obscured by the torus. Only a small proportion, less than 10%, are 'radio loud', that is, strong enough radio sources to be classed as radio quasars. The core is the origin of the jets emitted along the polar directions; where these jets exist, the core and the jets may be regarded as continuous.

Figure 13.13(a) shows an optical view of the elliptical galaxy M87, one of the first galaxies to be identified as a powerful radio source (it was then known as Virgo A). Near the centre the photograph is completely crowded out with star images. A short-exposure photograph (Figure 13.13(b)) shows a jet extending radially from the centre of the galaxy. In radio observations of the same high angular resolution, the core and the jet completely dominate the centre of the galaxy; farther out, the two lobes of radio emission are seen to be fed from the central core. There is ordered structure in the jet, with very similar appearance optically and in the radio map (Figure 13.13(c)). Only one jet is dominant; this is a common situation, which demands explanation in terms either of an intrinsic asymmetry in the core, or of the effects of Doppler brightening, to which we now turn.

For the brightest quasars only the core can be observed optically. According to the model, the AGN is now seen very nearly pole-on, so that any jet is close to the line of sight. Blandford *et al.* (1977) added a vital ingredient to the model: the jet material is streaming out with high relativistic velocity, so that radiation from any source within it will be beamed in the direction of the jet, with an increase in intensity because of *Doppler boosting*. A jet pointing away from the observer would be faint, or unobservable. This proposal allowed other known objects to be included in the same scheme; the continuum radiation from two classes of star-like objects, the OVVs (optically violently-variable objects) and the BL Lac objects (quasar-like, but having faint or no line emission, named after the type object), could also be generated by jets directed closely in the line of sight. The Doppler boost of the radiation from the jet enhances it over the line radiation from the core; the evidence in favour of this remarkable geometrical effect is presented in Section 13.10 below.

Supporting evidence for the line-of-sight model is provided by the Laing–Garrington Effect, in which one of a pair of jets can be demonstrated to lie behind the other. Both jets

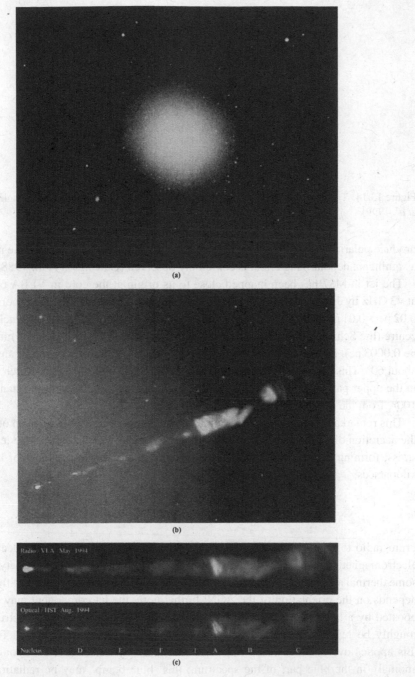

Figure 13.13. (a) The elliptical galaxy M87. (b) The central region in a short-exposure Hubble Space Telescope (HST) photograph. (c) Radio and optical images of the jet. The radio image (upper panel) was observed with the Very Large Array at wavelength 2 cm and resolution 0.15 arcsec. The optical image (lower panel) was observed with the HST with resolution 0.03 arcsec. Each image is 21 arcsec wide (Sparks *et al.* 1996).

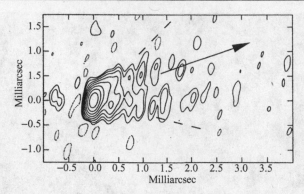

Figure 13.14. The nucleus and the base of the jet in M87, mapped with the VLBA at 43 GHz (Junor *et al.* 1999).

produce polarized radio emission, but where the far-side jet is seen throuh the near-side jet a significant depolarizing effect is commonly observed (Garrington *et al.* 1988).

The jet in M87 has been mapped close to its origin at the core in VLBA observations at 43 GHz by Junor *et al.* (1999). The beam of only 0.33 mas $\times$ 0.12 mas corresponds to 0.02 pc $\times$0.01 pc at the galaxy; this is only about 30 times the radius of the black hole at the centre (the Schwarzchild radius $r_s = 2GM/c^2$ for a mass of 3×10^9 solar masses would be 0.0003 pc). Figure 13.14 shows that the jet leaves the core in a cone of angular width about 60°. This is very wide compared with the narrowly collimated jet, which can be seen in the upper part of the same figure. Collimation apparently occurs at a distance of about $100r_s$ from the core.

This remarkable map of the base of the jet suggests that it originates as an outflow from the accretion disc, carrying out a toroidal magnetic field which confines the jet by a hoop stress, forming a corset that continues to confine the jet out to distances of hundreds of kiloparsecs.

13.8 Spectra of quasars and other AGNs

From radio to gamma-rays the spectra of many quasars extend over 11 decades of the electromagnetic spectrum. Within this range the radiation is emitted by a variety of sources, some thermal and some non-thermal. As we have seen, the relative strength of these sources depends on the orientation of the AGN; furthermore, the jet component may be Doppler boosted by a large factor. Despite this mixture, the overall effect is a spectrum that can roughly be represented as a simple power law $s = \nu^\alpha$, where $\alpha \sim -1$; furthermore, this applies over a range of up to 10^7 to 1 in intensity. There is a commonly observed anomaly in the blue part of the spectrum; this 'blue bump' may be radiation from the accretion disc. At the highest photon energies, in hard X-rays and gamma-rays, there is in addition a component from the inverse Compton effect; the interaction between the low-energy photons of the radio spectrum and the high-energy charged particles of the core

Table 13.1. *Approximate spectral indices of quasars, and corresponding proportions of luminosity, for various spectral bands*

Spectral range	Index α	Luminosity proportion
Infrared–optical (10–0.5 μm)	−1	30%
Optical–ultraviolet (500–100 nm)	−0.5	15%
Ultraviolet–X-ray (100–1 nm)	−1.3	30%
X-ray (1.2–100 keV)	−0.7	25%

and jet produces photons throughout the high-energy photon region, resulting in an overall smoothing and flattening of the spectrum. The main differences between the spectra of quasars lie in the radio region, where the *radio-quiet* quasars (the large majority) have radio luminosities well below the simple power spectrum extrapolated from the optical region.

For a continuous power-law spectrum with $\alpha = -1$, the total luminosity in any decade of frequency is the same. In more detail, shown in Table 13.1, we assign spectral indices to broad regions of a typical quasar spectrum. The relative importance of these regions is indicated by the proportions of the total luminosity shown for each wavelength range. Radio, which is the most variable part of the spectrum, is omitted; it is in any case only a small proportion of the total luminosity.

The radio spectra, seen in Figure 13.15, show great variety. When a compact core can be distinguished, the spectrum is flat ($\alpha = 0$); this appears to be the combination of several opaque sources with a range of optical depths. The jets and lobes are 'optically thin', with a spectral index of about $\alpha = -0.7$. The outer parts of the lobes have steeper spectra than the jets, as might be expected if the jets are feeding high-energy particles into the lobes, where the energy spectrum of the particles becomes progressively steeper as they diffuse outwards.

13.9 The radio brightness temperature of the core

The angular diameter of the core of radio-loud quasars is frequently measurable by VLBI. Typically, the core has an angular diameter of the order of one milliarcsecond, corresponding to a linear size of a fraction of a parsec. This small angular size implies a very high brightness temperature, which commonly lies in the range 10^{11}–10^{12} K. A typical spectrum shows self-absorption, giving a peak at a radio frequency ν_c; the peak is not ideally sharp, since the emitter is not uniformly bright. The brightness temperature of 10^{12} K would be a firm upper limit for a static, relativistic electron gas, giving a very simple relation between the angular size θ (milliarcsec) of the smallest component and its flux density S_m (Jy) at ν_c (GHz):

$$\theta \simeq S_m^{1/2} \nu_c^{-1}. \tag{13.1}$$

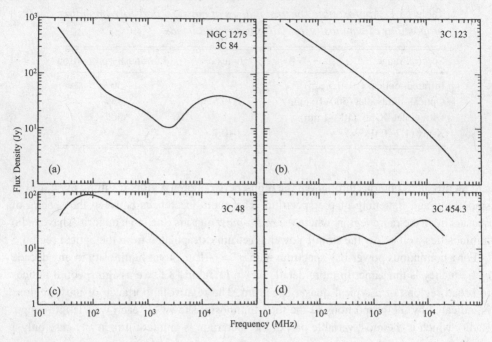

Figure 13.15. Typical spectra of radio galaxies (3C 84 and 3C 123) and quasars (3C 48 and 3C 454.3) (Kellermann and Owen 1988).

This upper limit on the brightness temperature is directly related to the effect of inverse Compton scattering. The ratio of particle energy loss by inverse Compton scattering, compared with energy loss by synchrotron radiation, increases rapidly with temperature, especially beyond 10^{12} K, at which temperature second-order scattering starts to dominate. The agreement between theory and observation in a number of sources provides the simplest confirmation that the radio emission is synchrotron, as well as demonstrating the important role of inverse Compton scattering.

Further confirmation of the action of inverse Compton scattering is found in an observed correlation between intensities at millimetre wavelengths and in X-rays; those quasars with higher radio intensities are expected to contribute more to X-rays through the scattering process.

The maximum brightness temperature to be expected because of the inverse-Compton limit has, however, been observed to be exceeded. Ordinary VLBI observations have difficulty in measuring the angular diameters of objects with such high brightness temperatures, because baselines are limited by the size of the Earth. Equation (13.1) implies that the limiting size will be inversely proportional to the critical frequency, that is, directly proportional to the wavelength. The fringe spacing, on the other hand, is also proportional to the wavelength, and it turns out that the critical test is dependent on having baselines greater than the diameter of the Earth. The first satellite-based observations were reported by Linfield *et al.* (1989); see Chapter 6 for details. At a baseline of 2.3 Earth diameters, interference fringes were observed for 24 of the 25 quasars and AGNs examined; furthermore, six of these sources clearly exhibited brightness temperatures that exceeded the self-Compton limit. This can be accounted for only by invoking Doppler boosting, which we discuss in the next section.

The observed radio intensity of many compact sources varies due to refractive scintillation (see Chapter 7), but there are for some sources intrinsic variations typically over timescales of some months (Ojha *et al.* 2004). Even making allowance for a compression of timescale due to the relativistic outward velocity of the jet, this short timescale indicates that the source must be small, typically no more than a lightyear across. A luminosity typical of a whole galaxy is concentrated in an object not much larger than our solar system.

13.10 Superluminal motion

The intensity variations of the core are accompanied by changes in angular structure on a similar timescale at the milliarcsecond scale observed by VLBI. The implications turn out to be dramatic. In observing the quasar 3C 279 with a single-baseline VLBI pair, Cotton *et al.* (1979) noticed that the visibility went nearly to zero at a particular baseline orientation, a pattern consistent with the Fourier transform of a double source. Some months later, the same deep null was observed, but for a different baseline length; the difference was consistent with the sources moving apart. Given the redshift and hence the distance of 3C 279, the angular motion implied a linear velocity at the source that exceeded the speed of light. Cohen *et al.* (1977) quickly verified that 3C 273 and others exhibited the same phenomenon. One could imagine that the apparent motions were the result of incomplete u, v coverage, but the effect is now well-established by multi-element VLBI observations. Figure 13.16 shows the classic example of 3C 273, which was mapped over several years by Pearson *et al.* (1981). In this case, the apparent velocity was $6c$ for a Hubble constant of $50 \, \text{km s}^{-1} \, \text{Mpc}^{-1}$; typical apparent velocities of VLBI (i.e. milliarcsecond-scale) jets fall in the range $(5–10)c$.

The superluminal-jet phenomenon, however, has an explanation that is consistent with conventional physics (Blandford *et al.* 1977). Doppler boosting, in addition to increasing the brightness of a source, can also create the illusion that the transverse velocity of a jet apparently exceeds the speed of light. We have already seen that the unified model of quasars and AGNs requires that the line of sight towards the nucleus of a radio-loud quasar must not be off the axis of the accretion disc by too large an angle, if the small proportion of radio-loud quasars is to be understood.

The geometry of superluminal velocities has already been presented in Chapter 11, in relation to radio sources within the Milky Way. Referring to Figure 11.22, we repeat the result for a source moving with velocity $v = \beta c$ along a line at angle θ to the line of sight. The apparent velocity is c times a 'superluminal' factor F, where

$$F = \beta \sin \theta (1 - \beta \cos \theta)^{-1}. \tag{13.2}$$

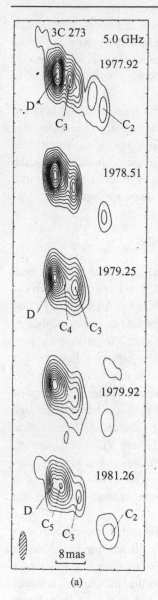

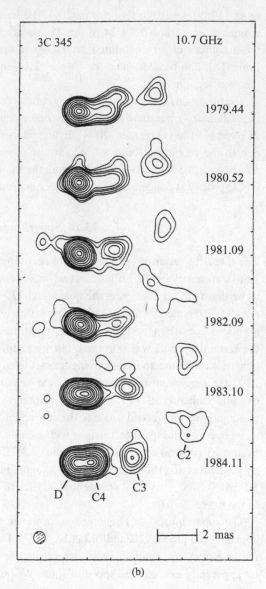

(a)   (b)

Figure 13.16. Superluminal motion in the cores of (a) 3C 273 (VLA at 5 GHz) (Pearson *et al.* 1981) and (b) 3C 345 (VLA at 10.7 GHz) (Biretta *et al.* 1986).

The maximum effect on the apparent velocity is at $\sin\theta = \gamma^{-1}$, where γ is the relativistic factor $\gamma = (1 - \beta^2)^{-1/2}$. For the typical observed values $F = 5$–10, the minimum required value of γ is about 7, and the largest effect is seen for line-of-sight angles $\theta = 5°$–20°. The effect is still important at $\theta = 45°$, where $\gamma = 7$ gives a superluminal velocity of $3c$.

The corresponding Doppler boost of intensity depends on the spectral index, but is approximately γ^3 when the velocity effect is maximum, and up to $8\gamma^3$ when $\theta = 0$. This increase by a factor of up to 1000 or more gives rise to a powerful selection factor, in which the large majority of observed bright quasars are those whose axes are nearly aligned with the line of sight.

Doppler boosting and superluminal velocities are not, of course, expected in jets pointing away from the observer; it is not surprising, therefore, that jets are seen leaving the core only on one side. It is, however, remarkable that the core and the emerging blobs have approximately the same luminosity. It seems possible that the core itself is the base of the jet, and is itself Doppler boosted by a similar amount. Some caution is necessary in this interpretation, and indeed in that of the emerging blobs of emission; it is not clear whether these are discrete moving material concentrations or shock fronts moving along a smoother stream.

13.11 The radio jets and lobes

The radio luminosity of powerful radio galaxies is dominated by the extensive double-lobe structure, seen for example in Cyg A (Figure 13.1). This structure extends to over 100 kpc from the active nucleus, deriving energy from the very narrow elongated jets. The jets are continuous from the core out to over 100 kpc, expanding only slowly from diameters of the order of 1 pc to 1 kpc before their energy is dissipated into the lobes. High-resolution observations by Krichbaum *et al.* (1998) show jets in Cyg A extending from 0.1 mas to 400 mas, with clear evidence of discrete concentrations leaving the nucleus with velocities of at least $0.4c$. The detailed 5-GHz map of the jet in 3C 390.3 (Figure 13.17) shows that the radio brightness varies very little over this astonishing distance. The energy of the radiating particles is evidently maintained by a flow of magnetic field energy and material down the jet; the small bright patches along the jet represent turbulent interactions with the surrounding interstellar and intergalactic medium, culminating in the major interactions at the 'hotspots' in the radio lobe. A high degree of polarization of the radio emission, reaching 50% in some cases, indicates that a well-organized magnetic field is involved in the stabilization of the jet; the field is directed along the jet for most of its length, and becomes transverse in the hotspots where the jet comes to a halt. Jets are generally straight; where large wiggles or bends are seen, this can be attributed to a projection effect in jets directed fairly close to the line of sight.

A remarkable double-jet structure is seen in 3C 75 (Figure 13.18), where jets emerge from two central nuclei (presumably two black holes in the same galaxy). Jets on both sides are swept in the same direction (towards the north-west), as in NGC1265 (Figure 13.5). The two separate jets on the south-west side are similarly swept and bent together, showing an interaction with some large-scale structure in the galaxy. The energy injected into the ISM by jets from active nuclei such as these may have significant effects, which are difficult to discern since the outer parts of the older jets are diffuse and their radio emissions

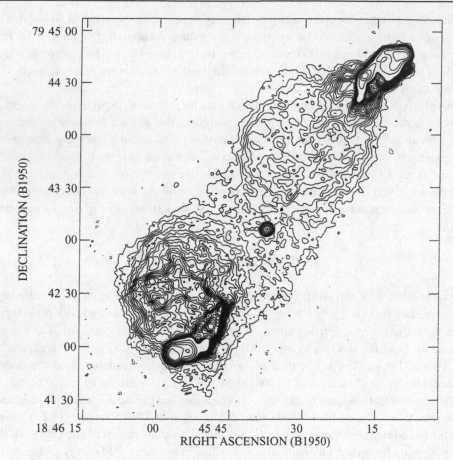

Figure 13.17. The radio galaxy 3C 390.3 mapped at 1450 MHz (Leahy and Perley 1995).

are expected to be measurable only at comparatively low frequencies. The development of large low-frequency arrays such as LOFAR and MWA (Chapter 17) may reveal the extent of this influence.

The *hotspots*, where the jets break up, are at the outer boundary of a cavity through which the jet bores its narrow tunnel. This cavity contains hot plasma flowing back from the hotspots; it still has enough energy to radiate synchrotron radiation, although it has fewer high-energy electrons and consequently a steeper radio spectrum. The cavity is therefore seen best at lower radio frequencies: Figure 13.17 shows a VLA map of 3C 390.3 at 1450 MHz (Leahy and Perley 1995), in which the jet is almost invisible but the cavities and the lobes are prominent. The radiation from the jets, hotspots and radio lobes is synchrotron, as shown by the high brightness temperature, the spectrum and the polarization. The total energy in the lobes can be found from the reasonable assumption of equipartition between particle energy and magnetic-field energy; it reaches over 10^{60} erg in the most luminous

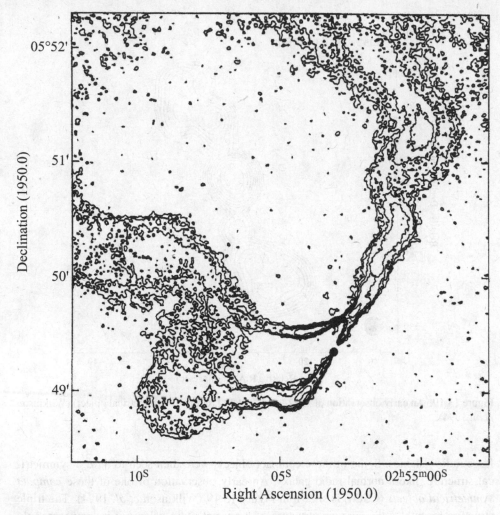

Figure 13.18. The radio galaxy 3C 75, showing two sources of radio jets within one galaxy (Owen *et al.* 1985).

radio galaxies. Typically, the radio emission is from electrons with relativistic energies $\gamma \sim 1000$, accelerated in a magnetic field of about 10 microgauss.

13.12 The kiloparsec-scale radio sources

As surveys of radio sources extended to higher frequencies and greater sensitivities, and as VLBI techniques provided improved resolving power, it became apparent that an increasing proportion of the fainter sources were physically very much smaller than the familiar

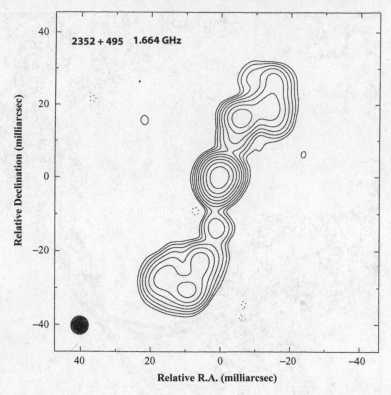

Figure 13.19. An early observation of 2352+495 – a CSO (compact symmetrical object) (Wilkinson *et al.* 1994).

large radio galaxies. Some of these compact objects were then seen to have a symmetrical structure, like a normal radio galaxy. An early observation of one of these *compact symmetrical objects* (CSOs) is shown in Figure 13.19 (Wilkinson *et al.* 1994). The triple structure resembles the core–jet structure of a normal radio galaxy; it is, however, only about 150 pc across, about 10^3 times smaller than a typical FR II radio galaxy. It has a flat radio spectrum, as might be expected from a compact source.

Many such sources are now known, with various sizes, shapes and spectra; these are usually designated according to their radio spectra as *compact-steep-spectrum* (CSS) or *gigahertz-peaked-spectrum* (GPS) objects. The GPS objects are less than 1 kpc in size, as shown in VLBI observations (Stanghellini *et al.* 1997); they are at the core of the active galactic nucleus. The CSS objects are also designated *medium-sized symmetrical objects* (MSOs); they are larger, up to 15 kpc, but contained within the host galaxy. In addition, some objects are one-sided, and are designated *core–jet* (CJ) sources. A comprehensive review by O'Dea (1998) contains many references to the observational data.

Apart from size, there appears to be a whole population matching the variety of the normal radio galaxies, but with sizes indicating that they are confined closely to the centres of otherwise normal galaxies. Are these an independent phenomenon, or are they to be regarded as the small-scale end of a continuous distribution of sizes? If the latter, are they the young progenitors which would evolve and expand into the larger normal radio galaxies?

An unequivocal answer came from the detection of expansion in several CSOs. A global VLBI array was used by Oswianik *et al.* (1999) to construct the image of the radio source 2352+495 shown in Figure 13.20. The structure is now seen to be similar to that of the much larger FR II radio galaxies. There are two well-defined hotspots, whose distance apart has increased from 49.1 to 49.4 milliarcsec during observations repeated over 10 years (Figure 13.20(b)). Dividing the scale by the expansion rate of 21 microarcsec per year gives an age of only 1900 ± 250 yr. The separation velocity is $0.3c$ (assuming $h = 0.6$). Similar ages and velocities have been found for several other CSOs.

There is a surprisingly large number of these young compact radio sources; Polatidis *et al.* (1999) found that about 8% of radio sources in a flux-limited survey at 5 GHz were CSOs. A simple evolutionary scenario in which CSOs evolve rapidly via MSOs into large radio galaxies (which might now be designated LSOs) suggests that a much smaller fraction should be in the short-lived CSO phase. The difference suggests that the expansion of some CSOs may be halted within the host galaxy, creating a class of frustrated radio galaxies.

13.13 Repeating and quiescent quasars

Active galactic nuclei, whether in quasars or Seyfert galaxies, are all powered by accreting black holes. Massive black holes are, however, known to exist in a quiescent state at the centres of many normal galaxies, our own Milky Way included. The circumstances in which the centre of a galaxy becomes a radio galaxy or quasar are as yet not understood. It is accretion, not the black hole itself, that supplies the energy for a quasar, and accretion may be the result of a large-scale merger with another galaxy. This may occur at any stage in the cosmological evolution of galaxies, but it may be significant that quasar activity is most frequently seen at redshifts of about $1.5 < z < 3$ (Wall *et al.* 2005). The duration of quasar activity is also unknown, but here there is a clue in several radio galaxies that exhibit *episodic* activity. Two examples are shown in Figure 13.21.

The radio image at 1.4 GHz of the giant radio galaxy J0116-473, shown in Figure 13.21(a), has a central unresolved quasar-like source and two superposed double-lobed sources, each of which has the appearance of a standard double radio galaxy. The larger double is more diffuse, with bright edges, and is not fed with any obvious jet, as for example in Cyg A (Figure 13.1). The smaller double appears to be connected with the central object, and is aligned on an axis that would be the axis of a jet, now non-existent, which might have powered the larger double. The radio galaxy B1545-321 (Figure 13.21(b)) also shows two double structures with the same alignment, but with no bright central source.

The obvious interpretation of these double–double radio galaxies is that the activity of the central source is varying dramatically within a timescale similar to the lifetime of the lobe structure, estimated at about 10^8 yr (Komissarov and Gubanov 1994; Safouris *et al.* 2008).

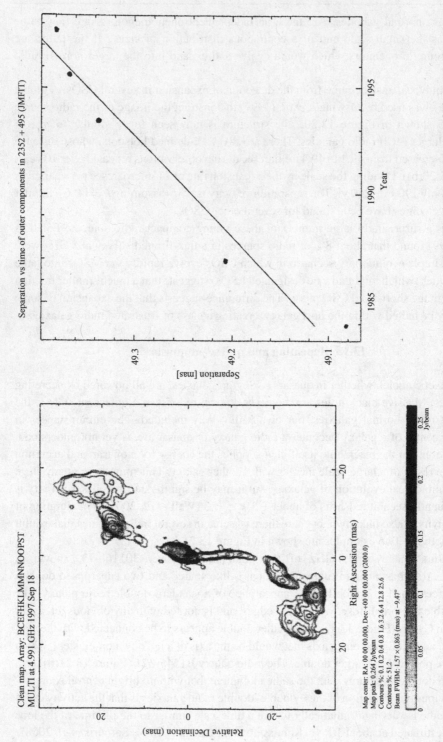

Figure 13.20. (a) A global VLBI image of 2352+495, a compact symmetrical object with z = 0.238. The three components of Figure 13.19 are now resolved into the core-and-double-jet structure closely resembling an FR II radio galaxy. In a series of similar observations over 10 years, the angular separation of the hotspots was found to increase linearly as shown in (b) (Oswianik *et al.* 1999, arXiv:astro-ph/9907120).

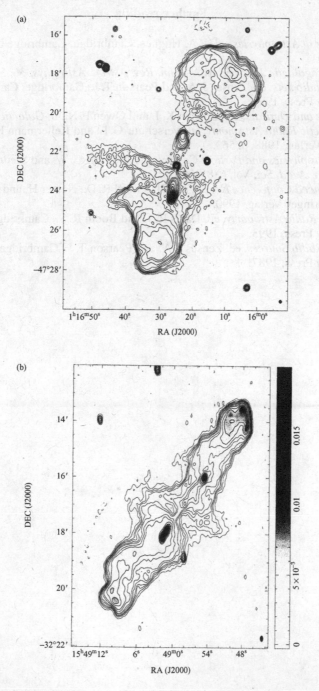

Figure 13.21. Two double–double radio galaxies imaged at 1.4 GHz, providing evidence for episodic activity. (a) J0116-473 and (b) B1545-321 (Saripalli *et al.* 2002, 2003, arXiv:astro-ph/0209570, 0303157).

Further reading

Beams and Jets in Astrophysics, ed. P. A. Hughes, Cambridge, Cambridge University Press, 1991.

Modelling Extragalactic Jets, Ferrari A., *Ann. Rev. Astron. Astrophys.* **36**, 539–598, 1998.

Parsec-scale Radio Jets, ed. Zensus J. A. and Pearson T. J., Cambridge, Cambridge University Press, 1990.

Radio Galaxies and Quasars, Kellermann K. I. and Owen F. N., in *Galactic and Extragalactic Radio Astronomy*, ed. Verschuur G. L. and Kellermann K. I., Berlin, Springer-Verlag, 1988, p. 563.

Relativistic Astrophysics and Particle Cosmology, Akerloff C. W. and Srednick M. A., *Annals N.Y. Acad. Sci.* Vol. 688, 1993.

Saas Fee Advanced Conference No. 20, ed. Blandford R. D., Netzer H. and Woltger L., Berlin, Springer-Verlag, 1990.

Sub-arcsecond Radio Astronomy, ed. Davis R. J. and Booth R. S., Cambridge, Cambridge University Press, 1993.

Superluminal Radio Sources, ed. Zensus J. A. and Pearson T. J., Cambridge, Cambridge University Press, 1987.

14

Cosmology fundamentals

For astronomers prior to 1925, the stars of the Milky Way seemed to form a Universe of immense size, but cosmology was transformed from philosophy to science in that year, when Edwin Hubble announced that he had identified Cepheid variables in the Andromeda nebula, M31. Through the known period–luminosity relation of Cepheids, he proved that M31, contrary to majority opinion, was a giant system of stars, comparable to the Milky Way. The stars of the Milky Way represented only a tiny fraction of a Universe populated by other galaxies fully as rich as our own. The climax of the Hubble revolution was reached in 1929 when, through measuring the Doppler shifts of the external galaxies, he demonstrated that the entire system of galaxies was expanding. The more distant the galaxy, the higher its radial velocity. This fitted nicely with cosmological models derived from solutions of Einstein's general theory of relativity (GR), whose cosmological interpretation of GR had puzzled theorists for over a decade, but it was now applied to Hubble's discovery and provided the intellectual framework within which a consistent cosmology could be developed. The 'standard cosmological model' of a Universe of galaxies, expanding from an initial state of high density, became the principal focus of theoretical cosmology for 20 years.

The second revolution in cosmology took place 35 years after the discovery of the expanding Universe, not from conventional optical observations, but from radio astronomy. Penzias and Wilson (1965), working at the Holmdel Research Station of the Bell Telephone Laboratories (the field station where radio astronomy had been born through the work of Karl Jansky), discovered that, at frequencies well above the domain of the Galactic synchrotron background, the radio sky was glowing with isotropic microwave radiation at a brightness temperature of approximately 3 K. The announcement of their discovery was accompanied by a theoretical treatment by Dicke *et al.* (1965) that explained the isotropic microwave radiation as the remnant of the thermal radiation from the 'big bang', the pejorative term invented by Fred Hoyle, who was advocating his rival 'steady-state' cosmology. The consequences of that discovery form the topic of Chapter 15. In this chapter we set out the geometrical framework of relativistic cosmology.

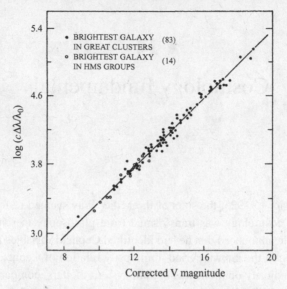

Figure 14.1. A Hubble diagram for galaxies with redshift up to $z = 0.5$, based on the brightest galaxy in clusters (Sandage and Hardy 1973).

14.1 Cosmology transformed

Hubble's law (more correctly the Hubble relation) states that the redshift, $z = \delta\lambda/\lambda$, of external galaxies is approximately proportional to their distance d. Uncertainties are imposed by the random velocities of individual galaxies; our nearest large neighbour, the Andromeda Galaxy M31, is coming towards us at a velocity of $300\,\mathrm{km\,s^{-1}}$ (at its distance of $600\,\mathrm{kpc}$, the collision may occur two billion years hence, unless its transverse velocity is large). As one extends the range of observations well beyond $z = 0.003$ (recession velocity $1000\,\mathrm{km\,s^{-1}}$, the value for the Virgo cluster, the closest large cluster of galaxies), the effects of random velocities become increasingly minor, and the Hubble relation is increasingly better determined:

$$cz = H_0 d. \tag{14.1}$$

The proportionality constant, H_0, is called the Hubble constant. Within the vicinity of our own cluster of galaxies, it is indeed a local constant to within the accuracy of measurement, allowing for random velocities among galaxies. This is not so on a cosmological scale, so the designation H_0 with the subscript will denote the local value, the Hubble 'constant'. More generally, with the subscript omitted, $H(t)$ designates the Hubble parameter, the measure of the expansion rate at a particular epoch t. Dimensionally, H_0 is an inverse time (the Hubble time T_H), but its units are traditionally given as $\mathrm{km\,s^{-1}\,Mpc^{-1}}$, and cz is, formally, a recession velocity in $\mathrm{km\,s^{-1}}$ (recognizing that this must not be taken literally for large redshifts, for which relativistic corrections must be applied). The linearity holds to redshifts of the order of 0.1 (Figure 14.1). Although the redshift data seemed to show that linearity

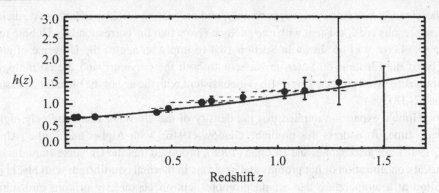

Figure 14.2. Deviation from Hubble's law for distant galaxies based on observations of Type Ia supernovae (Riess *et al.* 2007, arXiv:astro-ph/0611572). The plot shows $h = H/100$ as a function of z.

persisted at least to $z = 0.3$, this has turned out not to be the case. When supernovae of Type Ia were calibrated as distance indicators for galaxies at great distances a deviation from linearity was detected by two separate groups of observers (Riess *et al.* 1998; Perlmutter *et al.* 1997). Observations now extend beyond $z = 1$, and the deviation from linearity is shown by Riess *et al.* (2007) as a progressive change in slope, giving an evolution in the value of H (Figure 14.2). The implications for cosmology were fundamental: it appeared that a component of Einstein's equations, known as the cosmological constant, was not zero, as he had assumed, but in fact contributes the dominant contribution to the energy density of the Universe. The consequences were made clear by the WMAP mission, discussed in Chapter 15.

Determination of the Hubble constant requires measuring the average rate of expansion of the Universe to a redshift of 0.1, and conventional optical observations, based on Cepheid variables, cannot carry that far. The surveys of Type 1a supernovae carried the ladder further, beyond the linear 'constant' regime, but all methods are subject to systematic measurement errors and model assumptions. For these reasons, it has long been customary to write $H_0 = 100h_{100}$ km s^{-1} Mpc^{-1}, with the dimensionless scaling factor h_{100} appearing in both experimental and theoretical results. The subscript 100 is usually omitted, with the reference value 100 km s^{-1} Mpc^{-1} understood ($T_H = 10h^{-1}$ Gyr).

At present, the best value from optical observations, derived from combining data from the Hubble key project and the SN Ia results, is

$$H_0^{\text{opt}} = 73 \pm 3 \text{ km s}^{-1} \text{Mpc}^{-1} (h = 0.73). \tag{14.2}$$

The WMAP results presented in Chapter 15 lead to a pair of results, derived from the WMAP analysis of Dunkley *et al.* (2008), as discussed in Section 15.6:

$$H_0^{\text{WMAP}} = 70.1 \pm 1.3 \text{ km s}^{-1} \text{Mpc}^{-1},$$
$$H_0^{\text{WMAP}} = 71.9 \pm 2.6 \text{ km s}^{-1} \text{Mpc}^{-1}.$$

These results were derived independently, and are also model-dependent; nevertheless, all three results are consistent with one another. (Note that the corresponding Hubble time, close to 14 Gyr, will be shown in Section 14.4 to imply an age of the Universe of about 9.3 Gyr if the Einstein–de Sitter universe, with both the curvature and the cosmological constant zero, were correct. This result is inconsistent with the age of the oldest star clusters, about 12–13 Gyr.)

The Hubble expansion implied that the density of the Universe was extremely high at an early time. To address this problem, Gamow (1946), with Alpher and Bethe (Alpher *et al.* 1948), see also Alpher and Herman (1968), proposed that the Universe started with a primeval concentration of hot protons and neutrons in thermal equilibrium with blackbody radiation at a temperature that would promote neutron capture by protons and initiate nucleosynthesis. This contrasted with the steady-state universe proposed by Hoyle (1948), and by Bondi and Gold (1948), in which there would be no big bang.

The discovery by Penzias and Wilson of the *cosmic microwave background* (CMB), settled the matter; their brightness temperature of 3 K was remarkably close to the 5 K estimated by the Gamow group in their studies of the remnant radiation from the cosmic fireball, although their prediction had been forgotten. (Other evidence had also been forgotten; years earlier, McKellar (1941) had suggested that an interstellar absorption line of CN came from an unexpected excited vibrational state that might be explained by the presence of a universal heat source at 2.3 K.)

The CMB was remarkably uniform, apparently originating in an 'alabaster curtain' at redshift $z \approx 1000$ that masked observation of radiation from the earlier Universe. In succeeding years, the observations of Penzias and Wilson were extended to higher frequencies, showing that the radiation followed the Planck curve into the millimetre wavelengths. This work led to the launch of the COBE satellite (Section 14.2.2), which established that the spectrum was remarkably close to that from a 2.73-K black body. Ground-based radio searches for spatial structure in the CMB had also begun immediately, but it required the DMR radiometer in the COBE satellite to measure the effect, at the formidable level of 30 μK, 10^{-5} of the main cosmic background. These results led to the launching of a dedicated satellite, the Wilkinson Microwave Anisotropy Probe (WMAP) (Chapter 15) to examine the irregularities in greater detail. The results from that mission have had a profound effect on cosmology.

During the years when the experimental studies of the CMB were being conducted, there had been theoretical efforts to resolve several physical problems that plague a big-bang beginning for the Universe. This led to the proposal by Guth (1981), which was extended by Linde, Albrecht and Steinhart (Linde 1984), of the inflationary-universe scenario, joining the grand unified theory of fundamental physics with relativistic cosmology. The inflationary scenario demanded a universe with zero curvature, and the WMAP data led to that conclusion, together with the unexpected result that the largest contribution to the energy density of the Universe is in the form of dark energy, presumably the energy density of the vacuum. The process by which the reality of the CMB was established and linked to cosmology is examined next, in Section 14.2. Section 14.3 then presents the necessary background of relativistic cosmology to put the observations into context.

14.2 Observing the CMB

14.2.1 The CMB temperature

The discovery of the CMB was not the result of a carefully planned scientific campaign to test the big-bang hypothesis. Instead, it illustrates how exploratory science is done; 'good luck favours the prepared mind'. Penzias and Wilson joined a communications group at Bell Laboratories who had just developed the lowest-noise radio receiver ever built. It had been built as part of a satellite-communication project, designed to operate in a band centred at 2.4 GHz. The low-noise amplifier was a ruby maser and the total noise temperature of the system, after measuring the contributions of all components, including the Earth's atmosphere, was 18.9 ± 3.0 K. The measured sky temperature was 22.2 K, only slightly above the instrumental and atmospheric contributions.

When Penzias and Wilson joined the staff of the Holmdel laboratory, they recognized that the fully calibrated 2.4-GHz receiving system was ideally suited to investigating the Galactic microwave background in a new spectral region. As Arno Penzias, before their work was completed, remarked (to BFB, private communication) 'No one has measured the sky brightness at microwave frequencies, and we have the best equipment in the world to do it.'. The expected brightness of the sky at 2.4 GHz, including Galactic synchrotron radiation, was close to zero. The telescope was a 'sugar-scoop' horn, rotatable about its longitudinal axis. As emphasized in Chapter 4, the far sidelobes had to be as small as possible, to minimize the noise contribution from the ground, and the atmospheric noise contribution had to be determined by observing at a variety of zenith angles. To verify both contributions, the radiation pattern had been calculated in detail and measured on an antenna range. The ohmic losses in the waveguide system between the antenna's waveguide flange and the maser amplifier were measured, and the physical temperatures of these components were monitored, to determine the contribution to the system temperature. The reference temperature was determined by disconnecting the antenna and connecting the receiver to a cold load in a Dewar flask of liquid helium (4.2 K). The difference between the observed system temperature and the measured contributions of all components of the system, including the atmosphere, gave a cosmic background brightness of 3.5 ± 1.0 K. Once the result was known and had been verified by Dicke's group, the big-bang cosmology became the standard model.

14.2.2 COBE: the cosmic microwave background

The sky background is dominated at metre wavelengths by the integrated radiation from quasars and radio galaxies, and at infrared wavelengths both by extragalactic sources and by faint structure in the Milky Way. Happily, the microwave region between these contains the peak of the cosmic background radiation, whose discovery by Penzias and Wilson (1965) ranks equally in importance with Hubble's discovery of the recession of the nebulae (see Partridge (1995) for an account of the discovery and a more general review). The spectrum can be traced to wavelengths as long as 50 cm, but with decreasing accuracy as it becomes lost behind the Galactic synchrotron radiation (Sironi *et al.* 1990). It became clear that a space mission could measure the entire CMB spectrum with far greater accuracy than was possible from the ground, and in 1975 the COBE mission was selected by NASA.

The COBE satellite was furnished with three instruments: the far-infrared absolute spectrophotometer (FIRAS), the differential microwave radiometer (DMR) and the diffuse infrared background experiment (DIRBE). The satellite was launched into a circular orbit with an altitude of 900 km and an inclination of 99°; this orbit is 'Sun-synchronous', precessing one complete cycle per year in order to keep the same orientation with respect to the Sun. The instruments always point away from the Earth, and approximately perpendicularly to the Sun, thus ensuring a reasonably constant thermal environment, protected by their shielding and orientation from the powerful radiation from the Sun and the Earth. A general description of the satellite and its instruments has been given by Boggess *et al.* (1992).

The task of FIRAS was to measure the spectrum of the background radiation by comparison with radiation from a cold blackbody source whose temperature could be adjusted to give near equality. The instrument was a Fourier-transform spectrometer cooled by liquid helium and covering the spectral range from 1 cm to 0.5 mm. The detector was a bolometer and, strictly speaking, the instrument might be regarded more as a far-infrared than as a radio-astronomy device, but its Fourier-transform character and the close connection to other radio measurements can be cited to dismiss the point. The liquid helium for the FIRAS instrument lasted for nine months, establishing the isotropy of the background radiation, and measuring the background brightness temperature. The power spectral density is shown in Figure 14.3, where the experimental points have such small uncertainties that they seem to lie exactly on the theoretical blackbody curve. The comparison of data with the ideal Planck spectrum shows the deviation of the data from a 2.728-K blackbody curve as a function of wavenumber. The uncertainty of 0.010 K and the upper limits to the non-blackbody parameters (discussed in the following section) are derived from this fit. From the initial COBE data (and even from the early work of Penzias and Wilson) it was clear that the radiation was closely isotropic. Within a year, the first results of the DMR experiment, which had been designed explicitly to study the anisotropies, provided measurements of the solar motion through the background radiation of the Universe with great accuracy, and detected the feeble spatial structure that had been sought ever since the discovery of the CMB. This is such an important subject for cosmology that it will be discussed at length in Chapter 15.

The three main results from COBE were as follows.

1. The specific intensity and spectrum of the radiation fit closely to a black body at a temperature of 2.728 ± 0.002 K over the wavelength range 1 cm to 0.5 mm. The radiation is remarkably isotropic over a wide range of angular scale, and is strong evidence for a 'big-bang' cosmology.

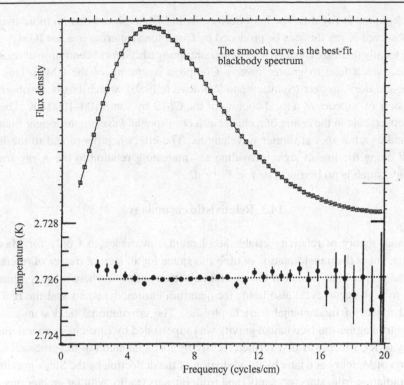

Figure 14.3. The spectrum of the CMB from spectrometer observations made by the COBE satellite (Mather *et al.* 1994).

2. There is a dipole term, of amplitude 3.343 ± 0.016 mK, in the direction $11^{h}15.6^{m}$, $-7.5°$, with an error circle of $0.5°$; this is interpreted as due to the motion of the Sun at a velocity of $364 \, \text{km s}^{-1}$ with respect to the coordinate frame defined by the CMB.
3. The CMB exhibits spatial temperature structure. On a scale of $7°$ (the beamwidth of the DMR), the measured ΔT has an r.m.s. value of $30.5 \pm 5.5 \, \mu\text{K}$ ($\Delta T/T \approx 10^{-5}$). The observed spectral index of the spatial spectrum is approximately $n = 1$, known as the Harrison–Zel'dovich spectrum. It is scale-free and consistent with the inflation scenario; see Chapter 15 for a detailed discussion.

The FIRAS result, with its close fit of the data to a blackbody spectrum, leads to a rigorous upper limit on possible distortions of the CMB spectrum by physical processes that might influence the radiation during its long journey to the observer. Hot electrons in the intergalactic medium can change the photon energy by Compton scattering; electrons with an energy kT that exceeds the CMB photon energy $h\nu$ will scatter the CMB photons to higher frequency, and the effect would show up at the high-frequency tail of the blackbody spectrum. The effect is characterized by the so-called Compton y-parameter (Mather *et al*. 1994). The observed upper limit to this distortion places severe limits on the temperature and density of the intergalactic medium (IGM). A hot ionized IGM might also be observed as a background X-ray flux; the limit derived from FIRAS shows that no more than 10^{-4} of the observed X-ray flux can be produced by Compton scattering in a hot IGM.

More locally there are clusters of galaxies containing a higher concentration of relativistic electrons, which lead to greater inverse Compton scattering of the CMB. This is the *Sunyaev–Zel'dovich effect* (Sunyaev and Zel'dovich 1980), which has been observed in rich clusters of galaxies as a local cooling of the CMB by some 100–1000 μK. The effect will be observable in the centre of a cluster as a cool spot at wavelengths longer than about 1 mm, and as a hot spot at shorter wavelengths. The effect is proportional to the integral $\int n_e T \, dl$ along the line of sight, providing an interesting relation to the X-ray free–free emissivity which is proportional to $\int n_e^2 T^{-1/2} \, dl$.

14.3 Relativistic cosmology

The special theory of relativity established Lorentz invariance, not only for Maxwell's equations, but as the transformation of time and space for all inertial frames of reference. It is also the proper transformation for quantum electrodynamics, where it is experimentally verified to high accuracy; it also holds for quantum chromodynamics and the rest of the standard model of fundamental particle physics. The gravitational field stands separate from electromagnetism. Newtonian gravity was superseded by Einstein's general theory of relativity (GR), which was first verified locally by its explanation of the precession of the perihelion of Mercury and later by its prediction of the deflection by the Sun's gravitational field of radiation from stars (optically) and from quasars (radio, with far greater precision). Gravity is a long-range force, and if it applies to the solar system it should apply to the Universe as a whole. On the cosmic scale, once Hubble had demonstrated the expanding Universe of galaxies, GR was applied immediately to the cosmological problems raised by his observations.

The new concepts that are introduced in relativistic cosmology are still rooted in the fundamental structure of physics, even though they are cast in geometrical terms for four-dimensional space-time. It is primarily a theory of gravitational interaction on a large scale. Locally, the laws of physics are Lorentz invariant under a coordinate transformation between inertial frames of reference. The local validity of the GR formulation works so well in the solar system that it should hold for the Universe at large, since gravity is a long-range force. Unifying gravity with the other force fields of physics is still an unmet goal, but this is not a barrier, since all work well locally. Relativistic cosmology is a complex study, but the main conclusions are reasonable and approachable. Only the basic results are given here; for understanding GR more deeply, consult texts such as Peebles (1993), Hobson, Efstathiou and Lasenby (2006), or Peacock (1999). A compact summary is given in Chapters 5 and 6 of Longair (2008), which will largely be followed here. A variety of conventions will be encountered in various authorities; here we adopt the conventions of Longair.

Some care must be exercised in defining the fundamental physical ideas of GR. The terms *coordinates*, *distance* and *time* have more than one definition, qualified by a descriptive adjective that specifies the exact meaning. First, we define *cosmic time* (equivalently, *proper time*) and *proper distance*. For the four-dimensional space-time of our Universe we postulate that the spatial coordinates are marked by a dense set of notional fundamental observers. Each fundamental observer has a location with specified spatial coordinates, and each has a clock that is synchronized with all the other observers, defining cosmic time. This emphasizes the basic role of observers in cosmology, although in a formal sense a complete coordinate system, labelled by both time and spatial coordinates, has been defined. By convention, the spatial coordinates of each fundamental observer are fixed for all time.

For most of the following discussion, the Universe is assumed to be isotropic, with no preferred direction, possessing homogeneous, average local properties that smooth out the fine structure that we observe around us. Thus, the proper separation between a pair of neighbouring observers can be taken as a radial proper distance Δr that will retain the same value as time progresses. (This description in terms of *co-moving coordinates* is a familiar concept in fluid flow.) Despite the fixed nature of these co-moving coordinates, in the real Universe the physical separation, $\delta x(t)$, does vary with time, dramatically so as the Universe expands. Thus a physical distance depends upon a scale factor $a(t)$ that in the limit for a pair of infinitesimally close observers, separated by dr, specifies the physical distance dx:

$$dx(t) = a(t)dr, \tag{14.3}$$

where $a(t)$ is dimensionless, with $a(t_0) = 1$ at the present epoch t_0 (following Longair and Peebles; note that not all earlier authors use the same convention). The dimensions are carried by the proper coordinate r, and with this normalization of $a(t)$ the dimensions of the co-moving coordinate system are those of the present epoch. Note that the time coordinate is cosmic time; other definitions of time will follow.

The general case for a non-homogeneous, non-isotropic universe is described in geometrical terms, a wholly new departure by Einstein in formulating GR. The Universe is described by Riemann's metric g_{ij}, which specifies the distance ds^2 between neighbouring coordinate points (dx^i, dx^j) in four-dimensional space-time, where $dx^0 = c\,dt$. The geometry of four-dimensional space-time is described by Riemann's metric

$$ds^2 = g_{ij}\,dx^i\,dx^j, \tag{14.4}$$

where the convention of summing over repeated indices from 0 to 3 is used. As in special relativity, the line element $ds^2 = 0$ describes the trajectory of a light ray. In Euclidean space-time, which is familiar from the treatment of inertial frames of reference in special relativity, $ds^2 = c^2\,dt^2 - dx^2 - dy^2 - dz^2$. Spherical coordinates are more appropriate for basic cosmology, where the Universe, to a first approximation, is homogeneous and isotropic; this gives the Euclidean metric form in spherical coordinates (r, θ, ϕ):

$$ds^2 = c^2\,dt^2 - dr^2 - r^2(d\theta^2 + \sin^2\theta\,d\phi^2). \tag{14.5}$$

This can now be generalized beyond the zero-curvature Euclidean case for geometries that both have curvature and are time-variable. Only isotropic geometries are being considered at this level, so the form of the metric tensor g_{ij} has only one parameter, the radius of curvature $\mathcal{R}$ of the three-dimensional spatial surface embedded in four-dimensional space-time. The Euclidean line element given above must now be derived for the case of a curved surface. The radial element dr^2 stays the same, but the angular element, $r^2\,d\Phi^2 = r^2(d\theta^2 + \sin^2\theta\,d\phi^2)$ has to be modified. The form is easy to derive for the case of a two-dimensional spherical surface in 3-space, where r is measured along great circles on the sphere. For an angle Φ between a pair of great circles the metric becomes $ds^2 = dr^2 + \mathcal{R}^2\sin^2(r/\mathcal{R})d\Phi^2$; if the curvature is negative, the term $\sin^2(r/\mathcal{R})$ is replaced by $\sinh^2(r/\mathcal{R})$. See Longair for a more detailed treatment.

With these conventions, the explicit form of the metric for a time-varying, homogeneous, isotropic four-dimensional space-time of curvature $\mathcal{R}$ was given first by Robertson and then independently by Walker, thus being known as the Robertson–Walker metric, RW metric for short. It follows by extending the case of the 2-sphere in 3-space to the four-dimensional space-time of GR:

$$ds^2 = \begin{cases} c^2\,dt^2 - a(t)^2[dr^2 - \mathcal{R}^2\sin(r/\mathcal{R})(d\theta^2 + \sin^2\theta\,d\phi^2)] & \mathcal{R} > 0, \\ c^2\,dt^2 - a(t)^2[dr^2 - r^2(d\theta^2 + \sin^2\theta\,d\phi^2)] & \mathcal{R} = 0, \quad (14.6) \\ c^2\,dt^2 - a(t)^2[dr^2 - \mathcal{R}^2\sinh(r/\mathcal{R})(d\theta^2 + \sin^2\theta\,d\phi^2)] & \mathcal{R} < 0. \end{cases}$$

Just as r gives the radial coordinate of a fundamental observer at the present epoch, so $\mathcal{R}$ is likewise the radius of curvature at the present epoch.[1]

So far, the definitions have been geometrical, containing no physics except for the intro-duction of fundamental observers and cosmic time. Now, GR dynamics can be introduced, since the metric tensor, g^{ij}, is set by the distribution of matter and energy in the Universe. Einstein's tensor field equation relates g_{ij} to the distribution of mass and energy in the Universe, as expressed by the stress tensor. The result is a set of ten non-linear differential equations that are far too complicated to solve without simplifications, for they describe the distortion of space-time for any distribution of matter and energy. There are simple cases; the commonly observed phenomenon of gravitational lensing of a distant radio galaxy or quasar by a foreground galaxy or cluster is an example. Schwarzschild's solution for a single point mass is a famous example, showing the potential existence of black holes. Inside a radial distance depending on the mass (the Schwarzschild radius), $R_S = 2GM/c^2$, (gravitational constant G and mass M) there can be no communication with the Universe outside. When matter falls within the Schwarzschild radius, it never returns. (The more general Kerr solution, for a rotating black hole, has a second region, the ergosphere, from which a return is possible; it is generally believed that black holes have angular momentum and are of the Kerr type, but Schwarzschild black holes are a good enough approximation for most discussions.)

---

[1] One will encounter other forms of the RW metric in the literature; both Longair (2008) and Peacock (1999) give summaries of the most common forms.

In relativistic cosmology, there are three fundamental assumptions that are made.

- The principle of equivalence, namely that the laws of physics are the same everywhere, for both inertial and accelerated systems, and that inertial mass and gravitational mass are identical.
- The Weyl hypothesis, namely that the fundamental observers follow separate, non-intersecting trajectories, except for a common origin in the finite or infinite past, the point at which the clocks were synchronized to cosmic time.
- The cosmological principle, namely that there is no preferred location in the Universe, which presents the same aspect to all fundamental observers at a given epoch.

The third assumption clearly needs amplification, for on the local scale we live in inhomogeneous surroundings. There has to be a qualification that all cosmological quantities must be averaged over a sufficiently large volume.

The physical implications of the RW metric, and the subsequent derivation of $a(t)$ from Einstein's tensor field equation, must now be examined. As in special relativity, the trajectory of a light ray is described by the condition $ds^2 = 0$. Several conclusions follow immediately from the RW metric: the total mass/energy density in the Universe must scale like $1/a(t)^3$, if mass/energy is conserved, and, since a length scales like $a(t)$, a wavelength of electromagnetic radiation also scales like $a(t)$. Therefore, the radiation density in a given region will scale like $1/a(t)^4$. The blackbody temperature, therefore, scales like $1/a(t)$. If radiation is emitted with wavelength λ_e when the scale factor has the value $a(t_e)$, and the observer, at time t_o, measures the wavelength λ_o, then, considering λ as a small interval Δr, it follows that the wavelengths and frequencies are related to $a(t)$ by

$$\lambda_o/\lambda_e = \nu_e/\nu_o = a(t_o)/a(t_e). \tag{14.7}$$

The relation between the scale factor $a(t)$ and the redshift $z = (\lambda_o - \lambda_e)/\lambda_e$ follows. Since $a(t_o) = 1$ for the present epoch,

$$a(t_e) = 1/(1 + z). \tag{14.8}$$

Longair emphasizes the importance of relation (14.8): 'Redshift is a measure of the scale factor of the Universe when the radiation was emitted by the source.'.

The co-moving radial distance from an observer at time t_o to a source emitting an electromagnetic signal at time t_r is given by taking $ds^2 = 0$. By integrating a ray path in time along the radial direction from the source at the time of emission t_e, received by the observer at cosmic time t_o, the co-moving radial distance to the emitting source is obtained:

$$r = \int_{t_e}^{t_o} c \, dt/a(t). \tag{14.9}$$

This formal expression shows that the co-moving coordinates depend upon the form of $a(t)$; it will appear in the treatment of distance below.

The field equations simplify when the universe is homogeneous and isotropic. Since the coordinates are co-moving, it is only the scaling that matters, and the result is a differential equation for $a(t)$. Three quantities are given by the real Universe: the mean mass density

$\rho(t)$, the mean pressure $P(t)$ (which is actually the energy density) and a third term, the cosmological constant Λ, which was invented by Einstein to give what he thought was a stable universe, constant in time. Hubble's expanding universe disposed of the idea of an unchanging universe, and Einstein disavowed his invention, but it is a perfectly reasonable term to include in the field equations, and it became firmly embedded in the literature. The result is an equation for $a(t)$ that looks like a force equation (remember that $a(t_o) = 1$ for the present epoch):

$$\ddot{a}(t) = -(4\pi G/3)[\rho(t) + (3/c^2)P(t)]a(t) + (1/3)\Lambda a(t). \qquad (14.10)$$

When P and Λ are zero, this becomes Newton's law for the force on a small test mass at the surface of a sphere of density ρ. There are two added terms: the time-dependent pressure P, which can be the pressure of gas and/or radiation pressure, and the constant cosmological term Λ. This also represents a force, albeit an unfamiliar one, since the force increases linearly as the distance increases. The force equation (14.10) can be integrated to give the GR analogue of the Newtonian energy; this is sometimes called the Friedmann equation, after the theorist who wrote it in 1922, before Hubble's discovery of the expanding Universe:

$$(1/2)(\dot{a}(t))^2 - (4\pi G/3)\rho a(t)^2 - (1/6)\Lambda a(t)^2 = -c^2/\mathcal{R}. \qquad (14.11)$$

The curvature term on the right is the constant of integration, written in terms of the radius of curvature, $\mathcal{R}$, which can be positive, negative or 0, corresponding to three geometrical families of solution for positive, negative and zero curvature (curvature will be designated κ, the inverse of $\mathcal{R}$, which will be used frequently). The solution with $\kappa = 0$ and $\Lambda = 0$ is sometimes called the Euclidean solution, even though ray paths are not straight lines because $a(t)$ is changing with time. For the RW metric given in Equation (14.6), when the radius of curvature goes to infinity (alternatively, a universe of curvature $\kappa = 0$) we should use the term 'flat geometry'.

In the Newtonian formulation, Equation (14.11) exhibits the physical principles at work in the GR Universe. For a unit test mass at the surface of a sphere (the sphere must be small compared with $\mathcal{R}$, a condition that is easily met), the first term is clearly the kinetic energy of the test mass, while the second term is its gravitational potential energy (ρ contains both the mass and the energy density of the cosmic medium). The third term looks like the potential energy of a harmonic oscillator; since the cosmological Λ term represents a force that is linearly proportional to distance, as in a harmonic oscillator, that interpretation is correct. A positive Λ represents a repulsive force, so the corresponding potential is negative; the test mass accelerates increasingly as the separation increases. Current results, discussed in detail in Chapter 15, indicate that Λ is indeed positive and is a major influence on the expansion of the Universe.

The curvature term on the right, following the Newtonian interpretation, is the total energy of the test mass. For a flat universe, therefore, the total energy is zero, and the expansion must therefore be determined by the initial conditions, with the expansion gradually slowing to zero as the age of the universe tends to infinity. A negative curvature corresponds to a net positive energy for the expansion, and such a universe would continue to expand for ever. Conversely, a positive curvature represents a universe whose total energy is negative; in that case, that universe would expand to a maximum and then fall back to a singularity at the origin, the case popularly known as 'the big crunch'. The data from the satellite mission indicate persuasively that the curvature, $1/\mathcal{R}$, is close to zero, a result that is discussed in the following sections.

Advances in cosmology have generally come through observations, but the ideas of GR set the basic framework within which models of the Universe are discussed. A homogeneous, isotropic universe, characterized by the scale factor $a(t)$ in the RW metric, is only an approximation, for it is obvious that the Universe is inhomogeneous, out to a scale well beyond ten megaparsecs. Nevertheless, the assumption of homogeneity is a useful and successful approximation. The task is now to make the connection between the abstractions of GR and the realities of the observations.

14.4 Connecting GR cosmology with observations

Cosmology is the study of our Universe, and is necessarily based on observations, setting aside its occasional definition as the study of all possible universes (e.g. by Bondi (1960) and Harrison (2000)). General relativity (GR) provides cosmology with the theoretical framework for the action of gravity, supplemented by the basic physics through which we understand the phenomena. The three fundamental assumptions given above, namely the principle of equivalence, Weyl's hypothesis and the cosmological principle, are all subject to experimental and observational tests.

The principle of equivalence has been verified to high accuracy; the equivalence of inertial and gravitational mass by delicate experiments that now reach an accuracy of three parts in 10^{13}, while the uniformity of the laws of physics throughout the Universe has been demonstrated by observing spectra from high-redshift objects that have not yet been in contact with one another. The ratio of the calcium H and K lines to the spectral lines of hydrogen establishes the constancy of the fine-structure constant to better than 1 part in 10^4 out to redshift $z = 3$, and comparing the redshift measured from the 21-cm hyperfine radio line of hydrogen with the redshift measured from the optical hydrogen lines establishes the constancy of the proton magnetic moment with similar accuracy. Molecular lines of carbon monoxide, which depend upon the mass of the nuclei and the fine-structure constant, give a more complex check on nuclear forces, but the ratio of their frequencies to those of atomic lines also adds to the evidence that the laws of physics are constant throughout the accessible Universe; Will (2006) gives a review of the subject.

The Weyl postulate, that on the average the matter of the Universe is streaming from a singular point in the past, is demonstrated with reasonable certainty by the observed expansion of the Universe. A proviso has to be added that the origin seems unlikely to be a singularity; rather, the Universe started from a compact initial state, with the inflationary scenario being strongly favoured at the present time.

The cosmological principle, namely that the Universe presents the same aspect to all observers at a given epoch, has been well verified by observations both of extragalactic radio sources and of optical galaxies, which are spread randomly across the sky. This is only a partial verification, since the high-z objects are being observed at a much earlier epoch, but, for a given redshift range, in all directions we observe the same populations of galaxies, obeying the same laws, as evinced from their spectra. Presumably they will on the average evolve to the same state as our local neighbourhood when they reach our present epoch. Galaxies tend to occur in clusters, and there are larger-scale voids and superclusters, on a scale of tens of megaparsecs or more, which complicate the argument, but the CMB demonstrates that the initial state, from the time of decoupling, was highly uniform. The observed patterns of galaxy clustering were shown by Zel'dovich (1970) to be those expected as a result of the primitive medium condensing into galaxies.

The cosmological constant Λ was for many years regarded as an unwanted guest, with no real evidence for its existence. Adding a repulsive force to the simple inverse-square law of Newton seemed to many to be inelegant. Nevertheless, there were enough discrepancies between theory and observation to justify keeping it as a possibility. In the field equations, Λ is a constant quantity, and can be regarded as the energy density of the vacuum. Parenthetically, it should be noted that the theoretical prediction of its value, derived from the 'standard model' of particle theory, is wildly wrong, by a factor of order 10^{120}, surely the largest discrepancy between theory and experiment ever achieved! This emphasizes the fact that gravity still seems to be outside the framework of quantum field theory.

14.4.1 The Einstein–de Sitter universe

In order to gain insight into the physical behaviour of a model universe in GR, a brief look at the Einstein–de Sitter universe is instructive. This simple model, which assumes that the Universe has flat geometry ($1/\mathcal{R} = 0$), with zero cosmological constant ($\Lambda = 0$), was favoured for many years, and in much of the literature prior to the mid 1990s it was taken as a tacit assumption. For the Einstein–de Sitter universe, the energy relation for $a(t)$, Equation (14.11), becomes

$$(1/2)(\dot{a})^2 - (4\pi/3)G\rho a(t)^2 = 0. \tag{14.12}$$

This has immediate applicability to physical measurements. Firstly, this is a statement that the total energy of the system is zero; the first term is the kinetic energy of a unit test mass at the surface of an expanding sphere of mass/energy density ρ, and the second is its potential energy, which could equally well be written in the form $GM(R)/R$, where $M(R)$ is the mass contained within the expanding sphere of radius R (which scales as $a(t)$ with mass/energy). Secondly, on solving for $R(t)$, which we identify with the scaling factor $a(t)$, it follows that

$$a(t)/a(t_o) = (t/t_o)^{2/3}. \tag{14.13}$$

The Hubble time, T_H, would be the age of the Universe if it were expanding at a linear Hubble rate, but Equation (14.13) shows that this model universe decelerates with time, because of the retarding effect of gravity. It follows that the actual age of the Universe would be 2/3 of the Hubble time. Unfortunately, the present value of $h \approx 0.7$ leads to an age slightly less than 10 Gyr. This is in sharp disagreement with the ages of the oldest stars, which have been derived from stellar-evolution models. There are several possibilities: the systematic errors in calibrating the distance scale foil the measurements; the stellar-evolution models could be seriously wrong; or the Einstein–de Sitter cosmological model of a simple, $\Lambda = 0$, flat universe does not describe our Universe. The discussion in Chapter 15 of WMAP measurements of the spatial power spectrum of the CMB shows that the latter is likely to be the case, accounting at least for the largest part of the discrepancy.

A useful cosmological parameter, however, emerges from the energy equation (14.12). The Hubble parameter, in terms of the scaling factor $a(t)$, is $H(t) = \dot{a}/a$. Rearranging Equation (14.11), and taking the Hubble constant and the mean density in our neighbourhood as a reference, defines a *critical density* ρ_c:

$$\rho_c = 3H_0^2/(8\pi G). \tag{14.14}$$

The numerical value of the critical density in terms of the Hubble scaling factor h is

$$\rho_c = 1.88 \times 10^{-28}h^2 \, \mathrm{g\,cm}^{-3} \quad (2.78 \times 10^{11}h^2 M_\odot \, \mathrm{Mpc}^{-1}). \tag{14.15}$$

The critical density, therefore, amounts to approximately 10^{-4} hydrogen atoms cm^{-3}, or one Milky Way per cubic megaparsec. If the actual mean density were greater than this, the Universe would have to be closed, with positive curvature, and, after its present expansionary phase, the Universe would eventually collapse. On the other hand, if the mean density were less than critical, the universe would be open, with negative curvature, and the expansion would continue for ever.

The value of ρ_c presented a problem for the Einstein–de Sitter universe. Most of the visible mass in the Universe appears to be in the form of stars, and the density of luminous mass in the Universe appears to be more than an order of magnitude smaller than the critical density. Furthermore, models of nucleosynthesis in the big bang predict a present-day density not far from the observed density of luminous matter, certainly within the range $(0.02–0.05)\rho_c$, in agreement with the observed present-day stellar mass density. There had been dynamical evidence from galactic velocities in clusters, first put forth by Zwicky (1933), that there was matter in clusters of galaxies whose mass far exceeded the observed masses of the stars in the cluster. Little attention was paid to this evidence for *dark matter* until studies of rotation in spiral galaxies, optically by Rubin and Ford (1970) and using 21-cm measurements by Burns and Roberts (1971), had demonstrated that the galactic rotation curves did not follow the distribution of mass in stars. For many spirals, at a large distance from the centre, the rotation curves were nearly flat. The 21-cm radio observations reach farther out in galaxies than the optical methods can reach, supporting the conclusion that the spiral structure is embedded within a larger sphere, a dark-matter halo, with a mass that exceeds the stellar mass by a significant factor. Thus, the evidence for dark matter in galaxies and clusters of galaxies, combined with the implications of primeval nucleosynthesis, led to the realization that it is necessary to distinguish between baryonic matter, with which we are familiar, and dark matter, which is non-baryonic but whose nature is still not understood. Its contribution to the matter density is uncertain, but it will be seen that the CMB evidence leads to an estimate of approximately $0.25\rho_c$. This is still not enough matter to give a flat universe.

The cosmological scale factor, $a(t)$, is related to the Hubble constant by $H_0 = \dot{a}/a$, evaluated at the present epoch, t_0. For any cosmological model, therefore, one can expect, for a given $a(t)$, that higher derivatives of $a(t)$ might be detectable in the redshift–magnitude observations. The dimensionless second derivative, $q = -a\ddot{a}/\dot{a}^2$, has been used to characterize this. When this parameter is evaluated for the present epoch, it is designated q_0, and is called the *deceleration parameter*. For the Einstein–de Sitter universe, its value is $1/2$, but its local effect is small, and during the period prior to 1980 or so, as the Hubble relation was being extended to fainter, more distant galaxies, the results were scrutinized for a possible effect, but with no definitive results. In recent decades, the deceleration parameter has seldom been referred to, for it is more convenient to describe higher-order effects in terms of the critical density, ρ_c, introduced in the following section.

14.4.2 The density parameters

It is convenient, therefore, to describe the actual mean density (the total of mass and energy densities) of the Universe at the present time in terms of the critical density, defining a total mass/energy *density parameter* Ω:

$$\Omega = \rho/\rho_c. \tag{14.16}$$

If $\Omega = 1$, the Universe has flat geometry. The total density parameter Ω is a sum of components including baryonic density, Ω_b, dark matter, Ω_{dm}, radiation, Ω_r, and Ω_Λ, the contribution of the presumed vacuum energy attributable to Λ; there may, of course, be other components of which we know nothing at present. Thus, for the present epoch in our local neighbourhood, where $a(t_0) = 1$, we can write

$$\Omega = \Omega_b + \Omega_{dm} + \Omega_r + \Omega_\Lambda, \tag{14.17}$$

with additional terms to be added if the need arises. Since baryonic matter and dark matter scale in the same way, exerting equivalent gravitational force, they can be combined into a single term, $\Omega_m = \Omega_b + \Omega_{dm}$. Note that the corresponding mass/energy densities scale differently with $a(t)$: the matter density scales as a^{-3}, the radiation energy density scales as a^{-4}, while the cosmological term, the vacuum energy density, is independent of a. The discussion of the structure of the CMB, in Chapter 15, shows that the curvature of the Universe is close to zero; $\Omega = 1$ to within 1%. Thus Equation (14.17) must be approximately

$$\Omega_b + \Omega_{dm} + \Omega_r + \Omega_\Lambda = 1. \tag{14.18}$$

This is the condition that satisfies the inflationary-universe scenario.

The concept of distance is fundamental to interpreting the observations, and, since there is more than one way of defining 'distance' in GR, some care has to be taken. In the present discussion, this will be limited to the case of the flat universe, where the curvature $1/\mathcal{R} = \kappa = 0$. For non-zero curvature, should it become of interest in the future, the cosmological texts can be consulted. For reference, the RW metric for flat geometry, repeating Equation (14.5), is

$$ds^2 = c^2 \, dt^2 - a(t)^2[dr^2 + r^2(d\theta^2 + \sin^2\theta \, d\phi^2)]. \tag{14.19}$$

14.4.3 Distances and angular diameters

Physical measures of distance depend upon knowledge and calibration of *standard candles* or *standard measuring rods*. In the first case, there is a sequence of standard candles of increasing luminosity, starting with nearby clusters of stars of known age, measurable by direct parallax (the Hyades, for example), transferring the standard to Cepheids, which are rarer but also more luminous and with a well-calibrated period–luminosity relation, and eventually extending to the rare but highly luminous supernovae of Type Ia, which have now been calibrated against the Cepheids and are visible at redshifts of 3 or more. The distance measured through the luminosity of understood objects is defined as the *luminosity distance*, D_L. There are no well-calibrated standard measuring rods as yet, but an *angular size distance* D_A appropriate to phenomena such as gravitational lensing can be defined. This is emphatically so in the study of the power spectrum of the spatial structure of the CMB, which will be discussed in Chapter 15.

A standard rod of proper length d, transverse to the line of sight, will subtend an angle $\Delta\theta$ that is used to define the angular size distance, D_A:

$$\Delta\theta = d/D_A. \tag{14.20}$$

Inspection of the RW metric above, and use of Equation (14.19), gives $\Delta\theta = (1 + z)d/r$, so the angular size distance for a flat universe is

$$D_A = r/(1 + z). \tag{14.21}$$

The *luminosity distance* D_L is defined by the observed bolometric flux S_{bol} compared with the bolometric luminosity of the source, L_{bol}, such that $S_{bol} \equiv L_{bol}/(4\pi D_L^2)$. The derivation follows similar lines to that for D_A, with the result

$$D_L = (1 + z)r. \tag{14.22}$$

For the cases of non-zero curvature, the co-moving radial distance r is replaced by $\mathcal{R}\sin(r/\mathcal{R})$ or by $\mathcal{R}\sinh(r/\mathcal{R})$ for the positive- and negative-curvature cases.

There are other measures of distance; in particular, there is *affine distance*, which takes account of the possible non-uniform distribution of matter. D_A and D_L, as defined here, refer only to a universe that has a smooth distribution of matter; it appears that the dominance of vacuum energy, Ω_Λ, validates their use to a first approximation. For applications such as gravitational lensing this is not necessarily the case. Another caveat should also be heeded: prior to about 2005, analytic expressions that appear both in the journal literature and in the review literature sometimes assume that $\Lambda = 0$, without explicitly saying so. As in so much of the cosmological literature, due attention has to be paid to the assumptions and conventions.

The observed spatial power spectrum from WMAP observations, discussed in the following chapter, has clarified the question of the curvature of the Universe, κ, which is certainly close to zero, and the analysis has established that the cosmological constant Λ, rather than being zero, is the dominant contributor to the energy density of the Universe at the present epoch. Furthermore, if the dark matter of the Universe is cold (meaning that its peculiar velocities are low enough to keep the matter within galactic haloes), there is a better fit to the mass function of galaxies. The need to incorporate matter density, radiation density, dark energy and (possibly) a slight deviation from zero curvature leads one to rewrite Friedmann's energy equation for $a(t)$ in the form (using the various Ω terms):

$$\dot{a}^2 = \left(H_0^2/\Omega\right)[\Omega_{\rm r}/a^2 + (\Omega_{\rm b} + \Omega_{\rm db})/a + \Omega_\Lambda a^2 + \Omega_\kappa]. \tag{14.23}$$

Here we assume that the curvature κ is close to unity, that the dark matter is cold and that the cosmological constant is non-zero; this is the ΛCDM model for the Universe. There is no simple solution to this general Friedman equation, but the general course of variation of the scale factor $a(t)$ with time can be seen from inspection. For the earliest time, when $a(t)$ is very small, the radiation term will dominate, since it varies as $1/a^2$, and a simple integration shows that it varies as $t^{-1/2}$. As $a(t)$ increases, the matter term, varying as $1/a$, will dominate. The form of $a(t)$, in this approximation, is simply that of the Einstein–de Sitter universe, treated earlier, with $a(t)$ proportional to $t^{-3/2}$. From the time of decoupling until a redshift of $z = 3$ or so, this is a reasonably accurate representation of the cosmological expansion. The transition from the radiation-dominated Universe to the matter-dominated Universe occurs when the two terms are roughly equal, only shortly before the era of decoupling. Much later, the cosmological term, proportional to a^2, becomes the dominant term, having an influence in the epoch starting between $z = 3$ and $z = 2$; this is presumably the epoch in which we now find ourselves.

Since we assume that $\kappa \approx 1$, the final, curvature, term, will be small, but notice that in between the matter-dominated era and the Λ-dominated era it could have a strong effect, even if it appears to be small. This is the 'curvature problem', one of the motivations for inventing the inflationary-universe scenario.

When radiation and curvature can be neglected, and the evolution of $a(t)$ is influenced both by matter and by Λ, a solution exists. This solution was originally pressed by Lemaître as a way of explaining the discrepancy between the apparent age of the Universe and the age implied by the (old) Hubble constant. There is actually a triplet of solutions, in which the attractive gravitational force of matter is decelerating the expansion, trying to drive it to condense at a singularity, but opposed by the repulsive force of a positive Λ. If these just balance, the result is a static solution $a = $ constant (Einstein's first try at cosmology); this

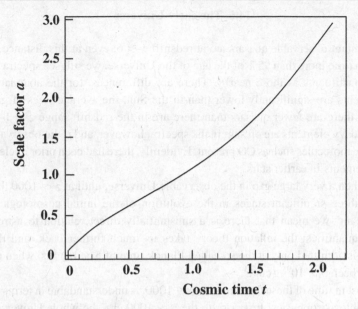

Figure 14.4. The evolution of the scale factor $a(t)$ (after Longair, *Galaxy Formation*, Figure 7.3).

solution is unstable against small perturbations, leading to either a collapsing solution, in which gravitational attraction wins and the universe shrinks to a singularity, or an expanding solution in which $a(t)$ tends to infinity with time. If the Λ-term is larger than the critical solutions, the attractive gravitational force of the matter induces an initial deceleration, but eventually the repulsive cosmological term wins, and, after an inflection at which the acceleration, $\ddot{a}$ goes to zero, there is perpetual acceleration thereafter. This was recognized by Lemaître as a way of stretching the expansion timescale. If the Λ-term is just above the critical value, the expansion nearly stops, and with a proper balancing this standstill can be lengthy.

One can also write a solution for an empty universe, but with non-zero Λ, (de Sitter recognized this solution at an early stage, causing Einstein great trouble). Inspection of the equation, with only the Λ-term on the right-hand side, shows that the solution will be a combination of exponentials; if $a(0) = 0$, the solution is

$$a(t) = A[\cosh(\Omega_\Lambda^{1/2} H_0 t - 1)]. \tag{14.24}$$

As the discussion of the WMAP analysis in Section 15.6 will show, the matter and cosmological Ωs are approximately $\Omega_{\rm m} = 0.3$ and $\Omega_\Lambda = 0.7$. The resulting variation of the scale factor with time is illustrated in Figure 14.4. The other solutions discussed above are currently ruled out, although, at the proper epoch at which they dominate, they still give physical guidance and useful approximate solutions for $a(t)$. The following chapter, Chapter 15, addresses the present problems of the CMB, both experimental and theoretical, on the basis of the GR considerations that have been discussed.

14.5 The early Universe

The most distant observable quasars are at redshift $z \approx 6$; even at this distance, where we look back in time more than 95% of the age of the Universe, we still see spectra with much the same constituents as those nearby. There are differences, for the abundances of the heavy elements are significantly lower than in the Sun, the average sizes of galaxies are smaller and there are fewer quasars than there are in the redshift range $z = 1$–2. Atomic lines from heavy elements are present in the spectra, however, and radio observations show that there are molecules such as CO present. Evidently, there had been prior nucleosynthesis of heavy elements in earlier stars.

There is then a very large gap in the observable Universe, until at $z \sim 1000$ there occurs the last of three significant stages in the evolution of the initial cosmological fireball. By 'significant' we mean that there is a substantially direct relation to astronomically observable quantities; the inflation theory takes us much further back, and the limit to speculation is admitted only at the so-called Planck time of 5×10^{-44} s, when the density would have been 5×10^{93} g cm^{-3}.

The closest in time of these three eras, at $z \sim 1000$, is understandable in terms of present-day observable astrophysics. Just prior to the $z \sim 1000$ era, the whole Universe had been a plasma, with the radiation energy density and the matter energy density approximately equal, but with the energy density falling like a^{-4} in contrast to the matter energy density falling like a^{-3}; the balance was changing. Superficially, the plasma resembled the H II regions discussed in Chapter 11, but with the important difference that there was no external energy source to sustain it. As the plasma expanded, it cooled to a temperature in the range 3000–4000 K, and, when the electrons and nuclei combined, the probability that they would be reionized diminished. Finally, the point was reached at which no reionization took place, defining the era of radiation decoupling, the so-called recombination region. This is the surface we observe as the CMB.

The next earlier stage to be considered is at a temperature of 10^8–10^9 K, which occurred 100–300 s after the big bang. This was the era of nucleogenesis, before which the energies of free protons and neutrons were too great for nuclear reactions to occur. The nuclear reactions which determined the cosmic abundances of the light elements occurred at this time, when the thermal energy was about 0.1 MeV and the density of the primordial Universe was somewhat less than that of water. The main reaction was the combination of a neutron and a proton to form deuterium,

$$n + p \rightarrow D. \tag{14.25}$$

Subsequent collisions of deuterium with protons and neutrons led via tritium to the synthesis of ^{4}He. These reactions, whose rates are well known from laboratory measurements, also led to traces of ^{3}He and ^{7}Li. The proportions of these elements are frozen at equilibrium values that agree remarkably well with those found today in the least evolved parts of the Universe. Almost all is hydrogen (77% by weight) and helium (23% by weight), but the small proportions (compared with hydrogen) of D (10^{-5}), ^{3}He (10^{-5}) and ^{7}Li (10^{-9}) are all in good agreement with the theory (Audouze 1987).

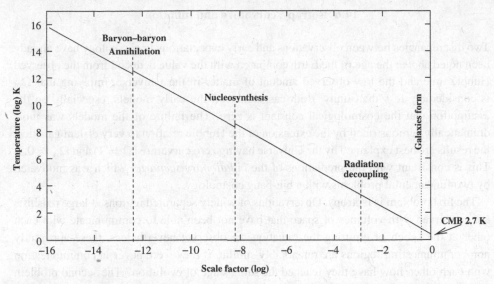

Figure 14.5. The major events in the early Universe and its thermal history (adapted from Longair, *Galaxy Formation*, p. 278).

The initial fireball of the big bang contained radiation and elementary particles effectively in thermal equilibrium, but at a rapidly decreasing energy density. As the energy decreased, elementary particles and their antiparticles annihilated, leaving gamma-rays and a small excess of normal baryons; this small excess of normal matter remains today, when the ratio of baryons to photons is about 3×10^{-10}. The interactions between the elementary particles changed fundamentally at the third of our three transition eras, at the very much earlier time of 10^{-35} s after the origin, when the electroweak interaction separated from the strong interaction. This is the more speculative era of the grand unified theories (GUTs), when the physical conditions were beyond any of our present-day experience; before this era we can consider only the overall geometry of the Universe, which seems to be determined by inflation.

These early events are related to the scale factor and the evolution of temperature in Figure 14.5. The temperature of the Universe diminishes as the scale factor $a(t)$ increases; if there are no additional energy inputs, the temperature varies as a^{-1}. Working back from the present day, the principal events consist of galaxy and star formation, out to an ill-determined $a \sim 0.1$ (i.e. redshift 10, with an uncertainty of perhaps ± 4). The Universe became transparent to photons at the era of recombination, $z \sim 1000$, with unknown events proceeding in the 'dark ages' between redshifts 10 and 1000. The earlier epochs are inferred from fundamental physics, relying on the 'standard model', which is based on quantum electrodynamics and chromodynamics. This framework is relatively secure, since it agrees with observations of phenomena ranging from quarks to the weak-interaction W-bosons; quantum electrodynamics, in particular, has been verified with extraordinary accuracy. The small upward 'glitches' in the temperature are caused by the energy input from electron–positron annihilation at $R \simeq 10^{-9}$ and baryon–antibaryon annihilation at $R \simeq 10^{-13}$.

14.6 Isotropy, curvature and inflation

Two discrepancies between observations and early expectations in cosmology have already been noted above: the age of the Earth compared with the value deduced from the observed Hubble time and the low observed amount of matter in the Universe, implying that Ω_m is considerably less than unity. Both cast doubt on the early models, especially on the assumption that the cosmological constant is zero. The failure of the models was most dramatically demonstrated by the extension of the Hubble relation to very distant galaxies; the results are best explained by the Universe having zero curvature ($\Omega = 1$) and $\Omega_\Lambda = 0.7$. This is consistent with the predictions of the *inflationary scenario*, which was motivated by two fundamental problems with a big-bang cosmology.

The first problem is isotropy. Observations of widely separated regions at large redshifts are concerned with volumes of space that have not been able to communicate with each other at the epoch of emitting the radiation we observe; nevertheless, these apparently non-communicating regions are remarkably similar. If they were never in communication with each other, how have they reached the same stage of evolution? The second problem is implied by the Friedmann equation (14.11). The integration constant, the total energy of the system, is proportional to the geometrical curvature $1/\mathcal{R}$, which evolves with time. Observations show that there is very little curvature in our neighbourhood of space-time; it follows that the curvature near the time of the big bang must have been very small indeed, i.e. the net energy per unit mass must have been very small indeed.

In order to address these fundamental problems, Guth (1981) proposed that grand unification theory (GUT) could provide a theory that was both consistent with modern theoretical ideas and also able to satisfy the uniformity and flatness problems. The idea of *inflation* quickly became the favoured cosmological scenario: a useful summary of its development is given by Peacock (1999).

The central idea of the inflationary scenario is that the matter described by GUT derived from a universe that had expanded exponentially through about 70 e-folding times. The origin is described as a fluctuation in the vacuum, giving rise to a 'false vacuum'; the subsequent expansion occurs in 10^{-35} s, followed by a decay into matter when the electroweak interaction separated from the strong interaction. At that time the effective temperature of the Universe was approximately 10^{14} GeV. The matter was a initially a 'quark soup'; after a further expansion, at an age of about 1 s, the quarks produced protons, neutrons, neutrinos and electrons. As the primeval gas expanded and cooled, element formation could proceed.

Although inflation theory addresses an era in cosmology that is inaccessible to direct observation, it does provide remarkably simple answers to the two problems outlined above: the communication problem is solved by the identical origin of all parts of the observable universe, and the overwhelming expansion at inflation ensures that the curvature is essentially zero. The convergence of theory and observation on the conclusion that the curvature is zero does, however, raise a different problem: the observable matter in our cosmological neighbourhood is insufficient to give $\Omega = 1$, the density that would give zero curvature. At most, baryonic matter can constitute only about 2% of the required mass;

furthermore, the total mass in baryonic form would be too large to account for the formation of galaxies and stars in the proportions and on the timescales that are required. Part of the discrepancy can be accounted for by dark matter, whose existence had already been inferred from the rotation curves of galaxies and from the dynamics of clusters of galaxies, but even the most optimistic estimates of the density of dark matter are insufficient to account for the zero-curvature universe (Coles and Ellis 1994), which is a requirement to support the inflationary scenario.

Further reading

An Introduction to Cosmology, Narlikar J. V., Cambridge, Cambridge University Press, 3rd edn, 2000.

Cosmological Physics, Peacock J. A., Cambridge, Cambridge University Press, 1999.

Cosmology, Bondi H., Cambridge, Cambridge University Press, 2nd edn, 1960.

Cosmology, Harrison E., Cambridge, Cambridge University Press, 2nd edn, 2000.

Galaxy Formation, Longair M. S., Berlin, Springer-Verlag, 2008.

General Relativity, Hobson M. P., Efstathiou G. P. and Lasenby A. N., Cambridge, Cambridge University Press, 2006.

Observational Cosmology, IAU Symposium 124, ed. Hewitt A., Burbidge G. and Fang Li Zhi, Dordrecht, Reidel, 1987.

Observational Tests of World Models, Sandage A., *Ann. Rev. Astron. Astrophys.*, **26**, 561, 1988.

Principles of Physical Cosmology, Peebles P. J. E., Princeton, NJ, Princeton University Press, 1993.

The Cosmological Distance Ladder, Rowan-Robinson M., New York, Freeman, 1984.

15

The angular structure of the CMB

The 'alabaster curtain' of the cosmic microwave background, the CMB, defines the limit of the directly observable Universe. Beyond that limit, the mean free path of photons is so short that the blackbody radiation field and the primitive plasma of ions and electrons are locked together in tight coupling. As expansion of the plasma proceeds, its density and temperature drop, and, when the Universe reaches an age that appears to be about 370 000 years, the era of decoupling is reached, the ions and electrons combine into atoms, and the photons escape through the nearly transparent neutral medium, to be observed as the CMB. Indirect paths of study of the Universe beyond the era of decoupling are few; one of these is the first era of nucleosynthesis, discussed in Chapter 14. There is the hope that neutrinos or more exotic particles might give another observation channel, reaching beyond the era of first nucleosythesis.

The Universe that we observe around us is clearly inhomogeneous, and that inhomogeneity had to grow from far smaller, unstable density fluctuations, urged on by the relentless force of gravity. After the discovery of the CMB, the search for spatial structure began almost immediately; it was generally recognized, by experimenters and theorists alike, that the testimony of those first perturbations must be present at some level in the apparently featureless façade of the CMB, motivating an intensive and difficult search. The search was successful, and this chapter describes how that search was carried out, and gives some theoretical background with which to understand the significance of those observations.

15.1 The coordinate frame of the Universe

The search for the irregularities of the CMB proceeded entirely experimentally. There was no credible theoretical guidance; in retrospect, all the theoretical estimates tended to be orders of magnitude too high. The only way to proceed was to build the lowest-noise, most sensitive state-of-the-art instruments possible, covering as wide a range of angular resolution as possible. As ground-based observations extended the data base following the initial discovery of the CMB, it became clear that thermal radiation from the atmosphere presented a major limitation. This led to the approval of the COBE mission, which included the FIRAS spectrometer described in Chapter 14 and a differential microwave radiometer (DMR) designed to measure the spatial structure of the CMB. The DMR experiment faced two

340

observational obstacles. Both the Galactic synchrotron radiation, at low frequencies, and interstellar dust, at high frequencies, presented a bright foreground with angular structure that would have to be corrected for, as had been done for the measurements of average sky brightness. This was done by selecting a suitable suite of observing frequencies, coupled with a substantial software effort to analyse the data. The second obstacle would be the distortion caused by our motion with respect to the Universe as a whole, which is a more tractable problem.

The Planck blackbody curve is necessarily Lorentz invariant. Thus, an observer travelling at velocity $\mathbf{v}_0$ through a radiation field at temperature T_{CMB} will observe, in direction $\hat{\mathbf{r}}$, a Doppler-shifted Planck distribution with an apparent temperature shift ΔT_{CMB} (the non-relativistic formula is sufficiently accurate):

$$\Delta T_{CMB} = T_{CMB}(\mathbf{v}_0 \cdot \hat{\mathbf{r}})/c. \tag{15.1}$$

This Doppler-induced temperature shift caused by the solar motion through the CMB (duly corrected for the Earth's rotation and orbital motion about the Sun) was first measured by a differential microwave spectrometer carried at high altitude in a U2 research aircraft. The experiment was successful; the solar motion was detected at a level of one part in 10^3 of the CMB background (Smoot *et al.* 1977). This also set a secure upper limit on the background fluctuations of the CMB. The DMR subsequently improved the measurement, determining that the Doppler shift due to the Sun's velocity had an amplitude of 3.343 ± 0.006 mK, corresponding to a velocity of 363 km s^{-1} towards Galactic coordinates $\ell, \textit{b} = 264°, 48°$ (Fixsen *et al.* 1996). This is a composite of the solar motion with respect to the local standard of rest, the motion about the centre of the Milky Way, the motion with respect to the local group and (probably) the motion of the local group with respect to the local supercluster. This represents our motion within the coordinate frame of the Universe. The motion of the local group within the local supercluster is not known, but our net motion with respect to the local Universe is about 600 km s^{-1}, substantially greater than our motion with respect to the frame defined by the Universe.

15.2 COBE and WMAP: the Wilkinson Microwave Anisotropy Mission

The DMR experiment (Figure 15.1), the precursor of WMAP, carried two exponential horns, each with a half-power beamwidth of 7.5°, comparing the sky brightness in two patches 60° apart; the rotation of the satellite scanned the sky. The choice of observing frequencies was governed by the considerations illustrated in Figure 15.2, which shows the perturbing Galactic contributions as a function of frequency (a major Earth-atmosphere absorbing band at 55 MHz from oxygen is not shown). The three radiometer frequencies were 31.5 MHz (dual linear polarization), to provide information on the synchrotron and H II contributions, 90 GHz (dual linear polarization), to monitor the dust, and 53 GHz (dual circular polarization), near the optimum for CMB observations. The comparison radiometer, performing a relative measurement of the difference between two beams on the sky, subtracts out the average CMB brightness and as a result is less subject to systematic errors. Careful

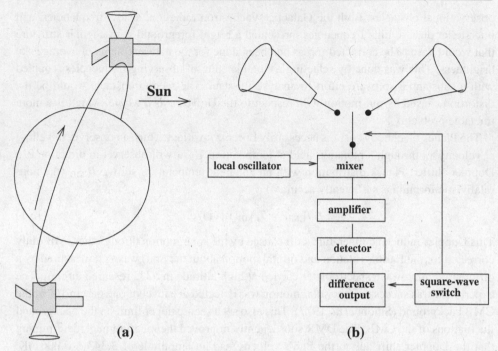

(a) **(b)**

Figure 15.1. (a) The COBE satellite in orbit. The telescope axes were maintained perpendicular to the direction of the Sun, so that a full scan of the sky was achieved in one year. (b) The DMR detected the difference between signals from two horn antennas, using a Dicke-switch system (Chapter 3).

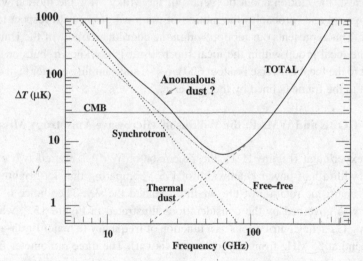

Figure 15.2. The spectrum of the Galactic components contributing to the structure of the microwave background at angular scales greater than $\sim 7°$, showing the contributions from synchrotron and free–free radiation, and thermal radiation from dust. A possible so-called 'anomalous dust' component may be responsible for the increase in total radiation between 20 and 90 GHz, as shown. The angular structure of the CMB at $\Delta T/T = 10^{-5}$ is shown for comparison. After Banday *et al.* (2003).

calibration of the radiometer was still required, since the horns, the comparison switch and the electronic systems are not precisely symmetrical. The surrounding spacecraft and the thermal radiation in the horn sidelobes from the Earth and the Moon also have to be evaluated as the spacecraft rotates. The sky is scanned completely in one year.

The DMR results from four years in orbit are given by Bennett *et al.* (1996). After reduction of the data to calibrated sky maps, with the Galactic foreground and the solar motion removed to first order, the resulting brightness-temperature maps, $T_{B,C}(\ell, \&)$ represented in Galactic coordinates, $(\ell, \&)$, consist of three principal terms:

$$T_{B,C}(\ell, \&) = T_F(\ell, \&) + T_{S,R}(\ell, \&) + T_{G,R} + T_N. \tag{15.2}$$

The background spatial structure, or 'fluctuation', T_F, is the signal of interest; while any residual solar-motion term, $T_{S,R}$, has the form of a dipole and causes no difficulty. Correction for the residual Galactic foreground term, $T_{G,R}$, depends on the accuracy of the foreground model. The final term, T_N, is the random noise contributed by the receivers and is always present; it also includes imbalance effects within the comparison radiometer, which must be regarded as a source of system noise. Together with the foreground term $T_{G,R}$, the noise term is the major limiting factor in the measurement accuracy, and may be sufficiently large to dominate the fluctuations in the map.

The final COBE map was smoothed from the antenna resolution of 7.5° to 10°. Even so, there was not a sufficiently high signal-to-noise ratio in the map of $T(\ell, \&)$ to show the CMB fluctuations directly; the speckle pattern was a mixture of system noise and the cosmic signal, the two being of comparable amplitude. There are approximately 1600 independent pixels in the map, however, and the cosmic signal can be extracted by an averaging procedure related to the process for extracting a spectral line such as the 21-cm H I line from the noisy time series that a radio spectrometer produces (see Section 3.8). In a spectrometer the spectrum is derived by autocorrelating the signal amplitude, and Fourier transforming the result to obtain the power spectrum, S_ν (Appendix 1). Similarly, the fluctuation map can be regarded as a two-dimensional spatial spectrum, with a spectral density in wave number $k = 2\pi/\lambda$, where λ is the angular measure (or the linear measure if the distance is known).

It is natural to represent $T_B(\ell, \&)$ as an expansion in spherical harmonics Y_{lm}, the proper orthonormal base functions to use. These are the associated Legendre polynomials (normalized); note that, since $\&$ is measured from the celestial equator, whereas the customary polar angle θ in the usual spherical polar coordinates is a co-latitude, $\cos\theta$ has become $\sin\&$ in the expressions below:

$$Y_{lm}(\theta, \phi) = \left[\frac{(2l+1)(l-|m|)!}{4\pi(l+|m|)!}\right]^{1/2} P_{lm}(\sin\&) e^{im\phi} \times \begin{cases} (-1)^m & m \geq 0, \\ 1 & m < 0. \end{cases} \tag{15.3}$$

The expansion in spherical harmonics takes the form

$$T(\theta, \phi)/T = (T(\theta, \phi) - T_0)/T_0 = \sum_{l=0}^{l_{max}} \sum_{m=-l}^{l} a_{lm} Y_{lm}(\theta, \phi), \tag{15.4}$$

where the coefficients are derived from the data by calculating

$$a_{lm} = \int (T(\theta, \phi)/T) Y^*_{lm}(\theta, \phi) d\Omega. \tag{15.5}$$

The integral is taken over the entire celestial sphere; Y^*_{lm} is the complex conjugate of Y_{lm}. As a result, the spatial map of the brightness fluctuations, $T_F(\ell, ɓ)$, is represented as a finite sum of spherical harmonics, with l_{max} being determined by the angular resolution, which is usually of the order of the half-width, half-power effective beamwidth. Although a two-dimensional autocorrelation could be performed on $T_F(\ell, ɓ)$, followed by a spherical harmonic transform, it is unnecessary since one already has decomposed the brightness distribution into the spherical harmonic (l, m) space that is dual to the angular $(\ell, ɓ)$ space (analogous to the dual time and frequency spaces of signal analysis). The coefficients a_{lm} represent the wave amplitude for a given spherical harmonic (lm). Multiplying by the complex conjugate a^*_{lm} then gives the power in a given mode:

$$C_{lm} = a_{lm} a^*_{lm}. \tag{15.6}$$

The final step in the analysis relies upon the isotropy of the Universe. All directions are equivalent, so each of the $2l + 1$ coefficients for a given multipole l can be summed to include waves travelling in different directions on the sphere. This set of l-coefficients constitutes the *angular power spectrum, C_l*:

$$C_l = 1/(2l + 1) \sum_m a_{lm} a^*_{lm}. \tag{15.7}$$

Despite removing the Galactic foreground through the multi-frequency model, the Galactic residual term in Equation (15.2) can still cause trouble, so in reducing the DMR results the low-latitude portion of the map was masked, and the spherical-harmonic expansion used only the results above latitude $ɓ = |20°|$. The terms in the spherical-harmonic expansion were derived from the unmasked portion of the map.

Since it turns out that the resulting spectrum follows a power law over several orders of magnitude, it is convenient to multiply the coefficients by $l(l + 1)/(2\pi)$ to keep the spectrum within a convenient range when plotting the results, as in Figure 15.4 below. The ordinates are then in units of temperature squared.

The DMR results demonstrated that the angular power spectrum of the CMB was clearly detectable at spatial frequencies corresponding to $l = 2$ to 20, with an r.m.s. of 30 μK (Fixsen *et al.* 1994). The dipole term is not included in C_l, since it is a local term expressing the motion of the Sun with respect to the Universe. The quadrupole term for the CMB gave a value of 13 μK in the spherical-harmonic expansion, representing the lowest fluctuation mode of the Universe: there can be no cosmic dipole term, since the centre of mass is fixed (assuming that the concept has any meaning). The value of the quadrupole term C_2 has to be regarded as more uncertain than the stated error; correction for the foreground radiation from the Milky Way, which is clearly anisotropic, involves subtracting a large quadrupole term from the raw data, and this could lead to a serious error in the derived value of C_2. The same could well be true for the next few higher terms in l, but to a lesser extent.

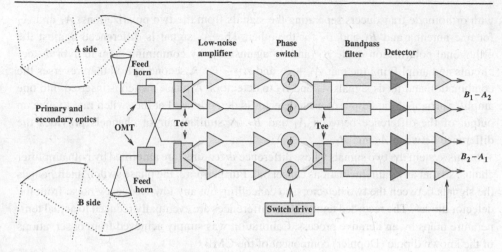

Figure 15.3. The WMAP radiometers.

The DMR experiment had successfully determined the magnitude of the CMB fluctuations, and had determined that the angular power spectrum was approximately a power law, with a spectral index of 1.1 ± 0.5, consistent with the conjecture of Harrison (1970) and independently Zel'dovich (1972) that a power-law spectral index of 1 (the Harrison–Zel'dovich spectrum) might be observed. The promising results encouraged the immediate planning for a successor mission, specifically aimed at determining the background CMB fluctuations over a larger range of spatial frequencies and with higher accuracy; no need was seen for a successor to the FIRAS experiment, for its measurement of the frequency spectrum had been definitive. The Microwave Anisotropy Probe (MAP) mission was approved by NASA in 1995 and launched in 2001. Major compilations of integrations over one, three and five years were published by Hinshaw *et al.* (2003), Spergel *et al.* (2007) and Hinshaw *et al.* (2008). As the first data were arriving, the death of David Wilkinson, an investigator on the mission and one of the pioneers of CMB studies, led to the mission being renamed the Wilkinson Microwave Anisotropy Probe (WMAP).

The WMAP mission built on the legacy of the DMR experiment. Thermal noise from the Earth, and to a lesser extent from the Moon, leaking into the sidelobes of the DMR antennas had contributed to the measurement uncertainty; to minimize sidelobe contamination, the WMAP satellite was placed at the L2 solar Lagrange point, 1.5 million kilometres from the Earth, where it was maintained by small thrusters. A pair of $1.6\,\text{m} \times 1.4\,\text{m}$ telescopes were mounted back-to-back to compare the values of the sky brightness temperature in opposite directions. These were furnished with comparison radiometers (schematically illustrated in Figure 15.3) for each of five different frequency bands, chosen to enable the construction of an accurate model of the Milky Way foreground contributions. Dual-polarization feeds were included to measure the polarization properties of the CMB fluctuations.

The receivers used a development of the hybrid switching technique briefly described in Section 3.6. Figure 15.3 shows the feed horns for the pair of oppositely facing antennas, with orthomode transducers separating the signals from the two polarizations, A_1 and A_2 for one antenna and B_1 and B_2 for the other. The A_1 signal is differenced against the orthogonal polarization mode B_2 (and A_2 against B_1) by combining them in hybrid 'tee' circuits and amplifying the pair $A_1 + B_2$ and $A_1 - B_2$. A second hybrid then reverses the combination and feeds signals A_1 and B_2 to detectors. A phase switch is inserted into one amplifer arm, and the output of each detector is demodulated at the switch rate to give an output of the difference between A_1 and B_2. A similar pair of channels provides the difference between A_2 and B_1.

In this system the two signals whose difference is required are amplified by both amplifier chains, so that any gain fluctuations cancel out. Furthermore, the phase switch interchanges the signals between the two detectors, so cancelling out any low-frequency noise from the detector diodes. The recorded temperature differences are eventually reduced to actual temperature maps by an iterative process. Calibration was simply achieved from observations of the known dipole (Doppler) component of the CMB.

The raw sky maps from the five observing bands, as expected, showed considerable differences due to the different frequency dependences of the three main components of the foreground contamination, namely synchrotron, free–free and thermal dust. The angular resolution was determined by the telescope gain patterns, ranging from $0°.88$ at K-band to $0°.22$ at W-band. The lowest-frequency band, at 22 GHz (K-band), shows a strong Galactic component from synchrotron radiation. The 33-GHz and 41-GHz maps (K- and Q-band) contain a decreasing Galactic synchrotron contribution, coupled with an increasing component from the free–free bremsstrahlung of H II regions. The fourth map, at 61 GHz, is close to the frequency at which all foreground contributions are at a minimum, while the fifth map, at 94 GHz (W-band), includes the foreground from thermal radiation by interstellar Galactic dust. From these maps, a model for the separate components of the Galactic foreground emission was constructed, and then this was subtracted from the data. The resulting 'clean' maps of the CMB were then combined, by suitable weighting, to give the 'concordance' map shown in Figure 15.4. The speckle pattern of the background is necessarily mixed with noise from the receiving system and from imperfections in the foreground-construction process, but the investigators estimate that there is a probability of about 0.99 that any given fluctuation is real. From this, the procedure outlined schematically above was followed with meticulous attention to detail and examination of possible sources of error, including the use of a more sophisticated mask to exclude Galactic residuals and assessment of the contribution of discrete extragalactic radio sources, to derive the angular power spectrum.

The resulting power spectrum is shown in Figure 15.5, where C_l has been multiplied by the customary scaling factor $l(l + 1)/(2\pi)$. The prominent peak at $l = 200$, followed by the lesser peaks at higher l, was not unexpected. Observations of the CMB had already been made with ground-based and balloon-borne instruments, which had the advantage of larger size and therefore better angular resolution than the spacecraft, but had the disadvantage of contending with the added thermal radiation of the atmosphere. The first results, from balloon observations, were mixed, although the expected 'acoustic peak' (discussed in the following sections) shows clearly in data from the balloon flights of Page (1997).

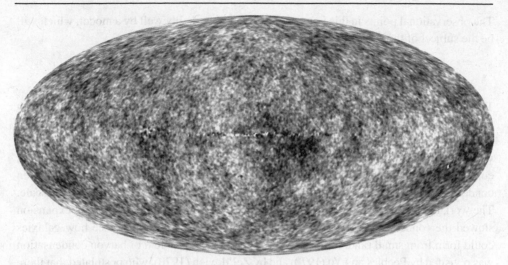

Figure 15.4. The 'concordance' map derived from the five WMAP observing frequency bands (Hinshaw *et al.* 2008, arXiv 0803.0732).

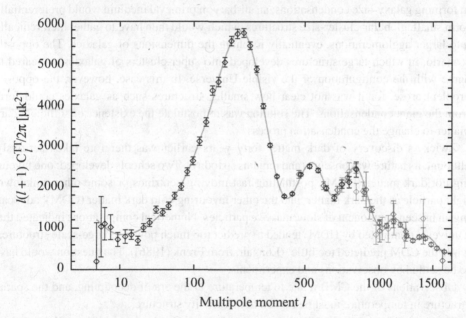

Figure 15.5. The angular spectrum of the CMB. Points up to $l \approx 600$ are derived from the WMAP concordance map. Points at higher l values were obtained from ground-based and balloon-borne observations (Nolta *et al.* 2008).

Much improved data, included in Figure 15.5, are now available from the ground-based experiments ACBAR (Kuo *et al.* 2004), CBI (Readhead *et al.* 2004) and VSA (Dickinson *et al.* 2004), plus the balloon project BOOMERANG (Jones *et al.* 2006). These extended the range of reliably determined l-values to $l = 1600$, leading to an improved best-fit spectrum.

The observational points in this figure can be fitted remarkably well by a model, which will be the subject of the rest of this chapter.

15.3 Baryons and cold dark matter

Theoretical speculation had begun early, examining the development of condensations in the primitive medium as they became galaxies in an expanding universe. The classic Jeans theory for condensation does not apply in an expanding medium, and it was Lifshitz (1946) who solved the Jeans instability problem for the cosmological expansion. Classical Jeans condensation develops exponentially, whereas in an expanding medium gravity-driven condensation grows only algebraically, leading to a drastic lengthening of the timescale. The work of Lifshitz was particularly significant, for it showed that the effect of expansion slowed the condensation process to the point where it was difficult to see how galaxies could form from small random fluctuations. The physical problem of baryon condensation was revisited by Peebles and Yu (1970) and by Zel'dovich (1970), who postulated that there were primeval fluctuations that would grow in the early Universe to a sufficient amplitude at decoupling to develop into the galaxy-condensation process. A problem arose, however, in forming galaxy-size condensations; an all-baryon primeval medium would preferentially form small, globular-cluster-size structures, which would then have to gather hierarchically into larger agglomerations, eventually reaching the dimensions of galaxies. The opposite scenario, in which large structures developed into super-clusters of galaxies, appeared to agree with the configuration of the visible Universe. In this case, however, the opposite problem arose, for it was not clear how smaller structures such as galaxies could form from the giant condensations. The solution was to postulate the existence of sufficient dark matter to change the condensation process.

Zwicky's discovery of dark matter forty years earlier was therefore taken seriously, although its nature was obscure (and remains so today). Two schools developed: one favouring hot dark matter (HDM), postulating fast-moving neutrinos or some other, unknown, light particle as the dark matter; and the other favouring cold dark matter (CDM), advocating an unseen component of slow, massive particles. Numerical computations indicated that a universe dominated by HDM 'tended to predict too much power in large-scale structures, while the CDM predicted too little' (Longair, from Frenk (1986)). The question would have to be settled by observations and experiments.

Observations of the CMB relate to temperature at the era of decoupling, and the spatial structure in temperature must be related to the density structure.

The era of decoupling took place over a finite interval of redshift, as illustrated by Figure 15.6. This can be taken as the interval $z = 1200$ to 900, defining the *last scattering layer*. Its thickness, in the co-moving radial interval $\Delta r = -c\,\Delta z/a$ (i.e. the thickness that it would have for our present epoch), is 70 Mpc, or, when scaled by $a(t)$, a physical thickness at the epoch of decoupling of 65 kpc. A sphere of this diameter would contain about $7 \times 10^{15} M_\odot$ of matter, one-sixth of which would be baryonic matter, a mass comparable to that of a rich cluster of galaxies.

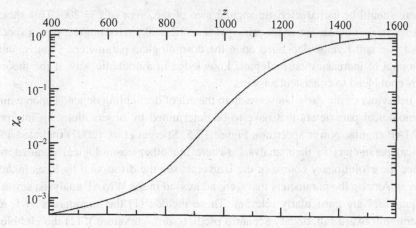

Figure 15.6. The development of the ionization fraction $\chi_e = N_e/N_H$ in the decoupling zone, shown as a function of redshift z (Chluba and Sunyaev 2006).

We now consider the physics of acoustic waves in the primeval medium, which reach the transition as the spatial structure observed in the CMB. Well before the WMAP mission was launched, the assumption that cold dark matter was a major constituent of the matter of the Universe had become the standard cosmological assumption (the CDM model), although hot dark matter lingered as a possible HDM model, in addition to intermediate variations. The model assumptions were further brought into question in 1995–96, when supernovae of type Ia were calibrated as distance indicators (Section 14.1). The most straightforward interpretation was that Einstein's cosmological constant Λ, rather than being zero, was so large that it implied that *vacuum energy*, or *dark energy*, was the dominant contribution to the energy density of the Universe. Furthermore, the data were consistent with a flat Universe, with curvature $\kappa = 0$ (although the uncertainty was large). For no reasonable model universe could one assume that $\Lambda = 0$. The CDM model became the ΛCDM model universe, with a large Λ and tight constraints on the curvature.

Following the demonstration of primitive fluctuations from the COBE DMR angular power spectrum, a strong rationale for an improved study of the CMB fluctuations led to the development of the WMAP mission, extending the power spectrum to much higher values of l. The structure measured by DMR was on such a large angular scale that it must have originated at a primitive state of the Universe. The theory of the fluctuations had to take account of the existence of two species of matter, *baryonic matter* that could interact through strong forces, electromagnetic forces, weak interactions and gravity, and *dark matter* that appeared to interact only gravitationally (although weak interactions could also be present). This led to the conclusion that, since the two forms of matter interacted differently, the baryonic matter could undergo a condensation process within the gravitational potential minima determined by the dark matter. By the time that the era of decoupling was reached, the physics of the process could be reasonably well understood. The baryon condensations would occur under quasi-resonance circumstances, and the calculations led to the conclusion that there should be a characteristic angular size of the order of $l = 200$. This should be visible as a peak (now known as the principal acoustic peak) in the angular power spectrum. Its exact size and l value depended upon the cosmological parameters. Clearly, this had the potential of increasing cosmological knowledge in a dramatic way, if the theoretical analysis could lead to consistent answers.

The behaviour of the early Universe up to the era of decoupling depends upon a number of cosmological parameters that have to be determined by observations. In interpreting the WMAP angular power spectrum, Figure 15.5, Spergel *et al.* (2007) included 15 cosmological parameters in their analysis. (There are other cosmological parameters that influence the evolutionary course of the Universe; see the discussion by Rees in *Just Six Numbers*). Among the parameters that were addressed in the WMAP analysis, seven principal quantities are particularly relevant. These include (1) the curvature $\kappa = 1/R$ (the most straightforward inflationary scenario predicts zero curvature.); (2) the Hubble constant $H_0 = 100h \, \text{km s}^{-1} \, \text{kpc}^{-1}$, with $h \approx 0.70$–0.75 (the Hubble parameter for our present epoch); (3) the cosmological constant Λ; (4) the baryon density parameter $\Omega_b = \rho_b/\rho_c = 8\pi G\rho_b/(3H_0^2)$, where ρ_c is the critical density that gives a flat universe for a given Hubble constant; (5) the dark-matter density parameter $\Omega_{dm} = \rho_{dm}/\rho_c$ ($\Omega_m = \Omega_b + \Omega_{dm}$); (6) the exponential spectral index, n, of the continuous part of the angular power spectrum (at low l); and (7) Q, the fraction of gravitational binding energy in the fluctuations, compared with their rest energy, serving as a measure of the amplitude of the fluctuations. If the inflationary scenario continues to enjoy success, then the curvature, κ, should be zero, and the data suggest that its value is, indeed, within 1% of zero. Otherwise, none of these parameters are predictable within current hypotheses; they are fundamental quantities imposed in the earliest epochs of the Universe, to be determined by observation.

The 'standard scenario', at least provisionally, starts with a flat ΛCDM universe in which primeval fluctuations have already been implanted, with $Q = (0.9 \pm 0.1) \times 10^{-5}$ (from the DMR results, reinforced by WMAP). As the Universe evolved from the first era of nucleosynthesis, electromagnetic radiation was, initially in the radiation-dominated era, the major contributor to the energy density. As noted in Section 14.3, the energy density of radiation diminishes as $a(t)$ to the fourth power, while the matter density diminishes only as the third power of $a(t)$, with the result that, shortly before the era of decoupling, the matter density became dominant. From the initial conditions of the Universe, the mass/energy density of dark matter was about six times that of the baryonic component, thus determining the first stages of the condensation process. The two matter components were initially locked together, with the baryons following the dark matter. As the era of decoupling approached, the difference in physical behaviour of the two components led to a gradual growth of baryon condensations within the gravitational potential of the dark matter, and the formation of the acoustic peaks could begin. The effect of the cosmological constant Λ is not important at this stage, since it scales as the square of $a(t)$; it would play a role much later. The scenario of matter condensation has been intensively analysed, and the relevant physical processes will now be summarized, to give guidance in interpreting the WMAP results.

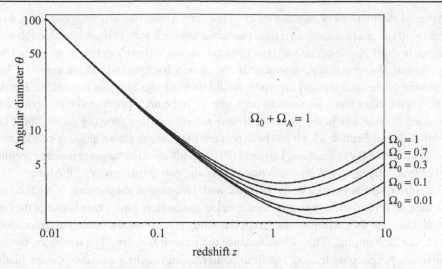

Figure 15.7. The variation of the angular diameter θ of a rigid rod of unit proper length with redshift z for world models with finite values of Ω_Λ and flat spatial geometry $\Omega_m + \Omega_\Lambda = 1$ (in the figure Ω_0 is the same as Ω_m). The ratio c/H_0 has been set to unity (Longair, *Galaxy Formation*, p. 222).

15.4 The geometry of the acoustic oscillations

The most striking, and the most significant, feature of the angular spectrum (Figure 15.5) is the peak at $l \approx 200$, with subsidiary peaks at roughly harmonic intervals. The linear scale of the acoustic peak depends upon local physical conditions such as the baryon density, the dark-matter density and the age of the Universe at decoupling, and upon the physical constants that regulate the recombination process. We observe the fluctuations as an angular spatial structure, and the linear and angular scales are related through the angular size distance, which in turn depends particularly on the key parameters curvature and Λ.

Since the CMB originates at $z \approx 1050$, we must relate observed angular diameters to distances with regard to the analysis of Section 14.4.3. The angle subtended by a standard measuring rod of proper length d was given in Equation (14.23), which can be written

$$\Delta\theta = (1+z)d/r, \tag{15.8}$$

where r is the proper distance for $a(t)$ (Equation (14.11)).

There is no elementary expression for r; the angular size distance depends on the curvature κ (assumed to be zero), the matter density index Ω_m and the Λ-index Ω_Λ; with the flat-universe assumption, $\Omega_m + \Omega_\Lambda = 1$ (the flat ΛCDM model). The calculated dependence of angular size on redshift is plotted in Figure 15.7 for various values of Ω_m.

The current value of Ω_m appears to be between 0.2 and 0.3, but the plot shows that for all values of $\Omega_m > 0.1$ there is a minimum angular size beyond $z = 1$. At larger redshifts, the angular size increases; in the apt description of Peebles 'the early galaxies loom overhead'; even though the geometry is flat, all ray paths converge towards the initial singularity.

Primeval fluctuations of matter density in the early Universe, which propagate as sound waves up to the era of decoupling, took place in a medium with two constituents following different laws of physics. The baryonic component, the minority constituent, carried with it by normal electromagnetic interactions the photon background, which carried a large component of the total energy; the whole would then act together gravitationally. The dark matter, on the other hand, interacts through gravity only, and appears to be dissipationless. The initial fluctuations in the primeval plasma are subject to a growing gravitational condensation, but in Section 15.3 it has been pointed out that the classical Jeans condensation is not appropriate for cosmology. Lifshitz (1946), in his solution for an expanding medium, showed that condensations grew algebraically as the two-thirds power of the time.

The subsequent behaviour of the baryonic and dark-matter constituents is not the same. For example, on sufficiently small scales the electromagnetic interaction between the baryonic medium and the CMB photons damps out acoustic oscillations, through the mechanism known as Silk damping. This occurs because the radiation density is a dominant factor in oscillations, so photons leaking out from compression maxima attenuate the oscillation; the effect is that baryonic clumping can persist only on a scale larger than the Silk damping scale (Silk 1968).

Two other characteristic lengths are related to the structure of the CMB: the *particle horizon scale*, r_H, given by (for the ΛCDM model at large redshift)

$$r_H = \left[2c/\left(H_0\Omega_m^{1/2}\right)\right]a^{1.5} = 2cH_{100}h\Omega_m^{1/2}(1+z)^{-1.5} \tag{15.9}$$

marks the radius out to which a light signal can travel since the beginning of time. No communication could have occurred with regions of the Universe beyond this radius. A second characteristic length, the *sound horizon* $\lambda_s \equiv c_s t_U$, where t_U is the age of the Universe at that epoch, marks the dynamical limit over which physical communication via sound can occur:

$$\lambda_s = c_s t_U = c_s\left[2z/\left(3H_0\Omega_m^{1/2}\right)\right]. \tag{15.10}$$

Here the sound speed, in the two-component medium of baryon density ρ_B and radiation density ρ_{rad}, in terms of the parameter $R_\rho = 3\rho_B/(4\rho_{rad})$, is

$$c_s = c/[\sqrt{3}(1+R_\rho)] \tag{15.11}$$

and t_U is for a matter-dominated universe (ignoring the era of radiation domination that occurred previously even though a lot of radiation lingers). The sound speed at $z = 1050$ is $0.43c$, close to the relativistic limit of $0.577c$. The sound horizon length λ_s has particular relevance for the evolution of the observed fluctuations of the baryon component. At the era of decoupling, the linear size of these acoustic fluctuations would be $\approx$100 Mpc co-moving (proper) distance; correcting for the scale factor $a(t)$, this gives a physical distance of about 100 kpc. On the sky, such fluctuations would therefore subtend an angle of about $0°.5$, close to the $l = 200$ peak observed by WMAP (Figure 15.5).

When the era of decoupling was reached, the principal condensations were of dark matter, carrying with it the baryonic matter. The radio brightness fluctuations of the CMB are related to the baryonic component of the matter density, not directly to the predominant dark-matter component. We now address the mechanisms of radiative transfer that determine the observed brightness structure at the era of decoupling, when the Universe had an age of only about 370 000 years. They depend upon three separate physical processes.

(i) If the fluctuations are adiabatic, the overdense, condensed regions will be hotter, and the emitted photons will be coming from a region of correspondingly higher temperature. Conversely, in an underdense region that has undergone expansion, the photons will be emitted from a medium that is cooler than average.

(ii) Photons from an overdense region will undergo a redshift, because they are climbing out of a gravitation potential well (the so-called Sachs–Wolfe effect; Sachs and Wolfe (1967)).

(iii) Photons emitted by matter in the fluctuating medium will undergo a Doppler shift as it moves towards or away from us in contraction or expansion.

The effect of the three perturbations listed above on the observed brightness temperature is given by an integral along the line of sight of the expression (where r is taken at the point of photon emission):

$$\int \delta T \, ds = \int [\phi(\mathbf{r}) - \hat{\mathbf{r}} \cdot \mathbf{v}(\mathbf{r}) + (1/3)\delta(\mathbf{r})]ds, \qquad (15.12)$$

where $\phi(r)$ is the local gravitational potential, expressing the Sachs–Wolfe effect, the second term is the Doppler term, while the last term expresses the effects of adiabatic compression, raising the local temperature. All three terms vary with the local wave number k. For the largest-scale fluctuations, the observed brightness-temperature fluctuation depends upon both the adiabatic term, $(1/3)\delta(\mathbf{r})$, and $\phi(\mathbf{r})$, the Sachs–Wolfe effect, which have opposite signs because the gravitational potential is negative, while the compression increases the temperature. The gravitational potential follows the local density fluctuations, so the two terms act in opposite directions. The net effect requires integration along the photon trajectory from last scattering to the observer. On the very largest scales, larger than a few degrees of angle, the net effect is to reduce the brightness temperature by $\phi/3$.

At an angular scale in the range $0°.1$ to $1°$, the fluctuations become coherent, as noted above, when the linear scale is within the travel time of a sound wave. The fluctuations have a degree of coherence, therefore, since oscillations can occur that result in stronger motion and larger temperature variations of the baryonic component. These influence the angular power spectrum markedly. This is the regime that gives rise to the maxima known as the acoustic peaks. Both the first and the third term of Equation (15.12) above become important: the Doppler term $\hat{\mathbf{r}} \cdot \mathbf{v}(\mathbf{r})$ because of the larger velocities and the temperature term because of the adiabatic compression of the baryons.

Although the dominant effect comes from the photons that are emitted last in the recombination process, the radiation leaving the CMB region will include scattered photons, which are polarized. There would be no net polarization if the medium were uniform, but the fluctuations introduce asymmetry, and the scattered photons, therefore, will show a net polarization. These are described as E-type scattering (from thermal baryon fluctuations, which will be curl-free) and B-type scattering (from fluctuations induced by gravitational waves, which will be divergence-free). The effect, first predicted by Zel'dovich, is an effective tool for determining the role of gravitational waves in the early Universe, and served as motivation to include polarization measurements in the WMAP mission.

15.5 Physics of the acoustic oscillations

The spectrum of the CMB in the region of the acoustic peaks (Figure 15.5) bore out the theoretical expectations in satisfying fashion. The physics needed to explain the observations is still an unfinished task, although the broad outlines seem soundly based. In the following discussion, the ΛCDM model will be adopted, together with the assumption of adiabatic processes for the oscillations of the baryonic medium. The review by Hu and Dodelson (2004) gives a detailed discussion with references, and the elementary exposition by Tegmark (1995) provides an excellent introduction to the physical ideas.

The acoustic oscillations appear before the era of decoupling because the matter condensations, mostly determined by the CDM, defined potential wells, and within the sound horizon the baryonic component could fall in and concentrate. The resulting compression could raise the temperature, through the electromagnetic interaction. A simplified dynamical model yields a harmonic-oscillator equation; if the effect of cosmic expansion is neglected, the local temperature fluctuation $\Theta_T = \delta T/T$ can be written

$$\ddot{\Theta}_T + k^2 c_s^2 \Theta_T = -(1/3)k^2 \Psi(k), \tag{15.13}$$

where the sound speed c_s is given by Equation (15.11) and is only slightly below the speed of light. The temperature variation is assumed to be adiabatic. The left-hand side is clearly the simple-harmonic-oscillator equation for wave number k, while the driving term on the right is in itself the solution of a damped-harmonic-oscillator equation involving the potential. If cosmic expansion were included, it would appear as a damping term, proportional to $\dot{\Theta}_T$. Thus, the density, and hence the temperature, can undergo oscillations. The region within the sound radius behaves something like a resonant cavity, with normal modes of oscillation. This is not strictly true, because the sound radius is changing with time, and cosmic expansion is damping the oscillations, so as a result the overtones should not be expected to follow a harmonic relationship exactly.

The frequency of oscillation for a given wave number will be $\omega = k c_s$, and so the expected variation of Θ_T turns out to be (given initial perturbations $\Theta_T(0)$ and $\dot{\Theta}_T(0)$ in amplitude and velocity)

$$\Theta_T(t) = \left[\Theta_T(0) + \Psi/(3c_s^2)\right]\cos(k\lambda_x) + \left[\dot{\Theta}_T(0)/(3c_s^2)\right]\sin(k\lambda_x) - \Psi/(3c_s^2). \tag{15.14}$$

The observed temperature fluctuations would correspond to choosing the era of decoupling as the time at which λ_x is evaluated.

The driving term Ψ has no Newtonian analogue (it is the scale-invariant gravitational potential for a given wave number λ, generated by the dark-matter fluctuations). It is derived from a harmonic-oscillator equation analogous to the semi-Jeans instability equation referred to at the opening of this section. To a first approximation, however, one can take $\phi_c \approx -\phi$, the negative of the Newtonian potential.

Care has to be taken in constructing approximate solutions to these equations. The combination of all three perturbations in Equation (15.12) can result in the acoustic peaks vanishing due to mutual cancellation! The time evolution of the Doppler term and the correct expression for the sound speed (rather than assuming that it is c) must be taken into account in order to display the brightness-temperature perturbations from the acoustic peaks. Equation (15.14) for $\hat{\phi}(k)$ has been solved approximately by using the WKB approximation. The separate contributions of the Doppler and the adiabatic perturbations, the second and first terms of Equation (15.12), are shown; the second term, the Sachs–Wolfe effect, imposed by the radiation origin in a gravitational potential well, has been included as a correction to the results. The first acoustic peak, sometimes referred to erroneously as the Doppler peak, marks the time of maximum compression of the baryonic fraction. The maximum Doppler perturbation occurs in quadrature, since it follows the velocity.

The resulting spatial spectrum, as a function of wave number k, where $k = 2\pi/\lambda$, appears to the observer as a function of angle on the sky, and hence it must be transformed in the usual way through the angular-distance relation. Thus the effects of Λ are present, even though Λ has no appreciable influence on the oscillations in the layer of last scattering.

15.6 Deriving the cosmological parameters

The angular power spectrum from the five-year WMAP data (Figure 15.5) covers the range $l = 2$ to 1000. The uncertainties are largest at each end of the range, but in the middle of the range the signal-to-noise ratio is limited, not by the equipment, but by 'the noise of the Universe'. The observed values are derived from a finite number of independent data points; we would need another universe to improve significantly on them. One conclusion is clear; beyond the third acoustic peak, the amplitudes of the maxima are noticeably smaller, which is the result of Silk damping.

In order to understand the effect on the angular power spectrum when the cosmological parameters are varied, numerical experiments have been conducted by various authors; see, for example, Hu *et al.* (1995), Hu and Dodelson (2002), Challinor (2005) and also Peacock p. 449. Increasing the density of baryons, as might be expected, increases the magnitude of the first acoustic peak without changing its l-value much, because the potential well of the CDM largely determines the dynamics of the baryon fluctuations. Increasing the total matter density, in addition to strengthening the peak, shifts its maximum l-value higher, since the potential well will be deeper and hence the acoustic oscillations will have shorter wavelength. Increasing the curvature κ from 0 (flat) to positive values has little effect on the magnitudes of the peaks but shifts the entire pattern to lower l, since it is the angular size distance that is affected; in effect, one is changing the overall magnification wrought by the Universe. For negative curvature (an open universe) the pattern shifts in the opposite direction, to higher l-values. Decreasing Λ from a finite value to zero also shifts the pattern to higher l-values without changing the peak ratios, since its local effect is small at the era of decoupling, affecting the angular size distance only at a much later epoch ($z < 1.5$).

The WMAP team published a detailed analysis, using the first 3 years of observations and a range of sophisticated computer models (far more realistic than the heuristic approach outlined in Section 15.5) to derive cosmological quantities from the observed angular power spectrum. Their extensive analysis (Spergel *et al.* 2007) considered a list of 23 cosmological parameters that could influence the angular spectrum. They first considered what quantities could be derived by use of only the WMAP angular spectrum, with two restrictions: a ΛCDM model universe, and a power-law angular continuum spectrum (not necessarily Harrison–Zel'dovich). In the later, 5-year analysis, the optimum results were obtained by using six independent quantities that give the best fit to the observed power spectrum: (1) the density parameter for baryons, $\Omega_b h^2$; (2) the total matter-density parameter for baryons plus cold dark matter, $\Omega_m h^2$; (3) the dark-energy density parameter Ω_Λ; (4) the power-law spectral index n_s; (5) the optical depth τ for Compton scattering during reionization after the era of decoupling; and (6) the r.m.s. deviation of the power spectrum, $\Delta_R^2(k)$ from a mean power-law spectrum $P_R(k)$. (The authors define this as the curvature deviation, which is of course related to the observed matter concentrations.) Removing parameters diminished the quality of the fit; conversely, adding more cosmological variables gave no significant improvement to the fit. The rest of the following discussion is based on the results of the 5-year analysis, which used improved algorithms and programs that narrowed the formal errors.

The analysis of the 5-year data is described by Hinshaw *et al.* (2008), Dunkley *et al.* (2008), Komatsu *et al.* (2008), Page *et al.* (2007), Gold *et al.* (2008) and Nolta *et al.* (2008); for experimental details the WMAP Explanatory Supplement (2008) can also be consulted. Note that the fundamental six parameters used to model the angular power spectrum do not permit the direct determination of several quantities of interest. Neither the Hubble constant H_0 (the scaling parameter h) nor the present age of the Universe can be derived directly from the six-parameter model. It turns out, however, that even modest estimates of these quantities derived from current independent measurements of the Hubble constant serve to resolve the degeneracies in the reduction and allow remarkably accurate estimates of these and other 'derived' cosmological quantities by adding the priors from other observations.

The principal results for the 5-year analysis are presented in Table 15.1, which is derived from Table I of Komatsu *et al.* (2008). The data derived from the WMAP data alone assumed a flat ΛCDM mode and a simple power-law form for n_s. The column labelled WMAP + BAO + SN was derived from an independent analysis using two additional priors: the local optical Galaxy correlation observations (BAO) and the Supernova Type Ia (SNIa) measurements. BAO refers to the observed large-scale structure of the Universe, measuring Galaxy correlations in our neighbourhood; this is expressed as σ_8, the variance in local Galaxy counts at a scale of 8 Mpc compared with the variance on very large scales.

Mean values are given, with formal errors of 1σ; refer to Komatsu *et al.* (2008) and Dunkley *et al.* (2008) for the maximum-likelihood values (which are not significantly different).

Table 15.1. *Principal cosmological parameters derived from 5-year WMAP analysis*

	WMAP 5-year mean	WMAP + BAO + SNIa
Primary fitting parameters		
$\Omega_b h^2$	0.02273 ± 0.00062	0.02265 ± 0.00059
$\Omega_{dm} h^2$	0.1099 ± 0.0062	0.1143 ± 0.0034
Ω_Λ	0.742 ± 0.030	0.721 ± 0.015
n_s	$0.963 + 0.014/-0.015$	$0.960 + 0.014/-0.013$
τ	0.087 ± 0.017	0.084 ± 0.016
$\Delta_R^2(k_0)$	$(2.41 \pm 0.11) \times 10^{-9}$	$(2.46 \pm 0.0093) \times 10^{-9}$
Derived parameters		
σ_8	0.796 ± 0.036	0.817 ± 0.026
H_0 (km s^{-1} Mpc^{-1})	71.9 ± 2.6	70.1 ± 1.3
Ω_b	0.0441 ± 0.0030	0.0462 ± 0.0015
Ω_{dm}	0.214 ± 0.027	0.233 ± 0.013
$\Omega_m h^2$	0.1326 ± 0.0063	0.1369 ± 0.0037
$z_{reion} \times 10^{-3}$	11.0 ± 1.4	10.8 ± 1.4
t_U (Gyr)	13.69 ± 0.13	13.73 ± 0.12
Σm_ν (eV)	< 1.3	< 0.61

For the derived parameters in which only the angular power spectrum of WMAP was used, it was necessary to use a prior value for H_0, but the results were not sensitive to the prior value chosen. This was already noted by Spergel *et al.* (2007) in their analysis of the 3-year data. Their value for the present age of the Universe, 13.73 Gyr, is the same as that given above for the 5-year data with the addition of the BAO and SNIa priors. It is surprising that the improvement between the two sets of results is relatively small. Adding the Type Ia supernova results (the summary given in the ESSENCE survey of Wood-Vasey *et al.* (2007) combining the SNIa data from the Kirshner and Perlmutter groups, and others) also leads to improvements, particularly for H_0, Ω_b, Ω_{dm} and Ω_Λ, but the improvements are not dramatic, except for the limit on the sum of neutrino masses (the upper limit on neutrino masses is due to the damping effect which would be expected if they had greater mass). This leads to increasing confidence that the derived values are represented reliably in Table 15.1.

Further analysis on these lines will be possible when the observations for larger l values are included. The curvature of the Universe is of particular interest, since the inflationary scenario, at least in its most straightforward formulation, requires a flat universe. Spergel *et al.* show that the observations up to $l = 1600$ already imply $\Omega = 1 - 0.014 \pm 0.017$.

In addition to this support for a flat ΛCDM universe following an inflationary scenario, further conclusions were drawn from this analysis: (1) the cold dark matter is an important forcing term in determining the acoustic peaks; (2) adding a-priori information from other observations, such as the approximate value of H_0 (i.e. h), gives an accurate value for the age of the Universe and (3) establishes that the value of Λ implies that dark energy is the dominant component of the matter density in the Universe; and (4) a scale-free (Harrison–Zel'dovich) angular spectral index $n_s = 1$ does not give a good fit to the data. In their analysis, the authors modified the power law by adding a scale-dependent term, but its value was not significant. One must conclude that the index appears to be 2.5σ flatter than the Harrison–Zel'dovich value.

The energy, or Friedmann, equation (14.13), assumed conventional equations of state for all of the energy-density terms, but on a cosmological scale this must be regarded as an assumption that should be subjected to observational tests. The WMAP observations provide an opportunity to examine the possibilities. In Equation (14.13), the pressure term from Equation (14.12), $3P/c^2$, is lumped in with the density, since it is an energy density with an equivalent mass. Similarly, the Λ-term is taken as a constant in time, an assumption that comes from Einstein's definition of Λ in the field equations, but might not correctly describe its behaviour in time. Thus it has become customary to modify the various mass-equivalents by introducing a set of w-parameters, defined by

$$\frac{E}{c^2} \sim a^{-3(1+w)} \qquad (15.15)$$

with the following values of w for the various components:

density or pressure (non-relativistic) $w = 0$
photons: $w = 1/3$
vacuum energy (Λ) $w = -1$.

These are all reasonable assumptions, but the equation of state, particularly for vacuum energy, may well be different. The analysis of the 5-year WMAP data, however, found no significant deviation in w for the vacuum-energy equation of state.

The increased body of observations after 5 years improved the signal-to-noise ratio of the polarization data set. This has particular significance for separating the E-mode and B-mode components of the polarization pattern, setting limits on the importance of gravitational waves as a product of inflation and as an influence on the matter/energy density of the Universe. The B-mode (divergence-free) component gives the characteristic signature of gravitational waves (which are tensor waves, as opposed to the vector character of photons). The decomposition leads to the conclusion that the observed polarization is due to scattering by electron inhomogeneity only; there is as yet only an upper limit on the energy-density fluctuations of gravitational waves. The polarization data also have relevance, for they give evidence for the ionization history of the Universe after the age of decoupling. Out to a redshift beyond 5, the Universe has evidently undergone extensive reionization. At the age of decoupling, the ionization fraction dropped dramatically, as Figure 15.6 shows, with the matter becoming almost neutral, but the formation of stars and galaxies had to follow sometime during these 'dark ages'. When these first-generation stars (known as

Population III) formed, the medium would undergo reionization, but that epoch has not yet been directly observed.

The polarization data from WMAP described by Page *et al.* (2007) allow an estimate of the time when reionization occurred. If the reionization was a sudden event, the best fit to the data from the augmented data set favoured a redshift of $z_{reion} = 10.8 \pm 1.4$, as shown in Table 15.1. One clear limit can be set: a single reionization event at $z < 6$ appears to be most unlikely. Determining whether the actual scenario favoured a sudden event or a gradual turn-on of ionizing radiation will depend on more accurate polarization measurements.

The principal conclusions of the WMAP were as follows.

1. The observed properties of the Universe are consistent with the ΛCDM cosmological model.
2. The Universe has flat geometry, at least to the level of 1%.
3. The energy density of the Universe is dominated by dark energy.
4. The mass/energy density of dark matter exceeds that of baryonic matter by at least a factor of 5; together they make up only about 25% of the total energy density.
5. Many cosmological parameters can now be measured with an accuracy approaching 1%, introducing an era of precision cosmology.
6. Among these quantities, the age of the Universe is 13.7 Gyr, the mean age of the Universe at the time of decoupling was 370 000 yr and the best value of the Hubble constant is $70.1 \pm 1.3 \, km \, s^{-1} \, Mpc^{-1}$.

These general conclusions are based on the values of the cosmological quantities that followed from the reduction of the first five years of data listed in Table 15.1. There is a broader significance to the WMAP results. In much of the pre-WMAP work in cosmology, the concepts were clear, but their accuracy was far from quantitative. The value of the Hubble constant differed by a factor of two throughout the latter half of the twentieth century among reputable observers. The errors of the WMAP analysis, in many of the key cases, are of the order of only one or two per cent, and some of the other quantities of interest, such as the contribution of dark matter and baryonic matter to the mass/energy density are uncertain to only 10% or so, an amazing improvement over earlier determinations. Cosmology has now become a quantitative science, where the errors of measurement can be small enough for one to distinguish between different theoretical concepts. The WMAP work has established a new standard of accuracy in cosmology.

Further reading

Cosmological Physics, Peacock J., Cambridge, Cambridge University Press, 1999.
Galaxy Formation, Longair M. S., 2nd edn, Berlin, Springer-Verlag, 2008.
Just Six Numbers, Rees M., New York, Basic Books, 2000.
Proc. Enrico Fermi Course CXXXII, Tegmark M., Varenna, 1995.
Science with the Square Kilometer Array, Carilli C. and Rawlings S., Amsterdam, Elsevier, 2004.
Cosmic Microwave Background Anisotropics, Challinor A., in *The Physics of the Early Universe*, ed. Papantonopoulos E., Berlin, Springer, 2005.

16

Cosmology: discrete radio sources and gravitational lensing

Both quasars and radio galaxies have large radio luminosities, and the most luminous of these can be detected at very large redshifts. Indeed, one of the complications encountered in determining the distribution of their intrinsic luminosities (the *radio luminosity function*) is that the objects are so faint optically that the largest telescopes are needed to obtain the spectra from which redshifts can be derived. At the same time, this raises the expectation that radio sources can serve as probes of the large-scale geometry of the Universe. The first indication of cosmological evolution was provided by the statistical relation between numbers and flux density of radio sources, the *source counts*; a similar test of cosmologies, the *luminosity–volume test* was applied to visible quasars. These tests immediately ruled out the steady-state model of the Universe, but did not contribute to the precision cosmology which later emerged from the WMAP measurements of the CMB. The extension of the source counts in surveys of extreme sensitivity has, however, contributed dramatically to the astrophysics of radio-source evolution.

Another cosmological test, the relation between apparent source diameter and luminosity, emerges from the geometry outlined in Chapter 14. This again proves to be more of interest regarding source evolution than for cosmology itself. A further observational field opened by the geometrical theory is *gravitational lensing*, which was discovered as a radio phenomenon and is now observed through most of the electromagnetic spectrum. Its main significance is in the detection of *dark matter* and an independent method of determining H_0.

16.1 Evolution and the radio-source counts

In a Euclidean universe, the flux density of a standard source would fall with distance D as D^{-2}, while numbers of sources would increase as D^3. The total number N detected above a flux level S would then vary as $S^{1.5}$. Extending this relation to sources with appreciable redshift requires both a careful definition of distance and consideration of the effect of redshift on observed flux density, which depends on the spectrum of the source. However, unless the sources have a steeply inverted spectrum, such that a large redshift increases the apparent luminosity, the geometry of an Einsteinian universe can only reduce the index to below 1.5.

The concept seemed straightforward, but the earliest surveys arrived at an incorrect power law for the source counts. The trouble for the early surveys originated in the relatively large beams on the sky, relative to the large number of apparent sources, an effect known as confusion. There were so many weak sources that they could not be individually distinguished with the large beamwidths then available. The 3C Catalog was the result of the first large-scale survey to have a sufficiently small beamwidth to indicate that the source-count slope for the bright sources was significantly steeper than 1.5 (Ryle 1958), but even so there were enough spurious sources in the catalogue for it to require editing; the resulting Revised 3C Catalog (Bennett 1962) has few errors, and is a reliable source of strong-source statistics. The confusion problem was discussed critically by Scheuer (1957, 1974), but the full value of the source-counting technique emerged only with the large surveys which are now available. The large value of index is now known to be the effect of evolution, in which the space density of radio sources increases dramatically up to a redshift z of between 2 and 4.

A related technique is the *luminosity–volume* test introduced by Schmidt (1968) for visible quasars and by Rowan-Robinson (1968) for infrared sources. This is also known as the V/V_{max} test. This requires a distance for each object; for the quasars this was derived from a measured redshift. Then for each object the volume of space up to its distance is designated V, and the volume up to the distance at which the object would be just detectable is designated V_{max}. The average of this ratio V/V_{max} for many objects should, in a simple Euclidean universe, be 0.5. Both for the quasars and for the infrared sources the average ratio was found to be $\sim$0.7; again this is related to source evolution on a cosmological timescale.

There are several large and well-defined catalogues of radio sources, which serve a variety of purposes. There are single-telescope ('single-dish') surveys that cover a large fraction of the sky, interferometric surveys that examine small areas of sky to low flux levels and specialized surveys carried out with particular purposes in mind (the 3CR survey as revised by Laing *et al.* (1978) is a historic example). An overall view of the way in which radio sources cover the sky is presented in Figure 16.1, derived from the 87GB northern-sky survey (Gregory and Condon 1991) and the PMN southern-sky survey (Griffith and Wright 1993), both at 5 GHz. The Galactic sources, mainly H II regions, are seen as a thin line outlining the plane of the Milky Way; these are confined to within 5° of the plane; at higher latitudes nearly all the sources are extragalactic (with the exception of a few associated with stars). Their distribution appears to be isotropic, and no clustering structure has yet been found.

The single-dish surveys have the merit of detecting sources regardless of their angular size, but they do suffer from the effects of confusion, through the chance occurrence of more than one source within a given telescope beamwidth. This limits severely the number of sources that can be observed reliably, since confusion can affect not only the apparent flux density of a source but also its position. As the number of sources increases with decreasing flux density, there is an increasingly significant upward bias to the measured flux density. For example, there are 30 times as many sources per beam at a flux 10% that of

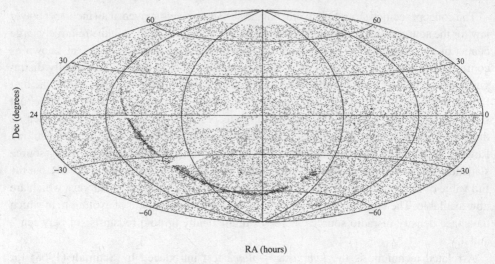

RA (hours)

Figure 16.1. An all-sky map of radio sources ≥ 80 mJy from the combined surveys of Gregory and Condon (1991) and Griffith and Wright (1993). Apart from the thin line of sources associated with the Milky Way, the distribution is essentially uniform over the sky.

the source being studied; therefore, if the number of sources per beam is 0.01 at the source minimum-flux limit, there is a 30% chance of a 10% flux enhancement being caused by confusion. The sources shown in Figure 16.1, at approximately one per 300 beamwidths, are free of confusion bias.

Interferometric surveys, with far smaller synthesized beams, can reach to lower flux densities, and hence include greater numbers of sources. There are three major interferometer surveys currently available: the NVSS survey (Condon *et al.* 1998), the FIRST survey (White *et al.* 2000) and the Westerbork Northern Sky Survey (WENSS; de Ruiter *et al.* 1998). The NVSS used the most compact array of the VLA to capture radio galaxies with large angular diameters; its resolution is 45 arcseconds, with a sensitivity limit of 2.5 mJy (at 5σ). The FIRST survey used a more extended array, with 5-arcsecond resolution and a sensitivity limit of 1 mJy; this survey obtained greater angular resolution on individual sources, at the cost of losing some sources with large angular structure. The WENSS survey covers 3.1 steradians at the comparatively low frequency of 327 MHz, with an angular resolution of approximately 1 arcminute and a limiting sensitivity of 15 mJy; a less extensive survey at 609 MHz is also included. All of these surveys can be accessed on the Internet.

The deepest surveys, covering smaller areas of sky, are necessarily interferometric, using aperture synthesis. Of these, the 'HDF' survey (Richards *et al.* 1998, Richards 2000) reaches particularly faint sources; it is centred on the Hubble Deep Field, allowing identification with exceptionally faint optical objects. For this survey the VLA was used in its most extended configuration at 1.4 GHz and 8.5 GHz, integrating for 50 hours to reach flux densities of 40 μJy at 1.4 GHz and 8 μJy at 8.5 GHz. The discussion of observing techniques by Richards (2000) is particularly illuminating as an example of the care needed when carrying out long

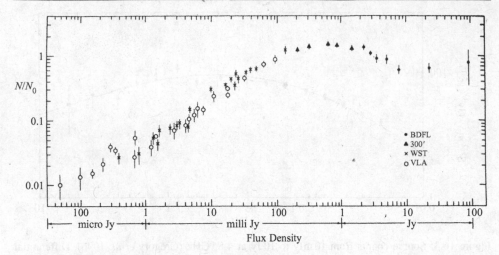

Figure 16.2. Counts (N) of radio sources, over a range of 10^6 in flux density, compared with those expected (N_0) from a static Euclidean universe with no evolution. The counts are presented as differentials dN against dS, plotting the ratio of observed number density to the number density expected in a simple Euclidean universe with constant number density (Kellermann and Wall 1987).

aperture-synthesis integrations; one must deal with confusing sidelobe effects both from strong sources outside the field of view and from millijansky sources within the field of view, and with differential effects across a wide frequency bandwidth, in addition to baseline calibrations. The 371 sources obtained in this survey in the 20-arcminute field provided a very useful extension of the statistics of faint sources.

The background of faint radio sources may also make an appreciable contribution to the sky brightness in observations of the CMB, especially if some of the sources already catalogued at centimetre wavelengths have an inverted spectrum. The angular resolution of the WMAP satellite is sufficient to identify discrete sources, and these have been catalogued by Wright *et al.* (2008). Their net effect has been included in the 5-year data reduction. Earlier ground-based surveys at lower frequencies, such as the 9C survey (Waldram *et al.* 2003) and the southern-hemisphere AT20G survey (Massardi *et al.* 2008) provide supplementary information, particularly on the incidence of inverted-spectrum sources.

Figure 16.2 shows counts based on a combination of surveys at 1.4 GHz and, for comparison, the 5-GHz source counts as well. The counts are presented as differentials, dN against dS; the differential counts are multiplied by $S^{5/2}$, giving the ratio of observed number density to the number density expected in a simple Euclidean universe with constant source density.

The early hopes that such plots could be related to the geometry of the expanding Universe proved to be unfounded. Although the counts in the top four decades of flux density all lie within a factor of two of a straight horizontal line, the deviations are systematic and increase at low flux densities. The problem is that the population is far from uniform. Instead of a

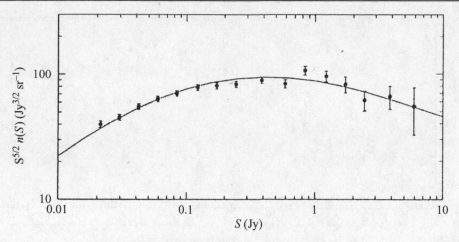

Figure 16.3. Source counts from 10 mJy to 10 Jy at 4.85 GHz (Gregory *et al.* 1996). Differential counts are plotted, showing the ratio to the counts expected in a Euclidean model with no evolution. The smooth curve is from an evolutionary model by Condon (1984).

single population of standard sources, there proved to be several major populations: radio-loud quasars, two types of extended radio galaxies (high and low luminosity, originally identified by Fanaroff and Riley (1974) and known as FR I and FR II types), compact steep-spectrum sources and, at the millijansky end of the catalogues, still another class of intrinsically low-luminosity galaxies. All differ in morphology, luminosity class and evolution in time. As a result, the counts were mainly determined by the evolution of the population rather than by the geometry of the space in which they were located. In this more complex situation it becomes essential to appeal to optical observations to obtain identifications and redshifts. Purely optical observations of quasars, which cover a more uniform class of objects and include a large number of radio-quiet quasars, evidently have some advantages. However, the radio surveys maintain an advantage in several respects; the surveys are almost free of contamination by objects in our own Galaxy, there is no Galactic absorption and it is relatively simple to carry out a survey that is complete and with uniform sensitivity over a large region of sky.

Apart from a small deficit of sources at the high-intensity end of the plot, there is a substantial range of flux density over which the differential counts follow a simple Euclidean law within a factor of two. Most sources in this range are known to be quasars, and the distances of many are known from optical redshifts. Below flux density 100 mJy (at 1.4 GHz) there is a steep fall in source numbers; Condon (1984) has shown that the deviations from the simple law in this part of the plot can be explained only by invoking a major dependence of luminosity on redshift. Figure 16.3 shows the comparison between models of this dependence and the observed differential counts (Gregory *et al.* 1996), from the catalogue of 75 162 sources. The solid line in this plot is the curve expected from a model in which the luminosity peaks between $z = 1$ and $z = 5$, but falls by a factor of 10 at $z = 1$ and at $z = 9$. The dramatic fall in numbers beyond this region, that is, below

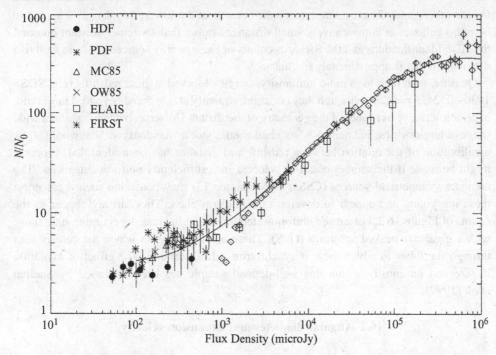

Figure 16.4. Source counts from deep surveys at 1.4 GHz (Richards 2000, arXiv:astro-ph/9908313). The counts are shown normalized with respect to the counts expected in a Euclidean geometry. The counts from observations of the Hubble Deep Field, extending to 40 μJy, are shown as filled circles; references to previous surveys may be found in the paper by Richards.

about 100 mJy at this radio frequency, is similarly explained in terms of source evolution, although the greater proportion of radio galaxies (rather than quasars) at low flux densities is a confusing factor.

Figure 16.4 shows the counts extending to very low flux densities, observed in a field extending 20 arcminutes from the centre of the Hubble Deep Field (Richards 2000). The data join reasonably smoothly to those at larger flux densities from surveys covering a larger solid angle, but with a pronounced levelling to a more nearly Euclidean distribution. This suggests that a different population of weak sources may be appearing at this low level of flux density.

The form of the evolution has been much debated. Simple analytical forms, in which, for example, all luminosities vary as a power of the distance factor $(1 + z)$, are clearly insufficient to explain the number counts; it seems necessary to adopt a more complex form in which evolution occurs mainly in the most intense sources, namely the quasars. The radio-source counts and the purely optical surveys here agree that the apparent population of quasars increases with distance up to a redshift of about 2, and decreases at redshifts greater than about 4. The populations at the high and low ends of the number-count curves are dramatically different. At the high-flux end, there are many high-luminosity objects.

On the other hand, the most numerous radio sources at the lowest flux densities appear to be radio galaxies at comparatively small distances rather than extremely distant quasars; the HDF identifications and the Richards counts of these nearby sources conform well to a distribution that is approximately Euclidean.

Quasars, with their high radio luminosity, can be observed at great redshifts (e.g. SDSS 1030+0524, $z = 6.28$), although they occur infrequently at $z = 5$ and beyond. They could, in principle, serve as probes of the geometry of the distant Universe, but they exhibit a wide range of luminosities, and neither a 'standard candle' nor a 'standard rod' that would allow a calibration of the relation between redshift and distance has been identified. Progress might be made if the sources could be selected into sufficiently uniform categories. The compact symmetrical sources (CSSs) (see Chapter 13) may constitute such a category; these are young and appear to develop as a uniform class. They already appear in the counts of Figure 16.2, but a more uniform selection might be made by counting only those with a gigahertz-peaked spectrum (GPS). These are the CSSs in which the core is seen through the lobes in which there is synchrotron self-absorption. The effect of evolution in size and luminosity within this well-defined sample has been discussed by Snellen *et al.* (1999).

16.2 Angular diameter and expansion velocity

Another and more direct method of exploring the geometry of the Universe is the measurement of the angular diameters of a series of standard-size objects at distances sufficiently large to show the nonlinear character of the Hubble relation, and especially the effect of the cosmological constant Λ. In standard cosmology, described in Chapter 14, angular size does not necessarily diminish indefinitely with distance; for example, in the simple Einstein–de Sitter model, a minimum angular size is reached at $z = 1.25$, and thereafter the angular size increases. For the ΛCDM universe, the same phenomenon occurs, although the exact minimum varies with the choice of parameters.

Hoyle pointed out in 1958 that the then recently discovered large angular size of the radio galaxy Cyg A at redshift $z = 0.06$ might provide such an opportunity, since the apparent angular size of a similar radio galaxy at redshift $z = 1$ might be measurable with only modest improvements in observational technique (Hoyle 1959). As so often occurs in observational tests of cosmological models, the situation proved to be far from simple. The techniques are certainly available, but it has been hard to find any set of objects that are sufficiently consistent to be used as standard measuring rods. As in the case of source counts, for which one required but did not find a standard population of sources with the same luminosity, there proved to be no standard size, even for double-lobed radio galaxies such as Cyg A.

Even though the compact source types CSS and GPS are short-lived, they are located at the centres of galaxies, which isolates them from any outside influence, and their physical properties might be expected to be independent of the larger-scale evolution we are attempting to measure. According to the model set out in Chapter 13, they should all be quasars in which the axis points towards us; this again limits the range of projected size, although the relativistic beaming will reduce the projected size for the closest alignments. Gurvits *et al.* (1999) analysed the VLBA observations of 82 compact sources of known redshift and showed that the angular diameters reached a minimum in the redshift range $z = 1$ to 2, although there was not sufficient precision to identify a particular cosmological model.

The superluminal jets at the cores of radio galaxies and quasars show rapid angular motion, and can be observed at high redshifts. This has led to the expectation that they might exhibit a dependence of apparent angular velocities on redshift similar to the angular-size–redshift test, with the additional factor that the velocity is reduced by a factor $(1 + z)$ due to time dilation. Little success has come so far, but the possibility remains viable.

A different class of angular measurement has been applied to the megamaser associated with NGC 4258, discussed in Chapter 13. The maser components exhibit proper motion and velocity, with sufficient precision to allow the acceleration of the maser components to be measured. This has yielded a distance to that galaxy to be measured by purely kinematic means, independently of the Cepheid distance scale. Herrnstein *et al.* (1999) determined, by two different methods, that the distance to NGC 4258 is 8.1 ± 0.1 Mpc, and, since Cepheid variables have also been observed there, they derived the definitive calibration for the Cepheid distance scale. With time, as more distant extragalactic megamasers are discovered, it may become possible to assess the cosmological effect on their angular diameters.

16.3 Gravitational lensing

The bending of a light ray in the gravitational field of a massive body was first observed at the solar eclipse of 1919, when it provided a crucial test of Einstein's Theory of General Relativity. In 1936 Einstein pointed out that, if a distant star were to be seen almost precisely behind a closer star, two ray paths would be possible through the gravitational deflection of the rays by the closer star, and it would be seen as a double image. Einstein dismissed the idea as being most unlikely. Nevertheless, in an extragalactic context, his conjecture was realized in 1979 when Walsh, Carswell and Weymann discovered that the radio source B0957+561 was identified with a close pair of star-like images on a sky-survey plate; these turned out to be quasars, leading to the colloquial name 'the double quasar' (Walsh *et al.* 1979). The two images, 6 arcsec apart, were found to have identical spectra, both with redshift $z = 1.41$; the separation of the two images was due to gravitational 'refraction' in a 'lensing' galaxy on the line of sight, at a redshift $z = 0.36$.

Although conceptually it is easier to think of gravitational refraction as the effect of gravity on a light ray, the correct analysis instead regards gravity as distorting space-time, in which light follows a re-defined straight line (a geodesic) with velocity c. Nevertheless it is permissible to state simply that the effect of a point mass is a deviation by an observable angle.

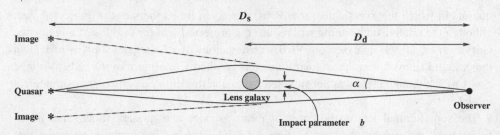

Figure 16.5. Ray paths in a gravitational lens in which the lensing galaxy is treated as a point mass. Two paths are shown; in the more general case there are three or five, depending on the geometry of the extended lensing galaxy.

There are two regimes in which such a deflection is observed.

(1) *Strong lensing*, with multiple images, distorted by the lensing process. Three categories of strong lensing can be distinguished.

(i)   Macrolensing, where a foreground cluster of galaxies creates many images of distant galaxies, on a scale of tens of arcseconds.
(ii) Mesolensing, where a foreground galaxy images a distant quasar or galaxy, on a scale ranging from arcseconds to a fraction of an arcsecond.
(iii) Microlensing, where the individual stars in a foreground galaxy, or in our own galaxy, image a more distant object, on a scale of microarcseconds. Here the images can be confused, with overlapping caustics, but the effect is manifested by scintillations of the lensed object.

(2) *Weak lensing*, when a background object shows only a single, but distorted, image.

The lensing phenomenon had been studied as a theoretical possibility some years prior to the discovery by Walsh *et al.* Zwicky (1937) predicted that galaxies were likely to act as lenses, and Refsdal (1964), among others, worked out the optics of a point-mass lens. The text by Schneider, Ehlers and Falco (1993) gives a comprehensive treatment of gravitational lensing, with a summary of the extensive literature and history of the subject.

According to general relativity, the angular deflection $d\alpha$ in a ray path dl due to a gravitational field g is

$$d\alpha = 2c^{-2}g\,dl. \tag{16.1}$$

Integrating along a ray path close to a mass M, with impact parameter b, as in Figure 16.5, gives a total angular deflection

$$\alpha = 4MG/(c^2 b). \tag{16.2}$$

The effect is very different from the geometry of a simple optical lens, in which the angular deflection is proportional to axial distance; here the deflection is *inversely* proportional to the axial distance, giving distorted images closer to those of a mirage rather than a lens. Note that the angle of deflection can be written $2R_S/b$, where $R_S = 2MG/c^2$ is the Schwarzschild radius for the mass M.

The lensing geometry is shown in Figure 16.5. The object, whether a quasar or a radio galaxy, is treated here as a point source. It is located in the object plane at distance D_s, and by convention an image plane is constructed centred on the deflecting mass, the lens, at distance D_d; the source would have been seen at an angle β from the centre had no lens been present. Two ray paths, on either side of the lens, give rise to two images seen by an observer; these are seen projected onto the image plane at an angle θ with respect to the axis joining the observer and the centre of mass. In this small-angle approximation, a simple geometrical construction, with the impact parameter $b = D_d\theta$, gives, using the point-mass deflection law of Equation (16.2),

$$\theta = \beta + \left(\frac{D_{ds}}{D_d D_s}\right)\left(\frac{4GM}{c^2}\right)\frac{1}{\theta}, \tag{16.3}$$

where D_{ds} is the lens-to-source distance (as noted earlier, angular-size distances apply to the lensing case). There is a particularly simple solution of Equation (16.3) when the point mass and the source are lined up ($\beta = 0$): the source becomes a ring, generally referred to as an 'Einstein ring', of angular radius θ_E,

$$\theta_E = \sqrt{\left(\frac{4GM}{c^2}\right)\left(\frac{D_{ds}}{D_d D_s}\right)}. \tag{16.4}$$

The Einstein-ring radius in physical terms can be written $R_E = (2R_s D_*)^{1/2}$, where $D_* = D_s D_d/D_{ds}$ is a 'corrected distance'. Equation (16.3) is a simple quadratic equation, and the solutions for non-zero β give one image outside the Einstein-ring locus (the principal image) and one image inside the ring. The images are magnified; the closer the source is to the observer–lens axis, the higher the magnification. The magnification of a small source is the ratio of the observed solid angle $\delta\theta_1\,\delta\theta_2$ in the image plane to the solid angle $\delta\beta_1\,\delta\beta_2$ that the source would have subtended in the object plane if there had been no lens, where the angles are now expressed in two dimensions. When the lens mass distribution is circularly symmetrical but extended, there will still be an Einstein ring for an aligned source, but the mass in Equation (16.4) will be the lens mass contained within the ring.

16.3.1 Imaging by extended lenses

The 6-cm radio map of the double quasar B0957+561, shown in Figure 16.6, demonstrates the complexities that arise in real cases. The northern source is the principal image, and it clearly has a core–jet structure, but no such structure is immediately evident for the southern image. The source near the southern image is definitely associated with the most prominent galaxy (designated G1) in the field, and it is that galaxy, the brightest in an extended rich cluster that covers the field, which exerts the largest lensing influence. There is, however, a considerable gravitational effect from the mass of the cluster as a whole, and from the dark matter that may be contained in it; finding the mass distribution from the details of such images is a complex task. Even an isolated galaxy acting as a lens is not a point mass, and can seldom be treated as spherically symmetrical. More general cases, therefore, must be considered.

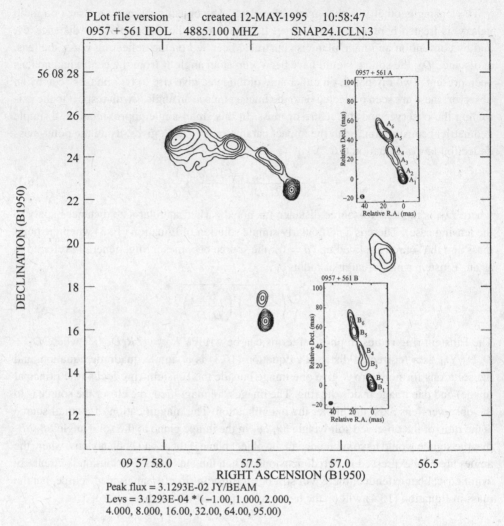

Figure 16.6. A radio map at wavelength 6 cm of the 'double quasar' B0957+561. The inset detailed maps of the two main components (Garrett *et al.* 1994) show similar, but not identical, structures.

The elementary point-mass case described by Equation (16.3) is easily generalized to the case of a mass distribution, which can, in the small-angle approximation, be taken as a surface density Σ in the image plane. The angles β and θ will now be vector quantities, and the scattering angle α will be a vector function of the impact parameter $D_\mathrm{d}\theta$. Thus, a brightness distribution in the object plane, $I(\rho)$, is mapped into a new brightness distribution, $I^*(\theta)$, given by solving the generalized form of Equation (16.3):

$$\theta = \beta + \left(\frac{D_\mathrm{ds}}{D_\mathrm{s}}\right)\alpha. \tag{16.5}$$

As noted earlier, angular-size distances are taken. All the angular variables are now two-dimensional, and solutions generally have to be found numerically. There are some general properties of the lensing equation, however, that aid understanding.

- For a mass distribution of finite density, there should be an odd number of images.
- At least one image, the principal image, has a flux at least as great as that of the unlensed source, and it must have positive parity (i.e. not a mirror image).
- Surface brightness is conserved in the lensing process.
- A low surface density in the lens gives a single image: an elementary solution of the circularly symmetrical case, in which there is an average surface density Σ within the impact parameter, gives a single value for θ if the average surface density is less than a critical surface density[1] Σ_c, given by

$$\Sigma_c = \left(\frac{c^2}{4\pi G}\right)\left(\frac{D_s}{D_d D_{ds}}\right). \tag{16.6}$$

As in the simple point-mass case treated above, the scalar magnification μ will be given by the ratio of the elements of solid angle in the image and object planes, and the more general magnification matrix follows by taking the Jacobian derivative of Equation (16.5) with respect to θ. On setting the derivative $d\beta/d\theta$ equal to the inverse of the magnification matrix $\mathcal{M}$ ($\mathcal{I}$ is the identify matrix[2]),

$$\mathcal{M}^{-1} = \mathcal{I} - \left(\frac{D_{ds}}{D_s}\right)\left\|\frac{\partial\alpha}{\partial\theta}\right\|, \tag{16.7}$$

we see that the Jacobian of the deflection angle is the critical quantity to be calculated. The elements of the inverse magnification matrix are conveniently expressed in terms of a deflection potential ψ; this is determined by the three dimensional distribution Φ of gravitational potential in the lens. The integral of $\Phi\,dl$ along the ray trajectory is related to ψ by

$$\psi = \frac{2}{c^2}\frac{D_{ds}}{(D_s D_d)}\int \Phi\,dl. \tag{16.8}$$

The second derivatives ($\psi_{1,2}$ etc.) of ψ with respect to the coordinates θ_1 and θ_2 in the lens plane determine the inverse magnification matrix, which can be written

$$\mathcal{M}^{-1} = \begin{vmatrix} 1 - \kappa + \gamma_1 & \gamma_2 \\ \gamma_2 & 1 - \kappa - \gamma_1 \end{vmatrix}, \tag{16.9}$$

where

$$\kappa = \frac{1}{2}(\psi_{1,1} + \psi_{2,2}), \tag{16.10}$$

$$\gamma_1 = \frac{1}{2}(\psi_{1,1} - \psi_{2,2}), \tag{16.11}$$

$$\gamma_2 = (\psi_{1,2} + \psi_{2,1}). \tag{16.12}$$

---

[1] Typically the critical surface density for a galaxy is of the order of $1\,\mathrm{g\,cm}^{-2}$.
[2] The identity matrix has 1s on the principal diagonal and 0s elsewhere.

The quantity κ, which is called the *convergence*, depends on the mass within the beam. The combination

$$\gamma = \left(\gamma_1^2 + \gamma_2^2\right)^{1/2} \qquad (16.13)$$

is called the *shear*, and depends on the mass outside the beam.

By choosing the normal coordinates, $\mathcal{M}^{-1}$ can be diagonalized, with elements $1 - \kappa + \gamma$ and $1 - \kappa - \gamma$. Given a source distribution, the magnification matrix $\mathcal{M}$ is usually the desired quantity; it can be derived from $\mathcal{M}^{-1}$ by standard matrix inversion. Starting from the diagonal form of $\mathcal{M}^{-1}$, the magnification matrix will also be diagonal, with elements $(1 - \kappa \pm \gamma)^{-1}$. This gives the transformation for a line element; the scalar magnification for a small element of solid angle will be $\mu = [(1 - \kappa)^2 - \gamma^2]^{-1}$. The magnification of a small source can be positive or negative; a positive magnification means that the parity of the image will be the same as that of the source, while negative magnification implies that the parity of the image is reversed. Refer to the text by Schneider, Ehlers and Falco (1993) for a more complete discussion of the detailed considerations of lensing properties. We now turn to several useful properties that are relevant to lenses as observed.

Given a complete solution of the lensing equation, one will always find that there is a locus that divides the object plane into an outer region in which a source has only a single image and an inner region in which it has three images. Nested within, there can be another locus that divides the regions for three and five images, and so on. These are the caustic curves, and in the image plane they will correspond to a set of critical curves, which mark the locus where multiple images will merge as the object approaches a caustic.

We now describe the lensing effect in more detail for a simple elliptical galaxy, describing the appearance of the images of a small test object at various angular distances and directions in relation to the axes of the lensing galaxy (Figure 16.7). The position of the source is shown in the object plane in relation to two caustic curves, which form an outer ellipse and an inner diamond. These caustic curves divide the object plane into three regions. A source outside both curves has a single image; between the curves it has three images; and inside the diamond it has five images. The appearance on the sky is described on an image plane, also seen in Figure 16.7; here the images of the caustic curves are the critical curves, dividing areas in which the various images appear.

We trace the development of the images of the test source as it is moved along the track shown in the object plane, moving inwards from outside both caustics. Initially there is an angular deflection with little distortion. As the source crosses the outer caustic, two new images, B_1 and B_2, appear, on either side of the corresponding (inner) critical curve; the original image A stays outside the critical curve, where it ends up on one of the principal axes of the elliptical galaxy. Similarly, when the source crosses the inner caustic, two or four detected images, C_1 and C_2, appear on either side of the outer critical curve; both pairs of images move towards the principal axes as the test source is moved closer to the centre of the galaxy.

The pairs of images have opposite parities; B_1 and C_1 have negative parity, that is, they are inverted as in a mirror. Surface brightness is always conserved, so the total intensity of

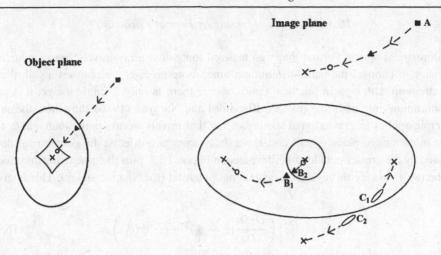

Figure 16.7. Object and image planes, showing caustic curves and critical curves for a small source lensed by a simple elliptical galaxy. The development of multiple images is traced as the source approaches the centre of the lensing galaxy. After Walsh (1993).

each image depends on its angular scale, that is, on its magnification. B_2 is usually strongly demagnified, and is unlikely to be seen; for this reason most systems have only two or four images. The images are distorted; close to a critical curve the two images of a pair are stretched towards each other, as shown for C_1 and C_2 in Figure 16.7. Near the inner critical line, that is, for an object close to the outer caustic, the stretching is radial; close to the outer critical line it is tangential.

As the test source approaches the centre of the lensing galaxy, the faint image B_2 moves towards the centre of the image plane, and the other four move to the principal axes of the elliptical galaxy, forming an 'Einstein cross'. If the source is sufficiently extended, covering the whole of the inner caustic, the cross is extended into a complete 'Einstein ring'. Both these formations have been observed.

The map of B0957+561 (Figure 16.6) illustrates the principles. There must be a critical line just above the southern quasar image, and there should be a faint third image of the quasar close to the lensing galaxy. The extensive jets associated with the northern source have no counterparts, so the outer critical line must pass close by the A image, with most of the jet structure falling into the single-image domain. Also VLBI maps of the two quasar cores are shown in Figure 16.6, and their similarity is a convincing demonstration that the lens hypothesis is correct. Moreover, they have reflection symmetry, and therefore opposite parity, as they must in any reasonable lens model. The images have different elongation; this shows that the magnification matrix for one is different from that for the other, and this difference aids in constructing a mass model for the lens. A comprehensive discussion of the lens model has been given by Falco *et al.* (1991); they show conclusively that the cluster has an important influence on the magnification matrix, and dark matter in the cluster is probably the main contributor, although the model is not unique.

16.3.2 Analysis of lensing by Fermat's principle

In geometrical optics, Fermat long ago realized that when a ray travels from a source to an image, it chooses the path of minimum time; more precisely, it chooses a path that is an extremum, although in practical optics, where there is only a single image, it is only the minimum-time path that matters. Blandford and Narayan (1986) show how the same principle applies to gravitational lensing. A ray that travels from a point source at θ_s to a point in the image plane θ_i will undergo a delay from two effects: the geometrical delay induced by the greater path length, illustrated in Figure 16.5, plus the time delay introduced by the ray's passage through the gravitational potential (the Shapiro delay). This is given by

$$\tau = (1+z) \left(\frac{D_d D_s}{2 D_{ds}} (\theta_s - \theta_i)^2 - c^{-3} \psi(\theta_i) \right). \tag{16.14}$$

Now, if the ray path from the source at θ_s is traced to all points θ_i in the image plane, a contour plot of time delay can be constructed. Without any gravitational lens there is a single minimum at the centre, where the image must appear. Weak lensing moves the minimum to one side. Multiple images, which result from lensing in an extended, more massive object such as a galaxy, appear as extrema in the delay plot; these may be minima or maxima, or saddle points. If there are three extrema, the first locates the minimum time delay for the principal image; the second is a maximum in the time delay, but still stationary with respect to variation in position, therefore an image will appear; and the third extremum is a saddle point where the curvature of the time-delay function is positive in one direction and negative in the other, so a third image will be formed.

One notes, therefore, that there are three classes of image. The first, where there is an image formed at a minimum in the time delay, occurs where the curvature is positive in both coordinate directions, so the parity in the image will be the same as for the source. At a maximum in the time delay, both curvatures are negative, so the parity of the image will also be even. For the saddle point, the two curvatures have opposite signs, resulting in a parity change in the image. The quasar images for B0957+561 show this effect; the VLBI results for the A and B images, shown in Figure 16.6, exhibit opposite parity. The third image is faint and merged close to the core of the imaging galaxy, and must be at a maximum in the time delay, but with the same parity as the A image.

When the surface density of the lensing galaxy is further increased, the delay plot becomes more complex, showing two minima and two saddle points, with a single maximum, and therefore five images appear.

16.4 Observations of lenses: rings, quads and others

Table 16.1 lists the clearly identified cases of gravitational lensing up to mid 1994, distinguishing those discovered optically (O) from those discovered by radio (R). There are more R entries than O in the table; this is at first surprising when the great majority of

Table 16.1. *Gravitationally lensed systems (Walsh 1993)*

Name	θ_{max}	Images	z_{source}	z_{lens}	O/R
B0957+561	6″.1	2	1.41	0.36	R
MGB2016+112	3″.8	3	3.27	1.01	R
B1115+080	2″.3	4	1.72	0.29	O
B0142–100	2″.2	2	2.72	0.49	O
MG0414+0534	2″.1	4	2.63		R
MG1131+0456	2″.1	Ring			R
MG1654+1346	2″.0	Ring	1.75	0.25	R
B2237+031	1″.8	4	1.69	0.039	O
MG1549+304	1″.7	Ring		0.111	R
B1413+117	1″.4	4	2.55		O
B1422+231	1″.3	4	3.62	0.64	R
0751+271	1″	4			R
PKS1830–211	0″.98	Ring			R
B1938+666	0″.92	4			R
B0218+356	0″.33	Ring		0.685	R

observable quasars are optical rather than radio objects, but it turns out that the greater angular extent of radio sources increases the chance that they will be seen through a lensing galaxy. It is essential to have observations with high angular resolution; although the first double-quasar system had a separation of 6 arcsec, most of the recent discoveries are of systems with sizes 1 arcsec and below, a domain in which the radio observations often have the advantage. Optical observations nevertheless now dominate the tally of observed lenses, since there are many large wide-angle cameras with high angular resolution. The Hubble Space Telescope in particular has produced a spectacular series of photographs of giant arcs, which are partial Einstein rings formed by compact galaxy clusters (Smail *et al.* 1996).

The radio sample contains enough examples of lensing to place lower bounds on the frequency of occurrence. The selection criteria have not been homogeneous, but it appears that every sample of 1000 radio sources will contain two to four lensed objects. Turner *et al.* (1984), when they addressed the statistical problem, concluded that unless every sample of 1000 quasars contained at least two examples of lensing, there would have to be something seriously wrong with the current ideas of cosmology. So far, the rate of occurrence of lensing appears to be well above this limit.

Table 16.1 shows that Einstein rings are not a rare occurrence. The first, MG1131+0456, was discovered by Hewitt *et al.* (1988); Figure 16.8(a) shows a later map at wavelength 2 cm (Chen *et al.* 1995). The ring is complete, and other sources are present. These are more compact, and lie just inside and outside the ring, as they should. Kochanek and Narayan (1992) developed an inversion algorithm that allows one to construct the appearance of the

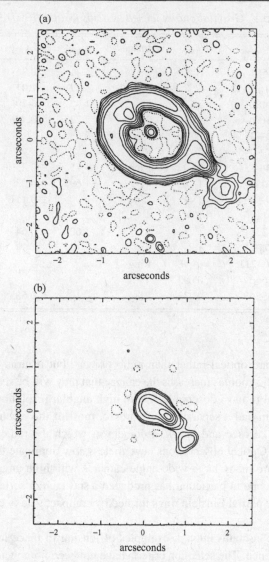

Figure 16.8. The Einstein ring MG1131+0456: (a) imaged at 2 cm and (b) the core–jet source derived from a modelling process (Chen *et al.* 1995, arXiv:astro-ph/9501031).

source in the object plane. Figure 16.8(b) shows the core–jet configuration of the source deduced by Chen *et al.*; the core is in the three-image domain (see Figure 16.7), while the jet crosses the caustic into the five-image domain.

Another ring, MG1654+134, is shown in Figure 16.9 superposed on an optical image of the lensing galaxy. The source here appears to be a single object, which, however, is one lobe of a triple source centred on the quasar at Q (Langston *et al.* 1990). The comparative simplicity of this ring provides a good possibility of measuring the mass-to-light ratio of the foreground galaxy.

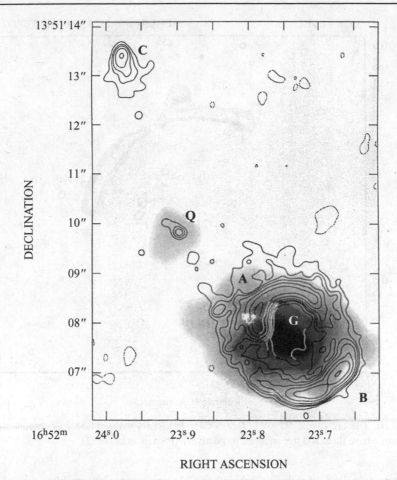

Figure 16.9. The Einstein ring MG1654+1346: an 8.4-GHz VLA map superposed on an optical image. G is the lensing galaxy, which is in front of one of the two radio lobes of the quasar Q (Langston *et al.* 1990).

The more complex object B1938+666, shown in Figure 16.10, is formed from a double source, with multiple images including two extended arcs. Finally, the infrared image of the lensed system B2045+265, shown in Figure 16.11, is an example of an 'Einstein quad', with four easily recognized point-like images and the lensing galaxy; a fifth component, too faint to be recognized but which is probably present in all quads, is expected at the position marked by a cross.

There are optically discovered examples of lensing that should be considered alongside the radio cases. The 'Einstein cross' is a quad discovered by Hewitt *et al.* (1988) in which the alignment of object and lensing galaxy is so close that it would be a ring, except for the compactness of the object quasar (Kochanek and Narayan 1992). There are now several striking examples of lensing on a larger scale, in which the mass of an entire

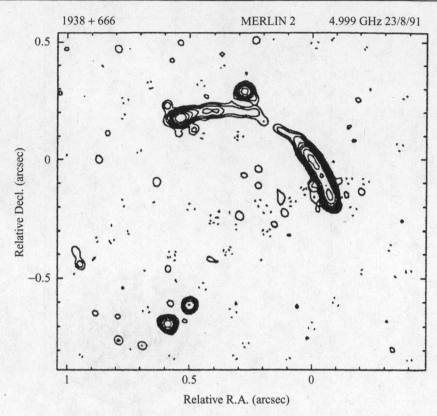

Figure 16.10. The arc source B1938+66.6 (5-GHz MERLIN map). The two components of the double source have three and five images, respectively (Patnaik *et al.* 1992).

cluster acts as a lens on galaxies beyond the cluster, producing arc-like images (Smail *et al.* 1996). Eventually, radio examples should also be found, but these phenomena will not be as common as the optical examples, since there is a richer background of distant optical galaxies to be imaged.

In practice it is often difficult to analyse an observed gravitationally lensed image and obtain both the mass distribution in the lens and the brightness distribution in the original object. Keeton *et al.* (2000) show that two stages are involved: a guessed model solution, using the general properties of the image such as multiplicity and parity of the image components, followed by a sophisticated optimization of the model to obtain the best possible match between computed and observed images.

16.5 Time delay

Except for the symmetrical system of an Einstein ring, the lengths of the ray paths corresponding to the various images in the strong-focussing regime will be different. Any

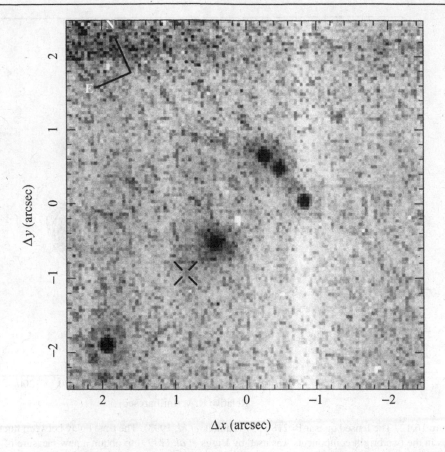

Figure 16.11. The fourfold image of the quasar B2045+265, seen in an infrared image from NICMOS. The lensing galaxy is near the centre; the cross marks the expected position of a fifth component (Fassnacht *et al.* 1999, arXiv:astro-ph/9811167).

variation in intensity of the source will therefore be observed at different times in the different images. As noted in Section 16.3.2, there are two components in the observed delay, one due to this excess geometrical path and the other due to the different gravitational time delays as the rays cross the gravitational potential well $\psi(\theta)$ of the galaxy (the Shapiro delay); in practical cases these two components are of comparable magnitude. Both must be scaled by the redshift z of the lensing galaxy, with the time delay expressed in Equation (16.14). Interpretation of an observed time delay therefore depends both on a measured value of the distances and on an evaluation of the potential ψ by modelling the mass distribution in the lensing galaxy.

There are several good candidates for this application of gravitational lensing. A delay of 10.5 ± 0.4 days has been observed in fluctuations of the two main components of B0218+357 (Biggs *et al.* 1999). The lensed image of this object also includes an Einstein ring (Figure 16.12), which greatly simplifies the construction of the geometry of the

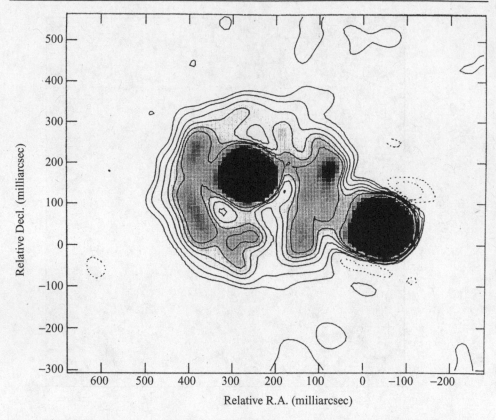

Figure 16.12. The lensed quasar B0218+357 (Patnaik *et al.* 1993). The time delay between fluctuations in the two bright components was used by Biggs *et al.* (1999) to obtain a new measure of H_0. Note the Einstein ring, which indicates a simple lens system directly on the line of sight.

lensing galaxy. Using the angular separation of the two components (335 milliarcseconds), the redshift of the quasar $z = 0.96$ and of the lensing galaxy $z = 0.68$, an independent distance scale can be deduced which gives a value of the Hubble constant $H_0 = 69^{+13}_{-19}$ km s^{-1} Mpc^{-1}, which is close to other recently determined values.

16.6 Weak gravitational imaging

We have so far considered gravitational lensing in the so-called *strong* regime, in which objects are close to or within the caustic curves and critical lines of Figure 16.7. Outside this area there is a regime of *weak gravitational lensing*, in which small effects may be observed over a larger area of sky. The effect of an individual lensing galaxy or cluster is to distort the spatial distribution of distant sources and to elongate their images: the problem of interpretation is that their original distribution and individual shapes cannot be known, and only a statistical effect can be observed. The importance for individual lenses is that the mass distribution may be observed out to larger radii than for the strong regime; as we have seen in Chapter 10, the dynamical behaviour of stars and interstellar gas suggests the existence of unobserved dark matter at large radii in many galaxies.

There is also a cosmological application for weak lensing. A distant background of galaxies distributed over a large area of sky may be seen through a statistical lens formed by random fluctuations in density. If many of the sources are at large redshifts, the statistical effects on spatial distribution and on source distortions may relate to distant mass distributions that are otherwise inaccessible. For such observations, as indeed for all applications of weak lensing, a large observational survey is required. Most such work will be done optically, where techniques for imaging some millions of galaxies are available, notably in the Sloan Digital Sky Survey (SDSS), where weak lensing has already been clearly demonstrated (Fischer *et al.* 2000). Radio astronomy, however, has a special importance because radio surveys such as those described in Section 16.2 select sources with high redshifts at which cosmological effects become especially interesting (see a comprehensive discussion of gravitational lensing by Mellier (1999)).

16.7 Microlensing

Microlensing, by individual stars in a lensing galaxy, is observed on rare but important occasions. Originally, the idea of individual stars acting as lenses was dismissed as too remote a possibility, but it was recognized by Chang and Refsdal (1979) that, if the object quasar was sufficiently compact, the individual stars of the lensing galaxy might complicate the lensing process. This is indeed the case, particularly at optical wavelengths, and the collective effect can induce complicated time variations due to proper motions. Wambsganss (1993) reviews the observational evidence and shows that Huchra's 'Einstein cross' exhibits strong microlensing. The radio observations are less likely to be subject to microlensing, since the object quasar tends to be more extended in the radio domain, but there are examples in which microlensing may have to be considered. Microlensing already is receiving an interesting application at optical wavelengths; a wide-angle camera with a CCD detector array can be used to monitor a large number of stars, searching for the transient lensing effect of any massive body passing in front of any one of them. Monitoring programmes of this kind are especially appropriate for the comparatively distant stars in the Galactic bulge and in the LMC, in a hunt for single dark stars (or other compact objects, sometimes referred to as MACHOs). The intention is to find out whether some or all of the dark matter known to exist in the Galaxy can be ascribed to compact stellar- or substellar-mass objects.

Further reading

Cosmological Applications of Gravitational Lensing, Blandford R. D. and Narayan R., *Ann. Rev. Astron. Astrophys.*, **30**, 311, 1992.

Gravitational Lenses, Schneider P., Ehlers J. and Falco E. E., Berlin, Springer-Verlag, 1993.

Probing the Universe with Weak Lensing, Mellier Y., *Ann. Rev. Astron. Astrophys.*, **37**, 127–189, 1999.

Radio Sources and Cosmology, Condon J. J., in *Galactic and Extragalactic Radio Astronomy*, ed. Verschuur G. L. and Kellermann K. I., Berlin, Springer-Verlag, 1998, p. 641.

The Deep Universe, ed. Sandage A. R., Kron R. G. and Longair M. S., Berlin, Springer-Verlag, 1993.

17

The future of radio astronomy

We have concentrated in this introductory text on the distinguishing features of radio astronomy, in the domains of both technique and astrophysics. In this concluding chapter we survey the prospects for astrophysical and cosmological advances through new and developing radio-astronomical observations, building on the attributes and techniques which are special to radio astronomy.

17.1 The Cosmic Century

In his masterly survey of the development of modern astronomy Longair (2006) gave this name to the twentieth century. Astronomy has been transformed both in terms of our conception of the Universe and regarding the ways in which we can observe it. Up to the middle of the century the term 'telescopes' needed no qualification: they are now 'optical telescopes', alongside infrared, X-ray, gamma-ray and radio telescopes. All have their own special techniques and their own particular domains of astrophysics and cosmology.

Radio astronomy provides unique information on phases of matter and radiation that are otherwise inaccessible. The low-temperature radiation of the cosmic microwave background has been studied from sub-millimetre to decimetre wavelengths, especially around its peak at millimetre wavelengths. Radio astronomy also encompasses radiation from electrons with cosmic-ray energies, often again in objects that are invisible or less accessible for observations at other wavelengths. Relativistic electron gas, detected by observing its synchrotron radiation, is characteristic of quasars, active galactic nuclei and many star systems. Magnetic fields in these synchrotron sources are revealed by the polarization of the synchrotron radiation. The dilute warm gas in our Galaxy and other galaxies is detected by observing its thermal radiation, and its electron content is measured by recording Faraday rotation. Molecular concentrations in the colder parts of the Galaxy and in star-forming regions are seen by virtue of their coherent line radiation. Pulsar radiation, which is almost entirely concentrated in the radio domain, provides our most direct observations of condensed matter, while precision timing of pulsars has already provided some of the most accurate tests of general relativity. Furthermore, radio astronomy suffers from hardly any obscuration by gas or dust in the Milky Way, and detects objects at the largest cosmological distances.

There are parts of astronomy that are inherently geometrical, and therefore independent of wavelength. The effect of a gravitational field on electromagnetic waves is the same for all wavelengths; only the techniques and the available accuracies may differ. Gravitational lensing, which was discovered in a double radio quasar, now produces prolific results both in the radio and in the optical domain; in particular, the Einstein rings which provide the most direct measurements of the mass content of lensing galaxies are observed in both, with angular resolutions of the order of 1 arcsec. Radio is here especially important in extending angular resolution to better than 100 milliarcsec. On the other hand, the vastly greater rate of data gathering available in wide-field optical imaging has so far provided the only way of searching for microlensing by dark stars in the Galaxy (Chapter 16): here is an interesting pointer towards the development of matching techniques in radio astronomy, in which wide-field imaging will be made possible by using massive computer power in multi-channel systems.

Positional astronomy gains from a remarkable complementarity between radio and optical domains, with radio providing the most accurate framework of positions of extragalactic objects (see Appendix 2), while the more comprehensive body of optical data links this framework to more local parts of the Universe. The Hipparcos satellite, which observed the comparative positions of over 100 000 stars during a continous period of over 3 years, yields parallax and proper-motion data with an accuracy of approximately 2 milliarcsec. The radio data gained through astrometric VLBI measurements have an accuracy that is about ten times better than this, but for far fewer objects. Monitoring Earth rotation similarly involves a combination of techniques: the fundamental frame is provided by radio interferometry of extragalactic objects, while variations in rotation rate and axial direction are in practice monitored by optical laser ranging on terrestrial satellites.

The complementarity of the various domains of astronomy is more obvious and more fundamental in investigations of the astrophysics of individual objects; here, as we have seen in the later chapters of this book, the long wavelengths of radio often play a special role. The following sections concentrate on some areas of particular importance for future research. As more sensitive techniques are developed, especially at millimetre wavelengths, these have increasingly come to include thermal sources, adding to earlier fields of observation where the radio sources are high-energy and non-thermal. We point first to the possibilities of better observations of the coldest and most precisely thermal blackbody source, the cosmic microwave background.

17.2 The cosmic microwave background

The beautifully precise demonstration of the blackbody spectrum of the CMB (Figure 14.3) was achieved with an infrared radiometer in the satellite COBE, using a Fourier-transform technique. The satellite WMAP followed with maps of the structure and polarization of the CMB, transforming our understanding of the physics of the big bang. The satellite Planck, using radio techniques, is continuing this remarkable series with more precise observations of polarization.

The CMB requires observations at millimetre wavelengths, a range that is on the borderline between radio techniques and infrared radiometry. A sensitivity in the range of some 10 microkelvins is required, and it is interesting to compare the radio and infrared techniques: radio typically measures the correlation between combinations of signals from individual horn antennas, whereas in the corresponding infrared radiometer technique each horn would have its own separate bolometric detector. The advantage of the radio technique is its inherent stability and repeatability. In a bolometric measurement, angular structure is seen as the difference between separately detected and amplified signals, whereas in the radio technique the signal amplitudes may be combined in common amplifier systems before detection. As we have seen in Chapter 5, these techniques are similar to the correlation techniques used in radio interferometers, with inherent advantages in stability. The full advantage of the radio technique is, however, realizable only if a very wide bandwidth is used, giving the same sensitivity as the inherently wide band of a bolometer; as we shall see, the development of such wide bandwidths forms a vital part of the technical development of our discipline.

Although in these CMB measurements satellites excel in avoiding atmospheric absorption and radiation, there is still a vital role for ground-based observations in extending observations to higher angular resolution. Furthermore, the structure in the CMB is confused, at some level yet to be determined, by foreground radiation from interstellar dust, synchrotron and thermal radiation from the Milky Way Galaxy, and radiation from weak but numerous extragalactic sources; ground-based radio mapping, at a range of longer wavelengths, may prove to be essential in evaluating these confusing sources.

17.3 The interstellar medium

In optical astronomy, the ISM is detected mainly through the absorption of starlight by metallic atoms. The ISM, however, contains a rich mixture of high- and low-energy gas, many species of which are detectable only at radio wavelengths. At the longest radio wavelengths the sky is dominated by synchrotron radiation from cosmic-ray electrons; here radio provides the only observations of the main flux of cosmic rays in the Galaxy. Our understanding of the complex structure of the lower-energy ionized and neutral gas, as described in Chapter 9, also depends largely on radio.

The richest area of radio research on the ISM is interstellar chemistry. This mainly concerns millimetre wavelengths, at which many possibilities are being opened up by recently constructed reflector telescopes with surface accuracies of about 50 μm. Low-noise detectors and wide-bandwidth correlation spectrometers are becoming available for these wavebands. The millimetre wavebands are relatively free of terrestrial radio transmissions and, as can be seen from Chapter 9, spectral lines may be observed at frequencies up to 900 GHz. A most interesting field is the measurement of isotopic ratios in a number of atomic species; if this can be extended to a range of locations, it may reveal the development of element abundances in the Galaxy.

The ISM also reveals the structure of the Galaxy. Through its dynamics, as measured in spectral lines such as those of neutral hydrogen and CO, we already have a detailed model of the dynamics and spiral structure of the Galactic plane. Furthermore, the concentration of high-mass molecular species in denser regions of the ISM is a useful indicator of features such as the bar of the Milky Way and the disc surrounding the central black hole. There is, however, still much to be learned about the dynamics of the high-latitude structure. The distribution of ionized gas is more complex. Here, observations of the dispersion measurements of pulsars are particularly rewarding, giving the integrated electron content on lines of sight out to large distances.

17.4 Angular resolution: stars and quasars

We group together the very different observational situations for stars and quasars in order to emphasize the complementarity of optical and radio astronomies. Stellar astronomy is, obviously, primarily optical; as for the Sun, the blackbody radiation from a stellar surface at some thousands of degrees is mainly in or near the optical range. When, however, a star has an extended atmosphere, such as the solar corona, this may radiate effectively only at radio wavelengths. It is then distinguishable from the star itself by the use of the very high angular resolution which is obtainable only in radio interferometry.

A similar complementarity exists in the study of quasars. Although the majority of discrete objects in a radio survey of the whole sky would be quasars, it turns out that most of these quasars are detectable optically. This is due to the sensitivity and huge data capacity of large-area optical detectors, especially CCDs and the wide-angle Schmidt cameras. Again, it is the higher angular resolution of radio interferometry that makes a unique contribution. Optical observations from ground-based observatories often achieve resolutions of better than 0.5 arcsec; at the best sites 0.2 arcsec may be achievable, and adaptive optics could improve this further. Radio, however, routinely produces maps of individual objects at resolutions better than 100 milliarcseconds from interferometer arrays such as the VLA and MERLIN; the longer baselines of the VLBA, the EVN and the world-wide array (WVN) now operate with angular resolutions better than one milliarcsecond. Of all the mysteries of quasars, the most fascinating concern their smallest components, and at present only radio can hope to delineate their innermost structures. The extension of radio interferometer baselines beyond the diameter of the Earth has been achieved in routine observations by the VSOP spacecraft, carrying a radio telescope to a baseline of many Earth diameters. The high angular resolution of VSOP has shown that the brightness temperatures of many quasars and AGNs are beyond the self-Compton limit (see Chapter 13). Space VLBI (sometimes called orbiting VLBI) can achieve the highest available angular resolution; furthermore it can achieve a high sensitivity when used with the large telescopes of the ground-based VLBI arrays.

We may summarize the requirements of astronomical observations in all wavebands as sensitivity, and resolution in frequency, angle and time. Radio has the advantage in angular resolution: X-ray telescopes achieve about 1 arcsec, optical about 0.1 and radio 10^{-3}–10^{-4} arcsec. (For comparison, note that a 1-pc structure in a galaxy at 200 Mpc subtends an angle of 1 milliarcsecond, and the angle subtended by the black hole at the centre of the Milky Way is 5 microarcseconds.) Radio also has the advantage in frequency resolution, this advantage stemming from the coherent techniques available both in single telescopes and in interferometers: there is no limit to frequency resolution apart from the need for sensitivity. Sensitivity is also central to the resolution of temporal variations; for example, only large radio telescopes can resolve the finest structure of individual pulses from pulsars.

17.5 Optical and infrared interferometry

Radio therefore has a vital and individual part to play in the range of observational astronomies, old and new, that are now available to astrophysics. From the perspective of this chapter, however, the major contribution of radio astronomy might now be regarded as the availability of angular resolving powers that improve on optical astronomy by several orders of magnitude. Is this a fundamental difference between the two regimes, or is it possible for optical techniques to follow the same route, perhaps using interferometers with resolutions measured in microarcseconds?

The terrestrial atmosphere is clearly the limiting factor for optical astronomy. Even at the best observatory sites, the wavefront arriving at the ground from a distant point source has a coherence width of order 0.5 m, corresponding to a best possible image size of about 0.2 arcsec. The situation improves rapidly towards infrared wavelengths, but even so the coherence width is smaller than the apertures of the current generation of 10-m-diameter telescopes. However, the original coherence in the wavefront from a point-like source, although confused by the atmosphere, is still present on a very short timescale, of order 50 milliarcsec. Within this short time the coherence may be measured, and a correction made by active optics; the situation is then not very different from that in radio interferometry. The difference is that in the radio case the relative phase of signals from individual interferometer elements is stable for at least a few minutes, giving a much greater integration time and hence a greater sensitivity. The second limitation also concerns sensitivity. In an optical interferometer only small individual apertures can be used, since each must sample a coherent section of the wavefront. At present, each must therefore be less than 0.5 m in diameter, even at the best site; again the effect is to limit sensitivity. Larger apertures must contain an assembly of such small elements, combining them by using active optics. The sensitivity limit which is imposed by the limited collecting power of each element can be overcome only by the use of artificial, laser-generated reference stars; much greater sensitivity should then be possible. In contrast, radio-telescope apertures are limited in size only by engineering possibilities, since the local coherence of the wavefront is almost unaffected by the atmosphere.

Even within these practical limitations imposed by the atmosphere, which might in principle be overcome by building an interferometer in space or on the Moon, optical interferometry suffers a fundamental disadvantage that stems from photon statistics. At optical frequencies, the laws of quantum mechanics require noise photons to be added by any amplifier, with dire consequences for the signal-to-noise ratio. A radio signal, on the other hand, is easily amplified, and can be split into any number of channels for spectrometry or interferometry (or a combination of the two) with little loss of sensitivity; this is a basic distinction between the two regimes. Although in many ways the various regimes of astronomy are merging, this technical distinction will remain. All radio telescopes of the new generation which is emerging depend on aperture synthesis, which is not available for optical telescopes.

17.6 New large radio telescopes

Reber's original inspiration in building a parabolic reflector, extending the concepts of optical telescopes to radio wavelengths, proved to be a model that has been followed by a succession of larger and more accurate single-dish telescopes, including the Lovell, Parkes, Effelsberg and GBT (see Chapter 4). These instruments continue to dominate observations in which a large collecting area is the main requirement, for instance observations of pulsars and much of radio spectroscopy. Extension of the same principle to even larger collecting areas is scarcely feasible while preserving the ability to steer over the whole sky, but there is nevertheless a need for larger telescopes at longer wavelengths. The Arecibo reflector, using a fixed reflector surface and movable feed, points to the possibility of some further development in this direction; investigations are under way for such a reflector, with a diameter of about 500 m, to be built in a remote area in China.

The installation of multiple feeds on the existing large reflectors has already given a great advantage in surveys, such as the survey of pulsars carried out at the Parkes radio telescope. Primarily, however, the large single dishes are used for the analysis of signals from individual sources, such as molecular spectroscopy from distant galaxies and the detection of weak pulsars in galactic clusters or other specific locations. The ultimate in sensitivity and angular resolution can, however, be achieved only by large aperture-synthesis arrays.

The outstanding opportunities for the development of radio astronomy are now concentrated on aperture synthesis, which seems to offer almost unlimited possibilities for greatly improved sensitivity and angular resolution over practically the whole range of wavelengths, from several metres down to less than one millimetre. As we have seen in earlier chapters of this book, the basic ideas have been understood for half a century, but their realization on a large scale has become possible only through the advent of massive computional power together with digital techniques, fibre-optic communication and broadband, low-noise amplifiers. A new generation of radio telescopes is now coming into being, covering the radio spectrum from wavelengths of 10 m to 300 µm.

Connecting large arrays such as MERLIN is now achieved by fibre optics, providing wider bandwidth and greater stability than the original radio links. Fibre connections now also allow VLBI arrays to be connected without the need for recording systems; on-line correlation has been demonstrated on a world-wide scale, and arrays such as the European VLBI can now operate in real time. The improvement in performance can be dramatic;

the improved eMERLIN has a sensitivity up to 30 times greater than that of its previous radio-linked version.

17.6.1 LOFAR and MWA

The Low Frequency Array for Radioastronomy (LOFAR) is being built to cover the wide band 30–240 MHz. At such low frequencies the basic antenna elements must be broad-band dipoles operating on the same principle as the 'bowtie' dipoles of Figure 4.8(c). One set of dipoles covers the range 30–80 MHz, another the range 110–240 MHz, abandoning the VHF radio range 80–110 MHz which is heavily used by FM radio broadcasting. There will be 15 000 such dipoles, grouped into 77 clusters, known as stations, distributed over distances up to 100 km, centred on Dwingeloo in the Netherlands. As with all such arrays, it will operate on an increasing baseline scale as more elements are added, eventually extending into several neighbouring European countries. Baselines of 1000 km are envisaged, providing an angular resolution of the order of one arcsecond (see the ASTRON website).

At these comparatively low frequencies the signals can be handled digitally throughout the system, the fibre-optic transmission from each station having a capacity of 10 Gbits per second. It will be essential to divide the frequency band into small elements, both to avoid the inherent problems of wideband inteferometry (Section 5.2) and to avoid the many sources of narrow-band radio interference that will be encountered. A massive computer system will handle correlations between all 77 stations simultaneously in these separate frequency bands.

The very wide beamwidth of the dipole elements will accept signals at all times from all the bright radio sources in the northern sky. A large dynamic range will be essential in extending the sensitivity down to the low signal levels which are theoretically available to this array.

Such low frequencies by modern standards were used in the early days of radio astronomy, and abandoned partly because of the greater angular resolution available at shorter wavelengths and partly because of the apparently overwhelming interference from the increased use of radio in communication systems. Apart from the newly developed techniques which have made systems like LOFAR possible, there is a pressing cosmological reason to build large telescopes in this relatively unexplored region. Following the era of reionization at about $z \approx 1000$, there is a 'dark age' in the evolution of the Universe until the era of the furthest observable galaxies at about $z \approx 7$. Before this time, when there was an *era of reionization*, due to ultraviolet light from newly formed stars, neutral hydrogen was a major constituent, and its '21-cm-line' radiation should be observable, redshifted to frequencies below 200 MHz.

A similar array, the Murchison Wideband Array (MWA) is under construction in a remote area of Western Australia. Here the level of radio interference is very low, and it should be possible to observe over the whole of the band 80–300 MHz, including the FM band. The full array will comprise 8000 dual-polarization broad-band dipoles, as illustrated in

Figure 4.10. They will be grouped into 'tiles' each comprising 16 dual-linear-polarization, wide-beamwidth antenna elements over a ground screen. These elements are arranged in a planar, 4×4 pattern with 1.07-m spacing corresponding to half a wavelength at 140 MHz, with low-noise amplifiers (LNAs) integral to each antenna element. An analogue RF beamformer combines the 16 signals of each polarization with appropriate delays to form a beam 15–50° wide (FWHM), depending on frequency. A massive digital receiver system will provide cross-correlation with bandwidth 32 MHz between all pairs of the 496 tiles, with full Stokes parameters.

17.6.2 *The Giant Metrewave Radio Telescope (GMRT)*

At shorter wavelengths it becomes impractical to use individual dipoles as elements in a large array, and steerable parabolic reflectors are used. The GMRT, built in India, comprises 30 telescopes, of diameter 45 m, at spacings of up to 25 km. The frequency range is 30–1500 MHz. At decimetre and metre wavelengths the reflector surfaces need not be solid; instead a wire mesh with spacing 10–20 mm is used, with a consequent large reduction in weight and windage. The GMRT offers the most sensitive observations in this wavelength range, which includes the 21-cm hydrogen band.

17.6.3 *The Atacama Large Millimeter Array (ALMA)*

The rapid development of molecular radio astronomy, particularly in the millimetre-wavelength domain, led to the construction of millimetre-quality radio telescopes such as the 15-m JCMT on Mauna Kea and the NRAO 12-m telescope on Kitt Peak, and to arrays such as the Hat Creek Observatory, the Plateau de Bure array, near Grenoble, and the Owens Valley Observatory. Their main object of study was the 'cool universe': the molecular and dust contents of the ISM and the dense clouds where star formation, attended by the creation of planetary systems, is taking place. By the early 1990s it was recognized that both higher angular resolution and larger collecting area were needed in order to investigate the complex processes that were being discovered. As a result, projects were started in the USA and in Europe to build large millimetre-wave arrays. The large expense, and the common interest, led to a merger, with the North American portion led by the National Radio Astronomy Observatory and the European share under the direction of the European Southern Observatory. The National Radio Observatory of Japan then joined ALMA, as did a number of other international partners. Chile, which provides the site, is an important principal member of the observatory.

A high and dry site was required, and the Llano de Chajnantur, at an altitude of 5000 m in the Andes of northern Chile, was chosen. At a lower altitude of 2600 m, a support base has been built, near the town of San Pedro de Atacama, where quality control of components can be done. The specifications for the array elements, of diameter 12 m, demanded an r.m.s. surface tolerance of 20 μm, in order to support the required wavelength band, 0.3–9.6 mm. There are two antenna designs under manufacture for ALMA, one of which is

Figure 17.1. The first antenna of the millimetre-wavelength synthesis telescope ALMA mounted on its transporter vehicle.

shown in Figure 17.1. The design is an on-axis Cassegrain, with the feed supported by a quadrupod designed for minimum blocking cross-section and maximum stiffness. The antenna structure is temperature-controlled by an air-circulation system, to maintain an accurate profile for the primary reflecting surface. The receiver bandwidths are 8 GHz, to maximize the sensitivity for continuum observations and to encompass as many spectral lines in one observation as possible. The array has baselines ranging from 15 m to 18 km, giving a maximum angular resolution of 5 microarcseconds. Thus, strictly speaking, the instrument is a millimetre/sub-millimetre array.

The key science goals are threefold.

1. The ability to detect line emission from CO and C II in Milky Way-class galaxies at a redshift of 3 in less than 24 hours of observation.
2. The ability to image the gas kinematics in protostars and protoplanetary discs around young Sun-like stars at a distance of 250 pc, enabling the study of their physical, chemical and magnetic-field structures and to detect the gaps created by planet formation in the discs.
3. The ability to provide images at an angular resolution of 0.01 arcseconds and precise images at an angular resolution of 0.1 arcseconds, where 'precise' means representing the brightness at all points having a brightness greater than 0.1% of the peak image brightness.

At the time of writing (2008) production antennas are being delivered to Chile for testing and transport to the site. Limited observations are to start in 2010, with completion of the 64-element array scheduled for 2013.

17.6.4 The Square Kilometer Array (SKA)

In every observing band, there is an abundance of problems that can be addressed by radio-astronomy observations, but many require greater sensitivity than current instruments provide. The instruments described above represent current steps at metre wavelengths (GMRT, LOFAR and MWA) and in the millimetre band (ALMA) to advance the frontiers. The Arecibo telescope is the largest monolithic telescope for the intermediate bands, while the VLA is the most sensitive array in existence, but both are several decades old (although they have both undergone major improvements in performance over the years). The development of a rationale for the next major advance began with the suggestion by Wilkinson (1991), who showed that the study of the early evolution of galaxies through observing their H I structure at high redshift required an order-of-magnitude increase in collecting area, roughly one square kilometre. Upon closer examination, a single parameter, $A_{\text{eff}}/T_{\text{sys}}$, specifies the necessary figure of merit, which should be at least $2 \times 10^4 \text{m}^2 \text{K}^{-1}$, if galaxies with an H I content of a few times $10^9 M_\odot$ are to be detected at $z = 4$ with 300 hours of integration. Improving the system temperature over current performance can yield no more than a factor of 2 or 3, so one is driven to the conclusion that a collecting area of the order of a square kilometre would be needed. It was immediately recognized that an instrument of such power could serve many other functions as well. A set of preliminary specifications was adopted at an international gathering in Sydney, in 1997, and an international study group was formed through the exchange of memoranda of understanding. Subsequently, the concept was named the Square Kilometer Array (SKA), and a study was held in Amsterdam in 1999 that examined the concepts and scientific rationale more closely (van Haarlem 2000). As the project became a more formal international undertaking, a more detailed study of the science rationale was prepared by a broadly based science working group under the auspices of the International Square Kilometer Array Steering Committee (ISSC). The conclusions were published under the editorship of Carilli and Rawlings as *Science with the Square Kilometer Array* (2004). Nine working groups examined a wide range of potential projects for the SKA, and gave detailed justifications for the need for such an instrument in order to address a broad variety of interesting problems. In addition, five key projects were identified, which served as a core set of objectives that the ultimate design must address.

1. **The cradle of life**. Molecular lines can be used to study the details of planet formation in early circumstellar envelopes, including the study of early terrestrial-planet evolution. In addition, the complex interstellar chemistry of amino acids, polycyclic aromatic hydrocarbons and other large molecules whose transitions fall at frequencies below the operating bands of ALMA can be studied. The search for extraterrestrial intelligence, SETI, should be possible as a side benefit of the large collecting area.

2. **Strong field tests of gravity**. Precision timing of pulsars shows general relativistic effects, and the tens of thousands of pulsars that can be observed will reveal interesting cases such as pulsars orbiting black holes. Thousands of millisecond pulsars will form a timing network through which the perturbing effects of gravitational waves will be revealed.

3. **The origin and evolution of cosmic magnetism**. Through polarization, Faraday rotation and the Zeeman effect the detailed topology of cosmic magnetic fields can be traced, both in our own galaxy and beyond, to a redshift of 3. This will reveal the relationship between magnetic fields and observed structures, and will cast light on the origin of the fields.

4. **Galaxy evolution and cosmology**. Galaxy structure and dynamics can be studied through time, since H I studies will be possible out to a redshift of 4 at least. A complete census of galaxies out to a redshift of 1 or 1.5 will allow the first precise studies of the equation of state of dark energy. Gravitational lensing will allow the study of the role of dark matter in galaxies and clusters of galaxies.

5. **Probing the dark ages**. The era of reionization can be studied in two areas: observing primeval H I as it is ionized by the initial formation of stars and quasars, and molecular-line emission from the most distant galaxies.

The preliminary work on defining the characteristics of the SKA, determining the site requirements and constructing technical specifications that will meet the largest and most vital objectives has been carried out under the auspices of the ISSC. The work has been a thoroughly international collaboration, and much of the fundamental groundwork has been laid. Part of the development effort for the SKA is to construct arrays with about 1% of the eventual size at the two locations which are candidates for the full SKA. These arrays are known as SKA pathfinder, in Australia, and MeerKAT in South Africa.

The protection of frequency bands for the SKA will be a vital part of future work; although some radio observatories have some degree of protection, only the Green Bank Observatory of the NRAO in West Virginia has a clearly defined radio quiet zone, and even that has exceptions. Satellite-generated radio interference will have harmful effects on all radio observatories, and great vigilance will be needed to enforce adherence to the ITU regulations by the operators of orbiting satellites.

17.7 The protection of radio frequencies in astronomy

Radio astronomy has always been closely related to its origins in communication engineering. Since Jansky's observation in 1932 of Galactic radio waves at 20 MHz, radio astronomy has led the continuous and dramatic extension of radio techniques upwards by more than four decades in frequency. This extension has transformed radio astronomy, but inevitably the parallel advances in radio technology and communications have raised interference levels everywhere and severely limited the scope of radio observations. At frequencies now regarded as comparatively low, that is, below 1 GHz, the main problem has been posed by the development of television and mobile communications, which occupy most of the available bandwidth. Between 1 GHz and 10 GHz ground-based and satellite-based navigation and communication systems, together with radar, are generating ever greater radio interference. It is impossible for the same frequencies to be used for radio astronomy, since the signal strengths necessary for communications and navigation systems are so large. The techniques described in Chapters 7 and 8 allow radio-astronomical observations to be made at signal levels of some 60 dB below receiver noise, while communications must operate at levels of 20 dB or more above receiver noise.

Doubtless this pressure to exploit the newly available radio frequencies will extend upwards as new techniques are developed for centimetre and millimetre wavelengths. Radio astronomy was, however, recognized by the International Telecommunications Union (ITU) in 1959 as a user of the radio spectrum, and there is a long history of collaboration between astronomers and the frequency-allocation authorities (Robinson 1999). In ITU terminology, radio astronomy is a 'service', and bands of radio frequencies are allocated to it. These allocations are always under review, and they must be preserved and extended with the support of careful and cogent argument, as in the ITU *Handbook on Radio Astronomy* (1995). There are two types of allocation need for radio astronomy, namely continuous broad bands spread throughout the spectrum and narrower bands covering specific spectral lines.

Continuous broad reception bands, free of terrestrial transmissions, are needed for observations such as the mapping of quasars and the discovery and study of pulsars. These allocated bands have existed in the lower range of frequencies, up to 22 GHz, since an ITU conference in 1979, although the pressure from other services meant that only 1.4% of the spectrum could be allocated to radio astronomy. These bands are usefully spaced at intervals of about an octave; examples of their use can be seen throughout this book. Specific line frequencies of the most important spectral lines, such as H at 1.4 GHz, OH at 1.6 GHz, H_2O at 22 GHz and CO at 115 GHz, also require protection. Often these separate requirements can be combined; for example the band 1400–1427 MHz is wholly allocated to radio astronomy and is used for both hydrogen-line and continuum observations.

The problem of making and preserving such allocations is illustrated by the example of the Russian satellite navigation system GLONASS. This uses transmitters on a network of up to 24 satellites, on a set of frequencies in the range 1602–1615.5 MHz. This range unfortunately covers a radio-astronomy allocation to an OH spectral line at 1610.6–1613.8 MHz: even worse, each individual transmitter uses a form of modulation that effectively spreads detectable power over some 150 MHz bandwidth, extending to the main OH lines at 1665 and 1667 MHz. Recognition of the disastrous consequences to radio astronomy has resulted in the retuning of the carrier frequencies of some of the satellites so as to avoid the 1612-MHz OH line, but a complete solution would also require the use of a modulation system that reduces the spread of the radiated spectrum. The flux level at which radiation inside the radio-astronomy bands becomes harmful to astronomical observations has been agreed internationally, and specified by the ITU (1998).[1]

The importance of making such allocations well in advance of the exploitation of the spectrum by a multitude of users was recognized at the ITU Conference in the year 2000,

---

[1] *ITU-R Recommendations RA Series (Radio Astronomy) 1998.* Radio Communications Bureau, International Telecommunication Union, Geneva.

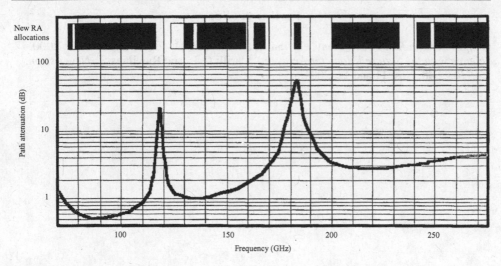

Figure 17.2. Frequency allocations to radio astronomy in the band 70–275 GHz, made at the ITU Conference in 2000.

which made allocations in the band 71–275 GHz. The allocations made to the radio-astronomy service are shown in Figure 17.2, together with a representation of atmospheric absorption at the zenith. Atmospheric windows occur in the ranges 70–115, 125–175 and 195–275 GHz. The new allocations extend over a substantial part of all three windows, covering many hundreds of molecular lines that have been detected in these bands. Parts of these bands are to be shared with other services, which may transmit from ground-based stations (but not from space). Provided that local arrangements can be made to place the transmitters well away from radio observatories, this should not be troublesome, since ground-level attenuation is high at these high frequencies.

Despite all these efforts it is inevitable that the wide bandwidths of the new large radio telescopes will cover frequencies in use for many communication systems. To some extent this can be tolerated by siting the telescopes in remote areas such as the Australian outback or the South Arican Karoo, where there are very few local radio transmitters, but sharing frequency bands is the only way to obtain the full advantages of the new systems. This is essential for LOFAR, which covers heavily used frequency bands: here the observing band must be divided into many sections, rejecting the signal contributions in the frequency bands affected by local radio transmitters.

The ITU 2000 Conference has given astronomers confidence that the community of radio communications and spectrum planners appreciates the need to share the electromagnetic spectrum as a vital natural resource. Provided that radio astronomers remain vigilant in the conservation and extension of their share, the world of astronomy can look forward to many more pleasant and surprising radio discoveries in the future.

Further reading

Perspectives on Radio Astronomy, ed. van Haarlem M. P., Dwingeloo, ASTRON, 2000.
Science with the Square Kilometer Array, ed. Carilli C. L. and Rawlings S., *New Astron. Rev.*, **48**, Nos. 11 and 12, 2004.

Appendix 1 Fourier transforms

A1.1 Definitions

A harmonically oscillating amplitude $a(t)$ can be represented by the complex quantity

$$a(t) = a_0 e^{i2\pi \nu t}. \tag{A1.1}$$

The complex modulus a_0 gives the absolute value of the amplitude $|a_0|$ and the phase offset ϕ:

$$a_0 = |a_0| e^{i\phi t}. \tag{A1.2}$$

The *phasor* a_0 rotates with angular frequency $\omega = 2\pi \nu$, and its projection onto the real axis gives the physical amplitude $a(t)$. The linear addition of various components of a quantity such as the electric field in a radio wave is conveniently done by adding the phasors as vectors. Adding two components with identical phasors but with opposite signs of ω gives the real quantity directly.

The amplitudes a_ν associated with components at various frequencies ν can be superposed to give general amplitudes that can represent any physically realizable form $f(t)$. Fourier theory relates $f(t)$ to the spectral distribution function $F(\nu)$ of these components by the dual transformation theorem

$$\text{if} \quad f(t) = \int_{-\infty}^{\infty} F(\nu) e^{i2\pi \nu t} \, d\nu, \tag{A1.3}$$

$$\text{then} \quad F(\nu) = \int_{-\infty}^{\infty} f(t) e^{-i2\pi \nu t} \, dt. \tag{A1.4}$$

The *spectral distribution function* $F(\nu)$ is the *Fourier transform* of $f(t)$. The frequency ν and the time t are *Fourier dual* coordinates. Another example of such dual coordinates is the relation between the radiation pattern of an antenna and the current distribution in its aperture (Chapter 5). The reciprocal relation between such pairs of functions may be written $f(t) \rightleftharpoons F(\nu)$. Figure A1.1 shows some particularly useful examples.

The Dirac impulse function $\delta(t)$ has an especially simple Fourier transform. Since by definition it is zero everywhere except at the origin, but has an integral of unity, its

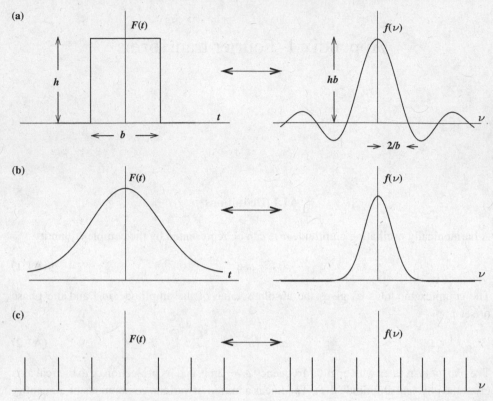

Figure A1.1. Useful examples of Fourier transforms: (a) a square step, or gating, function; (b) Gaussian; and (c) a periodic sampling function..

transform is

$$F(v) = \int_{-\infty}^{\infty} \delta(t)e^{i2\pi vt}\, dt = 1. \qquad (A1.5)$$

Thus, a unit impulse has a flat spectrum of amplitude unity. If such an impulse occurs at time T other than zero, the spectral distribution still has all frequencies present with unit amplitude, but with a frequency-dependent phase $2\pi vT$.

The square step or gating function, $\sqcap(t/T)$, with unit amplitude from $-T/2$ to $T/2$ and zero elsewhere, transforms into the 'sinc' function (Figure A1.1(a)),

$$F(v) = \sin(\pi vT)/(\pi vT) \equiv \text{sinc}(\pi vT). \qquad (A1.6)$$

The Gaussian function (Figure A1.1(b)) has the interesting property that its Fourier transform is also a Gaussian:

$$f(t) = e^{-(t/T)^2} \rightleftharpoons \frac{1}{\pi^{1/2}T}e^{-(\pi vT)^2} = F(v). \qquad (A1.7)$$

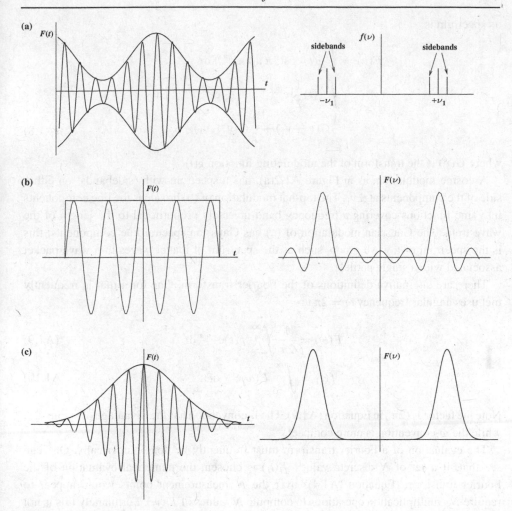

Figure A1.2. Fourier transforms of modulated waves.

Note that when $T = 1/\pi^{1/2}$ the Gaussian is its own transform. Another self-similar pair is the periodic sampling function $\mathrm{III}(t)$, named by Bracewell and Roberts (1954) after the Cyrillic character shah (Figure A1.1(c)). The Gaussian function illustrates a general principle: when the characteristic time T of a function is short, its spectrum is wide, and vice versa. Similarly, if a sinusoidal function is switched on and off at times $-T/2$ and $T/2$, its spectrum has a significant value over a bandwidth $\Delta\nu$ that is of the order of $1/T$.

Modulated waves, which are, of course, familiar to radio engineers, are encountered throughout physics. Typically, the amplitude of a cosinusoidal wave varies with time (Figure A1.2), either periodically as in the beat pattern (a), or aperiodically as in an isolated group of waves. If the modulating function is $g(t)$, so that the wave is $g(t)\cos(2\pi\nu_1 t)$, then its spectrum is

$$F(v) = \int g(t)\cos(2\pi v_1 t)e^{i2\pi vt}\,dt$$

$$= \int g(t)\frac{1}{2}\left[e^{i2\pi(v-v_1)t} + e^{i2\pi(v+v_1)t}\right]dt$$

$$= \frac{1}{2}G(v - v_1) + \frac{1}{2}G(v + v_1), \tag{A1.8}$$

where $G(v)$ is the transform of the modulating function $g(t)$.

A cosine modulation, as in Figure A1.2(a), has a spectrum with 'sidebands' on either side of the components at $\pm v_1$. The top-hat modulation of (b) broadens the line components into sinc functions covering a frequency band inversely proportional to the length of the wave train. The Gaussian modulation of (c) has Gaussian spectral line components; this is the spectrum of a wave group, such as the spectrum of matter waves in a wave packet associated with a single particle.

There are alternative definitions of the Fourier transform. One form that is frequently met uses angular frequency $\omega = 2\pi v$:

$$F(\omega) = \frac{1}{2\pi}\int_{-\infty}^{\infty} f(t)e^{-i\omega t}\,dt \tag{A1.9}$$

$$f(t) = \int_{-\infty}^{\infty} F(\omega)e^{i\omega t}\,d\omega. \tag{A1.10}$$

Note the factor $1/(2\pi)$ in Equation (A1.9). The v-convention has the advantage of symmetry, while the ω-convention is more compact.

The evaluation of a Fourier transform must frequently be done numerically. One can see that, if a set of N discrete values $f_i(t_i)$ is chosen, the numerical evaluation of the Fourier transform (Equation (A1.4)) over the N measurement points would appear to require N^2 multiplication operations to compute N values of $F_i(v_i)$. Fortunately this is not the case, since there exist *fast Fourier-transform* (FFT) algorithms that reduce the number of multiplication operations to the order of $N \ln N$. One of the most commonly used, the 'Cooley–Tukey' algorithm (previously developed by a number of authors, with Gauss as a forerunner), requires that the sample points be uniformly spaced; since in many practical situations, such as the aperture-synthesis technique (Chapter 6), the original data points are usually unequally spaced, an interpolation procedure is needed in order to obtain a regular grid of values. There are other FFT algorithms that do not require gridding, usually at the cost of greater demands on memory space.

A1.2 Convolution and cross-correlation

The *convolution theorem* plays an important role in practical applications of Fourier methods; for example in formulating the effect of smearing by the antenna beam on an aperture-synthesis map. For a pair of functions $g(t)$ and $h(t)$ with Fourier transforms $G(v)$ and $H(v)$,

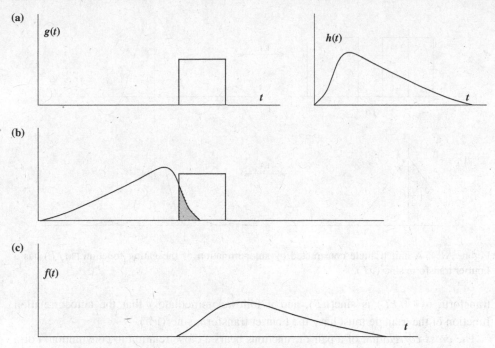

Figure A1.3. The convolution of two functions $g(t)$ and $h(t)$, shown in (a), evaluated (b) as their product (the shaded area) with $h(t)$ reversed. The convolution $f(t)$ is shown in (c).

the convolution is defined as

$$f(t) = \int_{-\infty}^{\infty} g(t')h(t - t')dt'. \tag{A1.11}$$

A convenient notation is $f(t) = g * h$, where the asterisk denotes the convolution process. Note the reversal of sign in $h(t - t')$; note also that the process is commutative, i.e. $g * h = h * g$. The convolution process is shown in Figure A1.3, which shows the functions $g(t')$ and $h(t)$ with the reversed and shifted function $h(t - t')$. Their product is the shaded area, which is $f(t)$ at the time shift t.

The convolution theorem states that the Fourier transform of $f(t)$ is simply the product of the Fourier transforms of the two convolved functions:

$$F(\nu) = G(\nu)H(\nu). \tag{A1.12}$$

The theorem is easily proven and has many practical applications. For example, a signal $g(t)$ may be an input signal to an amplifier or other linear device that has a response $h(t)$ to an impulsive input; the spectral distribution of the output is then the product of the spectra of the input signal and of the device. When a device is being considered in the frequency domain, $H(\nu)$ is known as the *transfer function*.

The convolution theorem may be used to facilitate the rapid calculation of Fourier transforms in many instances. For example, a triangle with unit height and base B is the autocorrelation function of the gating function $\sqcap(t/T)$, as seen in Figure A1.4. The

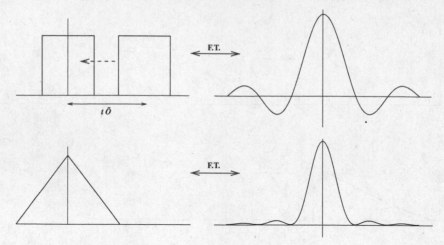

Figure A1.4. A unit triangle constructed by autocorrelation of the gating function $\sqcap(t/T)$ has a Fourier transform $\text{sinc}^2(\nu T)$.

transform of $\sqcap(t/T)$ is $\text{sinc}(\nu T)$, and it follows immediately that the autocorrelation function of the triangle must have the Fourier transform $\text{sinc}^2(\nu T)$.

The *cross-correlation* of a pair of functions bears a close relation to convolution. For a pair of functions $f(t)$ and $g(t)$ the operation is often designated $f \otimes g$, with

$$f \otimes g \equiv \int f(t')g(t' - t)\mathrm{d}t' \qquad (A1.13)$$

and thus

$$f \otimes g = f(t) * g(-t), \qquad (A1.14)$$

from which it follows that

$$f \otimes g = F(\nu)G^*(\nu). \qquad (A1.15)$$

Note the time reversal in Equation (A1.14); cross-correlation is not a simple commutative operation.

Cross-correlation has a particular significance in radio interferometry, where two signals $f(t)$ and $g(t)$ may be obtained from spaced antennas. The product fg^*, which has the dimensions of power, is then referred to as the *cross-power*, and the product FG^* is known as the *cross-spectral power density*.

The *autocorrelation function* $C(t)$ is defined by

$$C(t) \equiv \int_{-\infty}^{\infty} f(t')f(t' + t)\mathrm{d}t'. \qquad (A1.16)$$

Note that there is no time reversal; the function is multiplied by itself at a variable time shift. The autocorrelation function is related to the spectrum of $f(t)$; for zero time shift it is simply its square. The autocorrelation function has a particular relevance for signals and systems whose behaviour is often described in terms of *amplitudes* $e(t)$ (e.g. voltages, field strengths and currents), while the measurement process involves the *power* $p(t)$. Using a convenient choice of units, we can write

$$p(t) = e^2(t). \tag{A1.17}$$

The spectrum of a signal can be represented in Fourier terminology; the *spectral power density* $S(v)$ gives the power density in a given infinitesimal bandwidth dv. The total energy E_{em} emitted over time is then

$$E_{em} = \int e^2(t)dt = \int_0^\infty S(v)dv. \tag{A1.18}$$

Note that the distinction between positive and negative frequencies is no longer meaningful.

$S(v)$ is evidently proportional to $E(v)E^*(v)$, the product of the Fourier transforms of $e(t)$ and $e(-t)$. We explore this relation by writing the self-convolution $e(t') * e(-t')$, so that the emitted energy E_{em} may be expressed as

$$E_{em} = \langle e(t) * e(-t) \rangle_{t=0} = \int_{-\infty}^\infty e(t')e(t' - t)dt'. \tag{A1.19}$$

It follows from the definition of convolution (Equation (A1.11)) that

$$\int_{-\infty}^\infty E(v)E^*(v)e^{-i2\pi vt}\, dv = e(t) * e(-t) \tag{A1.20}$$

so that, for $t = 0$,

$$E_{em} = \int_{-\infty}^\infty e^2(t)dt = \int_{-\infty}^\infty |E(v)|^2 dv. \tag{A1.21}$$

This is known as *Rayleigh's theorem*, and is a generalization of the Parseval theorem for Fourier series. The integral over frequencies need be taken over positive frequencies only, since $|E(v)|$ is symmetrical in v, and so, for this physical case,

$$S(v) = 2E(v)E^*(v) = 2|E(v)|^2. \tag{A1.22}$$

The self-convolution $e(t') * e(-t')$ is symmetrical in the time offset t, and when written with the reverse sign becomes the autocorrelation function $C(t)$ as in Equation (A1.16) above, that is,

$$e \otimes e \equiv C(t). \tag{A1.23}$$

The autocorrelation function has a most important property, known as the Wiener–Khinchin theorem, which states that the Fourier transform of the amplitude autocorrelation is the power spectral density (with correction by a factor of 2 if positive frequencies alone are treated, as in Equation (A1.21)):

$$E(v) \otimes E^*(v) \rightleftharpoons C(t). \tag{A1.24}$$

This relation provides the basis for autocorrelation spectrometry.

The relation between autocorrelation and Fourier transformation may be summarized as

Signal $V(t)$ $\rightleftharpoons$ $V(\nu)$ **Spectrum**

autocorrelation $\downarrow$ $\downarrow$ **square**

$R(\tau)$ $\rightleftharpoons$ $S(\nu)$

In radio astronomy the autocorrelation function is obtained from the product of a digitized signal and the same signal subjected to a variable delay; in optical spectrometers light reaches a detector via two paths of variable path difference. In practice, many different delays may be used simultaneously, especially in radio astronomy, where the signal may be amplified coherently and split into many channels.

A1.3 Two or more dimensions

The definitions and examples in this appendix have all been presented in the one-dimensional case, but since the Fourier transform is a linear operation it can be generalized to Cartesian coordinates in many dimensions. Given a function $f(\mathbf{x})$ of the n-dimensional variable $\mathbf{x}$, it will have a Fourier transform $F(\mathbf{k})$. The Fourier dual coordinate $\mathbf{k}$ is called the *spatial frequency* by analogy with t and ν in the time/frequency case. A straightforward calculation shows that the fundamental Fourier inversion theorem, Equation (A1.3), becomes

$$f(\mathbf{x}) = \int_{-\infty}^{\infty} f(\mathbf{k}) e^{i2\pi \mathbf{k}\cdot\mathbf{x}} \, d^n\mathbf{k}, \tag{A1.25}$$

$$F(\mathbf{k}) = \int_{-\infty}^{\infty} f(\mathbf{x}) e^{-i2\pi \mathbf{k}\cdot\mathbf{x}} \, d^n\mathbf{x}, \tag{A1.26}$$

thus defining

$$f(\mathbf{x}) \rightleftharpoons F(\mathbf{k}). \tag{A1.27}$$

Further reading

The Fast Fourier Transform, Brigham E. Oran, New Jersey, Prentice-Hall, 1974.
The Fourier Transform and its Applications, Bracewell R. N., New York, McGraw-Hill, 2nd edn, 1978.

Appendix 2 Celestial coordinates, distance and time

A2.1 The celestial coordinate system

The Earth is a moving platform, and reference points in the Universe are not defined easily. There are no fixed stars in the Universe, and the closest approximation we have to a universal coordinate system is defined by the cosmic microwave background. This nearly featureless radiation background, discussed more fully in Chapter 14, appears to be the remnant of the cosmic fireball. The sum of all motions of the solar system and the Milky Way Galaxy in space causes a measurable Doppler shift in the background radiation, implying that our Milky Way Galaxy has a resultant velocity of approximately $340\,\mathrm{km\,s^{-1}}$ towards a point not far from the Virgo cluster. In theory, this could be used as the axis of a polar coordinate system, although the origin of the azimuthal angle would still have to be defined. In practice, the uncertainties in the measurement are so large that the idea is not feasible. The angular-momentum axis of the Milky Way Galaxy, and the line from the Sun to the Galactic centre, could be imagined as defining a nearly fundamental system, but neither direction is sufficiently well known as yet.

The positions and motions of bodies in the solar system thus provide the most practical coordinate reference system. The motions are complex, but are well defined and have been carefully analysed. One starts with the rotation of the Earth, whose axis defines the orientation of a polar coordinate system as shown in Figure A2.1. Instead of using the polar angle, it has been conventional to use the co-polar angle, measured from the *celestial equator*. This angle is the *declination*, usually symbolized by δ. The origin of the azimuthal angle is determined by the Earth's orbit about the Sun, which is given observationally by the apparent motion of the Sun along its path in the sky, known from ancient times as the *ecliptic*. The ecliptic has two intersections with the celestial equator. The *ascending node* or *vernal equinox* (Ω) (the point of the spring equinox, when the mean Sun crosses the equator in an upward direction) defines the origin of the azimuthal angle, known as the *right ascension*, symbolized by α, which increases in the counterclockwise direction about the celestial pole. Two further conventions can be noted: in celestial coordinates, north is taken in the direction of increasing declination, as in a normal terrestrial map, and east is taken to the left, in the direction of increasing right ascension. (This convention follows from mapping the terrestrial headings of north and east directly onto the celestial sphere.)

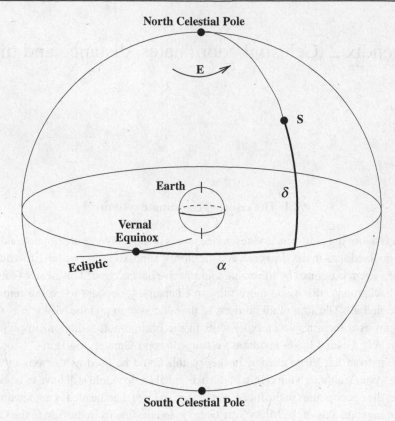

Figure A2.1. The polar coordinate system.

The *position angle* on the sky is defined from north through east, i.e. counterclockwise as seen from the Earth.

Care must be exercised in defining the two great circles, the celestial equator and the ecliptic, that define the celestial coordinate system. The Earth's pole precesses in space because of the torque of the Sun, Moon and planets, with a period of nearly 26 000 yr, or an angular rate of about 55 arcsec per year. There is also a nutational motion driven by periodic external torques with a more complex behaviour. The dominant term is a periodic motion of amplitude 9 arcsec and period 19 yr imposed by the Moon. The Earth's orbit, in turn, is perturbed by the other planets, so the ecliptic also varies in time. Even the Earth's rotation axis moves with respect to the Earth's crust, showing a 14-monthly variation (the *Chandler wobble*) with a fluctuating amplitude of amplitude approximately 0.1 arcsec, and a possible secular variation of about a milliarcsecond per year. Because of the large number of time-varying corrections, therefore, the right ascension and declination of a cosmic body have meaning only when the *epoch* is specified for the equinox. The equinox of 1950.0 is commonly used, but is progressively being superseded by 2000.0. Note that the equinox and pole determine the angular coordinate system, while epoch refers to time; the question of timescale is discussed more fully in Section A2.3. The International Astronomical Union has adopted a series of resolutions that took effect from 1 January 1984, adopting the equinox of epoch 2000.0 for future work, and this will surely supplant the old convention. The resolutions are summarized in *US Naval Observatory Circular* No. 163, in which illustrative examples are also worked out. A more accurately defined timescale, more accurate values of the astronomical constants and an improved theory of the Earth's nutation are included in the new system. To distinguish the old and new conventions the designations B1950.0 and J2000.0 have been adopted (B standing for *Besselian year*, the timescale defined for the former, and J for *Julian year*, the timescale for the latter epoch). Formal definitions and references are given in compact form in the *Astronomical Almanac*, jointly published by the UK and the USA (there are other national counterparts). Further material can be found in the *Explanatory Supplement* and in Smart's *Spherical Astronomy* (1977). One should note that the determination of the celestial pole had traditionally been made by observing stars, but now observations of extragalactic radio sources have become the method of choice. Similarly, the ecliptic plane has traditionally been determined by observing planets and asteroids, but radio and radar measurements of planetary spacecraft are also supplanting the optical methods.

The definition and realization of a reference frame depend on accurate astrometric measurements extending over wide angles. Flamsteed at Greenwich achieved an accuracy of 10″; nearly 200 years later most star catalogues based on optical observations were still accurate only to about 1″. The VLBI methods, which are more precise and more easily relate positions of widely separated objects, now cover the celestial sphere with an accuracy of the order of 1 milliarcsecond (mas), an improvement of three orders of magnitude. A new reference frame, the International Celestial Reference Frame (IRCF), which defines axes to 20 microarcseconds, is based on the positions of 212 compact extragalactic radio sources (Johnston and de Vegt 1999). The astrometric satellite Hipparcos produced a catalogue of 118 218 optical positions covering the whole sky and based on the IRCF, so there is now a grid of well-determined positions in this frame to an accuracy of about 1.5 mas.

A2.2 The astronomical distance scale

The ancient and still vexed problem of measuring the distance to celestial bodies must be addressed before physical quantities such as their size, mass and luminosity can be determined. If there were standard celestial yardsticks, a simple measurement of angle would do, but no such standards are known. Similarly, if some class of stars could serve as standard candles, a measurement of apparent brightness would give a distance, but this method also has its limitations. Both methods can be applied in a statistical sense, but the primary distance standards must start with triangulation – the measurement of parallax of stars as the Earth moves in its orbit about the Sun. Galileo first suggested measuring the differential parallax of bright nearby stars against the faint, distant stars as an application of his telescope, but larger effects such as aberration and atmospheric refraction defeated such efforts for more than two centuries. In 1838, Bessel was finally successful, when, instead of the bright northern stars, he chose 61 Cygni, a star known to have large proper motion and therefore likely to be close.

The Earth's motion about the Sun provides the base of the triangle for parallax measurements. (At some time in the future one might imagine space-borne instruments enlarging the baseline.) The semi-major axis of the Earth's orbital ellipse defines the *astronomical unit* (a.u.), which is best determined by radio echo techniques. Planetary radar observations and round-trip transponder measurements from planetary landers and orbiters are currently the methods of choice. The best current value for the astronomical unit is $1.495979(1) \times 10^{11}$ m, with the accuracy being limited primarily by current precision in determining the velocity of light (since the measured quantity is the round-trip echo time, which must be converted to a distance). The a.u., in turn, determines the most commonly used unit of distance, the *parsec* (pc), which is defined as the distance at which a body will exhibit a parallax of 1 arcsec ($1 \text{pc} = 3.08 \times 10^{16}$ m $= 3.26$ light-years). By coincidence, a velocity of 1 km s^{-1} is very nearly equal to 1 pc per million years.

At present, the error in measuring the orbital parallax of a given star is seldom less than 0.005 arcsec, equivalent to a distance of 200 pc, and nearly all stars are much further away. The centre of our Milky Way Galaxy is about 8 kpc from the Sun (see the discussion in Chapter 9), the great nebula M31 in Andromeda is about 80 times further removed, and some of the most interesting extragalactic objects are over three orders of magnitude more distant still. There are reasons to expect improvements in trigonometric parallaxes in the future. Some measurements by the radio interferometric techniques described in Chapter 5 have already reached the best available levels of precision, notably for pulsars (Brisken *et al.* 2002), and the extension of these methods to the optical domain also offers promise. Nevertheless, the general problem of distance determination requires the development of a hierarchy of distance measures, allowing one to proceed stepwise from direct parallaxes of nearby stars to the necessarily more uncertain distance criteria used for extragalactic objects.

The principal steps in the distance-measuring hierarchy are set out in many standard textbooks; see Jacoby *et al.* (1992) for a summary. In order of increasing distance, they involve galactic clusters (in particular the Hyades cluster), the Cepheid variables and a series of rather less precise indicators for the distances of galaxies, such as the brightness of components (H II regions, supernovae) as well as the integrated brightness and the fluctuations in brightness of the galaxies themselves.

We have already described in Chapter 13 an entirely independent method of measuring the distance of a distant galaxy, which depends solely on one set of radio observations, avoiding the previous long chain of observational steps. This is the observation of velocity and proper motion in the masers surrounding the black-hole nucleus of NGC 4258. The galaxy is found to be at a distance of 7.2 ± 0.3 Mpc, agreeing well with the value found by conventional methods.

A2.3 Time

Space and time are the coordinates by which we label events. Ever since Minkowski, in 1907, wrote about the geometrical interpretation of Einstein's special theory of relativity, the four-dimensional coordinate system of space-time has provided the framework for physics and astrophysics. Alternative schemes have been proposed from time to time, but no compelling experiments have yet dictated a change in this viewpoint. Time, in particular, can be measured with great precision using either gravitational clocks (the solar system) or atomic clocks (the frequency of photons from the ground-state hyperfine splitting of caesium-137). The rate of change of gravitational time versus atomic time is less than 10^{-12} per year; it is generally assumed, therefore, that time intervals are the same, regardless of the clock being used for the measurement.

The most accurate clocks in use today are probably the caesium atomic-beam devices, which achieve an absolute accuracy of at least one part in 10^{14} over a period of a day or longer. Hydrogen maser clocks have a *stability* of one part in 10^{15} over periods of hours to days, but are more sensitive to magnetic, pressure and thermal perturbation and thus may be less reliable as absolute standards. They have an important use in radio astronomy as stable frequency standards for VLBI, a subject treated in Chapter 6.

It now seems likely that some millisecond pulsars have a long-term rotational stability at least the equal of caesium clocks. If this is to be used as a check on the smooth running of terrestrial time standards, it will be necessary to monitor the rotation of a set of millisecond pulsars over a period of several years. This is the intention with the Parkes Pulsar Timing Array (PPTA), which is aimed to achieve an accuracy of the order of 10 ns in all necessary corrections for ephemerides and propagation phenomena (Hobbs *et al.* 2006).

Civil time is governed by the strong human desire to have noontime (1200 hours) occur at mid-day, and, at least on the average, for the year to start on the first day of January. There is the dual problem of specifying the time interval (the second) and the time (the elapsed interval since a specified origin for the time coordinate). Several conventions, agreed upon by international bodies, have been defined by reference to the motion of the Earth in its orbit. By definition, the *tropical year* is the interval required for the Earth to travel from vernal equinox to vernal equinox (i.e. for the mean Sun to travel 360° along the ecliptic). The tropical year, in turn, defines *Ephemeris Time* (ET). This requires accurate measurement of the celestial equator (the instantaneous axis of the Earth's rotation) and of the plane of the ecliptic (the apparent motion of the mean Sun), since the intersections of these great circles define the equinoxes. Since the mutual perturbations of the bodies of the solar system are appreciable, an accurate accounting must be made of the variations of the system with time. By international agreement, the length of the tropical year for epoch 1900.0, Jan 0 12^h E.T., was set at 31 556 925.9747 seconds. Currently, the repeatability of comparisons of intervals in this unit of time with atomic time is approximately one part in 10^{12} over several decades. It is generally accepted that atomic time and ephemeris (i.e. gravitational) time are identical; if they are not, the difference is evidently small.

Ordinary human activities are more strongly governed by the solar day, and the most common civil unit of time is *Universal Time* (UT), starting and ending at midnight, with respect to the transit time of the mean Sun at the location of the ancient transit-circle of the Royal Observatory at Greenwich. The most primitive definition, UT0, uses the instantaneous rotation of the Earth to define the time, and is the time that would be derived at any instant from observing the Sun and stars. This is an imperfect clock, since the Earth does not rotate uniformly about its axis: it is a multipiece gyroscope, exchanging angular momentum between the core and the mantle, mantle and crust, and crust and atmosphere, with dissipative tidal torques from the Sun, Moon and other bodies in the solar system. The seasonal and secular variations are known with some accuracy, and, when these corrections are applied, and successive improvements have been agreed on, better definitions of UT are adopted. Polar-motion corrections are applied to define UT1, which is then independent of the observer's location. UT2 is defined by applying corrections for seasonal variations in the Earth's rotation rate. The term UT is used when these distinctions are unimportant, and tables are given to relate the various timescales. Since caesium atomic-beam clocks are the most accurate devices currently in use, they provide the better time-interval reference. By international agreement, photons from the ground-state caesium-137 hyperfine transition are defined to have an average of 9 192 631 770 oscillations per second, thus defining *International Atomic Time* (TAI). Such an atomic clock runs about 300 ns s^{-1} slow with respect to UT (the offset was deliberately chosen in order to avoid confusion). In order to keep civil time in average agreement with solar time, and yet to have time intervals determined by the more precise TAI, the national time services generally use *Coordinated Universal Time* (UTC). Caesium clocks determine the short-term timescale, and 'leap seconds' are inserted at agreed times every few months to keep 0^h UTC in average agreement with UT. Corrections are listed in the *Astronomical Almanac* (AA).

For most purposes, the chosen origin of time is 12^h on day 0 of the tropical year 1900, defining an origin at noon, 31 December 1899. Time is sometimes specified in *Julian Days* (JD) with JD 0.0 starting at noon, on 1 January 4713 BC (Julian calendar). Conversion tables are given in the AA. The civil day 1 January 1990, for example, starts on JD 2 447 892.5. The choice of noon was made because the Sun was originally the best fundamental clock, but, as techniques developed, the Moon, planets and stars replaced the Sun as the best reference bodies, and the Time Commission of the International Astronomical Union agreed to define midnight as the start of the astronomical day as of 1 January 1923 and, ever since, this convention has been followed when the time of observation is referred to in year, month, day, hour, minute and second. Julian days, however, still start at noon.

Different timescales were adopted for the angular coordinate systems defined by B1950.0 and J2000.0: the Besselian year, adopted by convention, is defined in terms of UT0 and is thus defined by a clock (the Earth) whose imperfections are large and uncorrected; the Julian year, on the other hand, is defined as a tropical year and is therefore a sounder time reference. The fundamental constants describing the Earth's motion, such as the rate of precession, are also known to higher accuracy. For these reasons, adoption of the epoch of J2000.0 will certainly result in better positional measurements of celestial objects.

The details of time-keeping are spread through many references, with the AA being the starting point for the fundamentals. When one is interested in precision a careful search of the current literature, including the *Bulletin* of the Bureau International de l'Heure (BIH), should be made, combined with consultation with authorities in the field.

Further reading

First results from Hipparcos, Kovalevsky J., *Ann. Rev. Astron. and Astrophys.* **36**, 99–129, 1998.

The Hipparcos and Tycho Catalogues, European Space Agency Special Publication 1200, Vol. 1, Noordvijk, ESA Publication Division ESTEC, 1997.

Reference Frames in Astronomy, Johnston K. J. and de Vegt Chr., *Ann. Rev. Astron. Astrophys.*, **37**, 97–126, 1999.

Appendix 3 The origins of radio astronomy

> Thus the explorations of space end on a note of uncertainty. And nec-
> essarily so. We are, by definition, in the very center of the observable
> region. We know our immediate neighborhood rather intimately. With
> increasing distance, our knowledge fades, and fades rapidly. Eventually,
> we reach the dim boundary – the utmost limits of our telescopes. There,
> we measure shadows, and we search among ghostly errors of measure-
> ment for landmarks that are scarcely more substantial. The search will
> continue. Not until the empirical resources are exhausted need we pass
> on to the dreamy realms of speculation.

When Hubble wrote these words, he was probably not aware that Jansky had already completed work at the Holmdel station of the Bell Laboratories that would justify fully his ultimate optimism. Jansky's task had been to determine the levels of radio interference to be expected at wavelength 14.6 m. By 1932, he had identified a steady hiss component in the radio background and in 1933 he had concluded that this slowly varying background was varying with the sidereal day, and had to be associated with sources beyond the Earth (Jansky 1933a, 1933b). Furthermore, the maximum noise came from a well-determined location, in the constellation of Sagittarius. As Jansky continued his observations, he showed that the principal sources of radiation were distributed throughout the Milky Way, and that the Sun was such a weak emitter that the radiation could not be from stellar sources similar to the Sun (Jansky 1935). He also speculated that the noise might be generated in an ionized interstellar medium.

As Jansky's work came to an end, Grote Reber, a young engineer working in his spare time, constructed a 30-ft paraboloidal reflector in his back yard in Wheaton, Illinois, a story he recounted in some detail in 1958 (Reber 1958). After the antenna had been completed in 1937, Reber tried to observe Jansky's cosmic radiation at wavelength 10 cm, but even though his intentions were sound the electronic techniques available at that time were not up to the task. He worked his way gradually to lower frequencies, and finally succeeded in obtaining good results at 160 MHz in 1940. Otto Struve was editor of the *Astrophysical Journal* at the time, and in private conversation he described the difficulty that his editorial board had in deciding whether or not to publish this baffling paper, so unfamiliar in language and subject matter. Struve eventually took sole responsibility in deciding to accept the paper

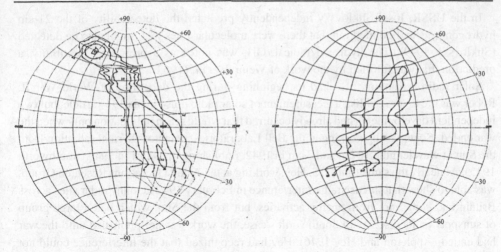

Figure A3.1. Reber's 1944 map of the radio sky.

(Reber 1940), and later agreed to publish Reber's succeeding paper in 1944 (Reber 1944). Both papers had interesting consequences. The 1944 paper contains a detailed map of the radio sky (reproduced in Figure A3.1) showing most of the salient features of the Galactic radiation. The concentration of the radiation towards the plane of the Milky Way and the maximum towards the Galactic centre are clearly shown. The other maxima can also be seen clearly, one at $(20^h, +40°)$ and the other at $(23^h, +60°)$. The first is principally the extragalactic radio source later to be known as Cygnus A, while the second maximum is caused by the most intense discrete radio source, the supernova remnant Cassiopeia A.

The 1940 paper had very little immediate influence on American astronomy, but it had an interesting consequence abroad. The Dutch astronomers at Leiden Observatory, under the leadership of Jan Oort, continued to hold a seminar on current astronomy despite the difficult wartime conditions and the burden of being subject to military occupation. In 1944, they received the 1940 issues of the *Astrophysical Journal*, and Reber's article was discussed intensively. Oort reasoned that the real need was to discover a spectral line in the radio spectrum. Galactic structure was a main focus of Oort's interest, and if radial velocity could be measured by observing a Doppler shift of the line frequency a powerful new tool would be available to Galactic astronomers. Interstellar dust obscures most of our Galaxy from view, but if the interstellar medium could be studied by radio waves, free of obscuration by the dust, the entire Galaxy could be studied. The challenge was taken up by Hendrick van de Hulst, who showed, within a few days, that atomic hydrogen, the most abundant constituent of interstellar gas, should have a measurable spectral line at wavelength 21 cm (van de Hulst 1945). At the close of the war, an intensive effort was mounted at Leiden to observe the line. The Leiden Observatory under Oort's leadership was among the first established observatories to pursue the new science, and in the following decade established a major presence among the leading radio-astronomy centres.

In the USSR, Josef Shklovsky independently predicted the detectability of the 21-cm hydrogen line, and pointed out that there were molecular lines that might also be detected (Shklovsky 1949). The hydroxyl radical, OH, was noted as a promising possibility, a prediction that was fulfilled by the work of Weinreb *et al.* (1963).

Solar radio astronomy also had its beginning during the dark time of World War II. Reber was the first to publish a measurement of solar radio noise in his 1944 article, but two independent developments had already occurred that would reach publication only when the war ended. Southworth, working at the Bell Laboratories, detected thermal radiation from the Sun at wavelengths 10 cm and 3 cm in 1942, publishing the work in 1945 (Southworth 1945). At nearly the same time, J. S. Hey, working at the Army Operations Research Group, was able to show that intense radio interference in February 1942 at many radar sites across Britain had come, not from enemy activities, but from the Sun, which had a large group of sunspots on its disc. As in Southworth's case, the work was not published until the war had ended (Appleton and Hey 1946). Hey had recognized that the interference could not have had a thermal origin, since it was far too intense, and the study of non-thermal solar radiation continues to be a rich and active field. (Hey's superior, Schonland, is reported by Edge and Mulkay (1976) to have recognized the irony that the discoveries by Jansky and Hey both owed their origins to the study of radio interference.)

The pace of radio-astronomy research quickened immediately after the end of the war in 1945. Observations of the Sun were vigorously pursued by the CSIRO Radiophysics group at Sydney and by the Cavendish radio-astronomy group at Cambridge. The radio bursts discovered by Hey had small angular sizes, and the filled-aperture arrays in general use had beamwidths of several degrees, an angular resolution that was totally inadequate. Both groups thereupon developed interferometric techniques to set more meaningful upper limits to the angular sizes of the regions responsible for the radio bursts. In Sydney, Joseph Pawsey's group mounted an antenna on the cliffs of Dover Heights looking out over the sea to detect the rising Sun and its reflected image, using the 'Lloyds mirror' effect to obtain a resolution of 8 arcmin at wavelength 3 cm. On the flat Cambridgeshire field that the Cavendish Laboratory had leased from Jesus College, no such method could be used, so Martin Ryle used a two-element interferometer with which a resolution of 20 arcmin could be obtained. As their work developed during the succeeding decade, both groups would gradually shift a significant part of their research efforts away from the solar system to the study of much more distant phenomena.

The new direction in their work was catalysed by the work of Hey and his group, who published a 64-MHz radio map of the sky (Hey *et al.* 1946a) (shown in Figure A3.2) that fully confirmed all the features of Reber's map. Shortly, they noticed that the maximum in Cygnus (approximately 20^h, $+40°$) showed fluctuations in intensity on a timescale of 10 s or so (Hey *et al.* 1946b). This implied source dimensions of a few light-seconds at most, comparable to the size of a star, and could not possibly be reconciled with an origin in the dilute interstellar gas, whose scale size would be many light-years. The cliff-top solar interferometer of the CSIRO Radiophysics group at Dover Heights was pressed into searching for the new source. Pawsey has privately related how his first

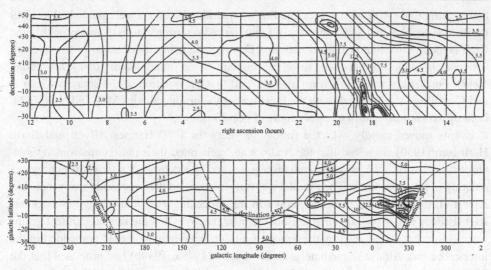

Figure A3.2. A 64-MHz map of the radio sky (Hey *et al.* 1946a).

attempts met with failure. He consulted an eminent optical astronomer, whose reaction was, in Pawsey's words, ' "Stuff and nonsense. Don't waste your time." This gave us great incentive to find Hey's source.'. The next observers scheduled to use the antenna were John Bolton and Gordon Stanley (1948), and for the first time the radio source Cygnus A was seen rising from the sea. Shortly thereafter, Martin Ryle and Graham Smith (1948) used the radio interferometer in Cambridge to observe the northern sky. In addition to observing Cygnus A, they found that the maximum in the constellation of Cassiopeia (near 24^h, $+60°$, in Reber's 1944 map) was the most intense radio source in the sky, and called it Cassiopeia A.

At this juncture, the literature shows clearly that there was a vigorous controversy over the question of whether the general Galactic radiation was a superposition of many 'radio stars', or noise from the general interstellar medium. Whipple and Greenstein had published the first analysis of the second possibility in 1937, and showed that the cosmic dust could not be the source of the radio noise. Reber's 1940 observations were immediately followed by a theoretical study by Henyey and Keenan (1940). They concluded that thermal bremsstrahlung from electron–ion collisions in a hot interstellar gas could be the cause, but only by ignoring Jansky's results at longer wavelengths. The definitive answer was given by Charles Townes (1947), who showed that bremsstrahlung from electron–ion collisions required a gas at a temperature of 150 000 K, a higher temperature than anyone would be willing to accept at that time. From Jansky's time, it had been generally recognized that stars like our Sun could not account for the observed intensity, so a new type of object was postulated. The matter took some years to settle, but in retrospect the vital information came quickly. By 1950, both the Australian and British radio astronomers had demonstrated conclusively that the fluctuations in brightness of the discrete radio sources were imposed by ionospheric scintillation, and were not intrinsic (Ryle 1950). Thus, the variable behaviour had nothing to do with source size. The first optical identifications were published in 1949 by Bolton, Stanley and Slee, who showed that the radio source Taurus A was probably the Crab Nebula, a well-known nebulosity that Duyvendak, using translations of the relevant Chinese chronicles, had deduced to be the remnant of the supernova of 1054 (Duyvendak 1942). Furthermore, Bolton's team also showed that the position of the sources Virgo A and Centaurus A agreed well with the locations of the bright galaxies M87 and NGC4528, both of which were known to exhibit optical peculiarities.

Events moved rapidly over the following years. In 1950 Hannes Alfvén and Bernt Herlofsen (1950) suggested that the 'radio stars' generated their radio emissions by synchrotron emission – the harmonic radiation generated by electrons of relativistic energy circulating in magnetic fields. Their model was not realistic, but the suggestion was prophetic. Albert Hiltner and John Hall had shown in the preceding year that starlight was polarized (Hiltner 1949), and the most reasonable inference was that the interstellar dust grains were elongated, and aligned by the action of a large-scale magnetic field embedded in the interstellar gas. Almost immediately, Enrico Fermi (1949a, 1949b) had proposed that the cosmic-ray particles were accelerated by collisions with the interstellar gas clouds in which the field was embedded. Only two months after Alfvén and Herlofsen had proposed their radio-star model, Kiepenheuer (1950) improved on their idea by suggesting that the Galactic radio background was synchrotron radiation from cosmic-ray electrons gyrating in the Galactic magnetic field. In contrast to thermal bremsstrahlung, the mechanism was decidedly *non-thermal* in character, and by this time it was clear that the discrete radio sources also had too high a surface brightness to be understood in terms of an equilibrium thermal mechanism.

Theorists in the USSR followed up the consequences of synchrotron radiation rigorously, culminating in Shklovsky's 1953 proposal that the optical and radio radiation from the Crab Nebula was mainly synchrotron emission from energetic electrons in the supernova remnant.

In 1951, the new initiative in radio spectroscopy set in motion by Oort came to fruition. The first observations came about almost simultaneously in Holland, Australia and the USA. Van de Hulst was visiting Harvard in the spring of 1951, where Edward Purcell and Harold Ewen were working on an experiment to detect the 21-cm line, as were van de Hulst's colleagues at Leiden. The original Dutch receiver was unfortunately destroyed by a fire during this period. They rebuilt their equipment, incorporating some of the features of Purcell's design, but, as luck would have it, Ewen and Purcell's detection came first. Oort and Muller had a successful confirmation within six weeks, followed shortly thereafter by the Australian group under Christiansen. The three teams reported their results in the same issue of *Nature* (Ewen and Purcell 1951, Muller and Oort 1951, Pawsey 1951).

During the same year of 1951, the problem of identifying radio sources and appreciating their significance was also reaching resolution. Graham Smith at the Cavendish Laboratory (1951) and Bernard Mills at Sydney (1952) were corresponding with Walter Baade and Rudolf Minkowski of the Mt Wilson Observatory, collaborating on an effort to use precise radio positions to allow faint optical objects to be identified securely (Smith 1951, Mills 1952, Baade and Minkowski 1954). The results were astonishing. Cassiopeia A was apparently associated with a compact region of faint nebulosity, with evidence of great agitation since the radial velocities varied by more than $4000\,\mathrm{km\,s^{-1}}$ from wisp to wisp. The wisps were evidently the remnants of a supernova explosion, but the character of the nebulosity was entirely novel. The radio source Cygnus A turned out to be associated with an even more surprising object: a faint galaxy having a peculiar emission-line spectrum, exhibiting a recession velocity of $16\,000\,\mathrm{km\,s^{-1}}$. Despite its great distance, Cygnus A was the most intense extragalactic object in the heavens. Even if it had been at ten times the distance, too far away to be detected optically, it would still have been a detectable radio source.

The results were reported at the 1952 IAU General Assembly in Rome and formally published two years later (Baade and Minkowski 1954). The significance of the new science of radio astronomy was secure. A broad new range of phenomena had been opened to study: known supernova remnants like the Crab Nebula, new supernova-related objects like Cassiopeia A, nearby peculiar galaxies such as NGC 5128 and M 87, and very distant radio galaxies akin to Cygnus A. In addition, the study of Galactic structure was transformed by the availability of the survey of 21-cm neutral hydrogen radiation by Kwee *et al.* (1954). A furious activity had come into being in England, not only in Cambridge under Ryle, but also in Manchester. Under Bernard Lovell a new branch of astronomy, the study of meteors by radar, was being created, and a broadly based programme, including innovative radio-interferometry studies of radio sources, was in progress. The CSIRO Radiophysics group, led by Bowen and Pawsey, had similarly built a radio-astronomy programme that was contributing to nearly every aspect of the new science. New surprises such as quasars, pulsars, interstellar masers and the cosmic microwave background lay in the future but, by 1954, when the optical identification work of Baade and Minkowski was published, the groundwork had been laid.

The 4th Symposium of the International Astronomical Union, on Radio Astronomy, held at Jodrell Bank in 1955, gives evidence of the changes that were under way in radio astronomy. The 250-ft steerable paraboloid was under construction at Jodrell Bank, while the Mullard Radio Observatory was in the midst of the 3C survey, and in the Netherlands the 25-metre telescope at Dwingeloo was nearing completion. Action of a major sort was finally stirring in the U.S. The National Science Foundation had decided on the foundation of the National Radio Astronomy Observatory, and had started to fund a radio-astronomy programme at Harvard, following Ewen and Purcell's discovery of the 21-cm hydrogen line. New programmes had begun with the establishment of the CalTech radio-astronomy programme (soon to become the Owens Valley Radio Observatory under the directorship of John Bolton), and in Washington, D.C., at the Department of Terrestrial Magnetism of the Carnegie Institution of Washington. In Australia, the CSIRO was about to receive a grant from the Carnegie Corporation to start building the 210-foot radio telescope at Parkes, Shklovsky's prediction of polarized light from the Crab nebula, a sure indication of synchrotron radiation, had been completely verified, and the resulting conclusion, that extremely high-energy phenomena were occurring in radio sources emphasized their unusual character. An active decade of radio-astronomy discoveries was well under way. A new low-frequency array, the 22 MHz Carnegie Mills Cross, (constructed by F. G. Smith and B. F. Burke) and the Mills Cross of B. Y. Mills had both gone into operation, and there were new astronomy initiatives in Germany, France, the Soviet Union, Japan, and Italy.

Planetary astronomy had been regarded as the domain of optical astronomers, but two new developments overturned conventional views. Observations with the Carnegie Mills Cross discovered the intense low-frequency radio bursts from Jupiter. Shortly after van Allen's 1958 discovery of the radiation belts of the earth, observations at the Owens Valley at decimetre wavelengths demonstrated that Jupiter possessed a strong magnetic field, that confined high-energy particles within an even more powerful radiation belt than that of the earth. As a result, all planetary spacecraft to Jupiter have had to be radiation-hardened; conventional electronics would not survive. Also in the period 1956–1958 observations of Venus at centimetre wavelengths showed that its temperature, at radio wavelengths, was of the order of 600 C, correcting the optical observations that claimed a cold temperature for the planet. Mercury, too, presented a puzzle, for it showed little variation of temperature with phase. This problem was cleared up by radar observations by Pettengill, who showed that the conventional belief, that Mercury kept the same face to the sun, from the same effect of tidal locking that keeps the same face of the Moon toward the Earth, was incorrect. Mercury rotates three times for every two revolutions, a 3/2 tidal lock that was completely unexpected.

The development of radio interferometry proceeded during the decade of the 1950s. It was clear that, in order to identify radio sources with optically-observed galaxies, accurate positions were needed, and interferometry could provide the accuracy if sufficient care were taken. Jennison, at Jodrell Bank, discovered the double-lobed character of the radio galaxy Cygnus A, using the technique of phase closure, which would later be an important tool in radio interferometry. Then, in 1961, Ryle formulated aperture synthesis, a technique that had been half-understood among radio astronomers, but which he demonstrated in its complete form. The One-mile aperture-synthesis interferometer at Cambridge, consisting of one movable and two fixed 25-meter dishes was constructed, and gave the first description of the radio galaxies now known as FR I and FR II, with the narrow, often one-sided jets ejected from the nucleus. Resolving the nucleus itself required interferometers with longer baselines, as shown in radio-linked observations at Jodrell Bank in 1955. The basis was laid for the next outstanding radio discovery, the quasi-stellar radio objects, generally called quasars.

In 1963, accurate positional measurements made at the Owens Valley Radio Observatory, coupled with optical observations at Mt. Palomar, showed that there were relatively bright, apparently stellar objects associated with the radio sources 3C48 and 3C273. Greenstein and Schmidt, carrying out spectroscopic observations at Mt. Palomar, announced in 1964 that these two objects were not stars, but exhibited excited optical lines with an astounding redshift; 0.16 for 3C273 and 0.37 for 3C48. This meant that these compact objects were emitting a hundred times more energy than the total stellar output of a galaxy. This opened an entirely new field of research, eventually leading to the realization that the energy came gravitationally, from matter falling into a central black hole of extraordinary mass. The final confirmation came from occultation measurements at the Parkes 210-ft telescope by Hazard, assisted by Bolton, in which the radio contours outlined the central condensed source and the emitted jet-like structure .

Two new discoveries soon followed. Penzias and Wilson, in 1965, discovered the Cosmic Microwave Background, sharing with Hubble's discovery of the expanding Universe the distinction of being one of the two most important cosmological discoveries of modern times. The 'big bang' cosmology, in which the Universe began as a hot, expanding plasma of fundamental particles and gamma-rays, was established as a credible model for the early stages of development in the Universe in which we live.

The second new revelation came the next year, in 1966, when Bell and Hewish found that there were regularly pulsing objects that they called pulsars. It shortly became clear that these were rotating, magnetized neutron stars, a class of object whose existence had been speculated upon (the earliest publication is that of Zwicky and Baade, in 1933). These would prove to be interesting physical laboratories, exhibiting relativistic phenomena that cannot be duplicated in laboratories on Earth. During this period, as the technique of aperture synthesis became thoroughly understood, ambitious aperture-synthesis telescopes were being constructed. The Westerbork Synthesis Radio Telescope (WSRT) was constructed near Groningen in the north of the Netherlands, an east-west linear array initially of 14 25-metre dishes, 10 fixed and two moveable, with a maximum extent of one metric mile (1640 m). The telescope went into operation in 1972, and has contributed widely to the advance of radio astronomy, including the understanding of the kinematics of spiral arms and the discovery of the high-velocity hydrogen clouds that accompany the Milky Way at high galactic latitudes. These successes led to the development of the Very Large Array (VLA) by the NRAO at Socorro, New Mexico, an array of 27 25-metre dishes, extending to a diameter of 35 km, giving an angular resolution of 0.1 arcseconds at 1.25-cm wavelength. The array was completed in 1980. Even longer baselines, up to 200 km, were used in the MERLIN aperture-synthesis array based on Jodrell Bank.

The next development had its origin in 1967, when groups at the NRAO, the Haystack Observatory of MIT, and at the NRC in Canada invented the technique of Very-Long Baseline Interferometry (VLBI). This allows interferometry to be carried out by radio telescopes that are too far apart to be connected by cables or radio links. Instead, a stable atomic frequency standard (usually a hydrogen maser) provides a time base, with the received signal being recorded, shipped to a central facility, where the signals are cross-correlated to give the fringe visibility. The size of the earth was the initial limitation (with larger baselines being made available by the Japanese HALCA mission, which placed a VLBI station in a high orbit in space). The first dedicated array, the VLBA of the NRAO, went into operation in 1990; this was followed by the European VLBA, a part-time array of existing radio telescopes, and by the World-wide Array, an organized consortium of all the VLBI stations of the world. Radio astronomy had become truly international.

Further reading

Astronomy Transformed, Edge D. O. and Mulkay M. J., New York, Wiley, 1976.
Classics in Radio Astronomy, Sullivan W. T. III, Dordrecht, Reidel, 1982.
Cosmic Noise: A History of Early of Radio Astronomy, Sullivan W. T. III., Cambridge, Cambridge University Press, 2009.
The Early Years of Radio Astronomy, ed. Sullivan W. T. III., Cambridge, Cambridge University Press, 1984.
The Invisible Universe Revealed, Verschuur G. L., Berlin, Springer-Verlag, 1987.

Appendix 4 Calibrating polarimeters

A radio polarimeter differs from an optical polarimeter by virtue of operating directly on the electric-field components of an incoming signal rather than on its intensity. In both, however, the polarization state of the signal is usually represented by the Stokes parameters introduced in Section 2.4. (Other representations in both optical and radio polarimetry, and a unifying discussion, are presented by Hamaker *et al.* (1996); see also Britton (2000)).

A4.1 Single-dish radio telescopes

The signal at the telescope may be represented by the vector

$$\mathbf{e} = \begin{pmatrix} e_x \\ e_y \end{pmatrix}. \tag{A4.1}$$

The radio polarimeter ideally operates on the field components to give voltage outputs (disregarding gain factors, and assuming a time average)

$$\mathbf{v} = \begin{pmatrix} v_{xx^*} \\ v_{xy^*} \\ v_{yx^*} \\ v_{yy^*} \end{pmatrix} = \begin{pmatrix} e_x e_x^* \\ e_x e_y^* \\ e_y e_x^* \\ e_y e_y^* \end{pmatrix}. \tag{A4.2}$$

The corresponding measured Stokes parameters may be represented as a column vector

$$S_m = \begin{pmatrix} I_m \\ Q_m \\ U_m \\ V_m \end{pmatrix} = \begin{pmatrix} v_{xx^*} + v_{yy^*} \\ v_{xx^*} - v_{yy^*} \\ v_{xy^*} + v_{yx^*} \\ i(v_{xy^*} - v_{yx^*}) \end{pmatrix}. \tag{A4.3}$$

In practice the Stokes parameters S_m as measured with a single telescope require correction for cross-coupling effects in the telescope and receiver systems, and for the parallactic angle P between the telescope axis and the sky coordinates. The measured Stokes parameters S_m are related to the Stokes parameters S_0 of the incoming signal by

$$S_m = M \times S_0. \tag{A4.4}$$

where M is the 4×4 Mueller matrix characterizing the total effect of the polarimeter and the parallactic angle.

The Mueller matrix may be written as the product of two separate matrices, namely M_P, representing the effects of the parallactic angle, and M_{pol}, representing the polarimeter; M_{pol} may in turn be decomposed into a set of independent matrices representing the polarization characteristics of separate elements in the polarimeter, such as cross-coupling in the telescope feed system, differential phasing in the detector and different gains in the output channels, noting that they must be combined in the right order (reverse order of occurrence) since matrix multiplication is non-commutative. (A further matrix M_{rot} can represent the progress of the signal from the source to the telescope, including Faraday rotation in interstellar space and in the ionosphere.) Each element of the combined matrix is an algebraic combination of terms representing the amplitude and phase of the cross-coupling and the differential gains.

The Mueller matrix M_P which accounts for the parallactic angle is

$$M_P = \begin{bmatrix} 1 & 0 & 0 & 0 \\ 0 & \cos(2P) & \sin(2P) & 0 \\ 0 & -\sin(2P) & \cos(2P) & 0 \\ 0 & 0 & 0 & 1 \end{bmatrix}. \tag{A4.5}$$

The Mueller matrix in Equation (A4.4) is then (again, note the sequence)

$$M = M_{pol} \times M_P. \tag{A4.6}$$

The elements of M now contain the variables $\cos(2P)$ and $\sin(2P)$. The calibration of a polarimeter is achieved by observing a partially polarized source over a range of parallactic angles, such as occurs over a period of some hours with an alt–azimuth-mounted telescope. The parallactic angle P for an observation of a source at declination δ at hour angle H from an observatory at latitude ϕ is given by

$$P = \arctan\left(\frac{\sin H \cos\phi}{\sin\phi \cos\delta - \cos\phi \sin\delta \cos A}\right). \tag{A4.7}$$

The terms of the combined matrix $M_{pol} \times M_P$ are functions of parallactic angle, and may be evaluated separately by measuring S_m over a sufficient range of P.

An example by Johnston (2002) is shown in Figure A4.1. Here the pulse peak of a highly polarized pulsar was observed with a polarimeter over a 180° range of parallactic angle. For an ideal polarimeter, the two measured Stokes parameters Q_m and U_m would yield equal-amplitude sinusoids in quadrature, while I_m and V_m would be independent of P. Analysis of the actual dependence of the components of S_m on P yields both M_{pol} and the Stokes parameters S_0 of the source.

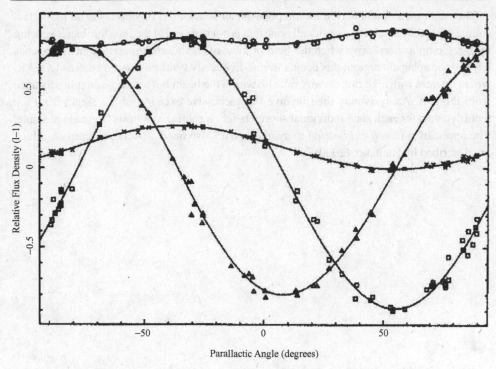

Parallactic Angle (degrees)

Figure A4.1. Measured Stokes parameters for PSR J1359-6038 as a function of parallactic angle. The sinusoidal lines represent the best fit, which yielded the calibration of the polarimeter and the Stokes parameters of the source (Johnston 2002).

A4.2 Polarization in interferometers

In an interferometric polarimeter two orthogonally polarized and separate antennas take the place of the polarimeter channels in Equation (A4.2) above. If both antennas A and B have orthogonal feeds, each feed sends a signal to a correlator that multiplies them in pairs; the outputs of the correlator channels are

$$
v = \begin{pmatrix} v_{Ax} v_{Bx}^* \\ v_{Ax} v_{By}^* \\ v_{Ay} v_{Bx}^* \\ v_{Ay} v_{By}^* \end{pmatrix}.
\tag{A4.8}
$$

Corresponding to the scalar Stokes parameters of Equation (A4.3), we now have *Stokes visibilities* $\mathcal{I}$, $\mathcal{Q}$, $\mathcal{U}$ and $\mathcal{V}$, which include the effects of angular resolution and may assume complex values. The correct Stokes visibilities are again related to the measured values by a *system matrix*, which acts like a Mueller matrix but whose individual terms may be complex. A detailed analysis may be found in Hamaker *et al.* (1996).

The calibration of an interferometric polarimeter is achieved by observing an individual source with known polarization, ideally one that is not resolved at the interferometer spacing in use. Complications arise when the field of view of the interferometer elements extends beyond the aplanatic region; this occurs in low-frequency synthesis arrays such as LOFAR, where sources with high flux density may lie within the beam but at a large angular distance from the axis. Analysis may then involve the successive excision of the signal from the recorded data for each such individual source before a routine synthesis becomes possible. The application of *self-calibration* to synthesis data also needs special techniques, which are described by Hamaker (2000).

References

Alfvén H. A. and Herlofsen N., 1950. *Phys. Rev.*, **78**, 616.

Alpar M. A., Cheng K. S. and Pines D., 1989. *Astrophys. J.*, **346**, 823.

Alpher R. A., Bethe H. and Gamow G., 1948. *Phys. Rev.*, **73**, 803.

Alpher R. A. and Herman R., 1968. *Nature*, **162**, 774.

Antonucci R., 1993. *Ann. Rev. Astron. Astrophys.*, **31**, 473.

Appleton E. V. and Hey J. S., 1946. *Nature*, **158**, 339.

Aschenbach B., Egger R. and Trümper J., 1995. *Nature*, **373**, 587.

Audouze J., 1987. *IAU Symp.*, **124**, 89.

Baade W. and Minkowski R., 1954. *Astrophys. J.*, **119**, 206.

Baade W. and Zwicky F., 1934. *Phys. Rev.*, **45**, 138.

Baars J. W. M., Genzel R., Pauliny-Toth I. I. K. and Witzel A., 1977. *Astron. Astrophys.*, **61**, 99.

Baars J. W. M., van der Brugge J. F., Casse J. L. *et al.*, 1973. *Proc. IEEE*, **61**, 1258.

Backer D. C., Kulkarni S. R., Heiles C., Davis M. M. and Goss W. M., 1982. *Nature*, **600**, 615.

Bahcall J., Kirhakos S., Schneider D. P. *et al.*, 1995. *Astrophys. J.*, **452**, L91.

Baldwin J. E., Beckett M. G., Boysen R. C. *et al.*, 1996. *Astron. Astrophys.*, **306**, L13.

Baldwin J. E. and Warner P. J., 1978. *MNRAS*, **182**, 411.

Banday A. J., Dickinson C., Davies R. D., Davis R. J. and Górski K. M., 2003. *MNRAS*, **345**, 897.

Barnes C., Limon M., Page L. *et al.*, 2002. *Astrophys. J. Suppl.*, **143**, 567.

Barnes J. E. and Hernquist L. E., 1992. *Ann. Rev. Astron. Astrophys.*, **30**, 51.

Bartel N., Bietenholz M. F., Rupen M. P. *et al.*, 1994. *Nature*, **368**, 610.

Barthel P., 1989. *Astrophys. J.*, **336**, 606.

Beck R., Brandenberg A., Moss D., Shukurov A. and Sokoloff D., 1996. *Ann. Rev. Astron. Astrophys.*, **34**, 155.

Bell A. R., Gull S. F. and Kenderdine S., 1975. *Nature*, **257**, 463.

Bennett A. S., 1962. *Mem. RAS*, **68**, 163.

Bennett C. L., Banday A. J., Górski K. M. *et al.*, 1996. *Astrophys. J.*, **464**, 1.

Bersanelli M., Bensadown M., De Amici G. *et al.*, 1994. *Astrophys. J.*, **424**, 517.

Beswick R. J., Riley J. D., Marti-Vidal I. *et al.*, 2006. *MNRAS*, **369**, 1221.

Beuermann K., Kanbach G. and Berkuijsen E. M., 1985. *Astron. Astrophys.*, **153**, 17.

Bhattacharya D. and van den Heuvel E. P. J., 1991. *Phys. Rep.*, **203**, 1.

Biggs A. D., Browne I. W. A., Helbig P. *et al.*, 1999. *MNRAS*, **304**, 349.

Bignami G. F., Caraveo P. A. and Mereghettis S., 1993. *Nature*, **361**, 704.

Binney J. J., 1992. *Ann. Rev. Astron. Astrophys.*, **30**, 705.

Biretta J., Moore R. and Cohen M., 1986. *Astrophys. J.*, **308**, 93.

Blandford R. D., McKee C. F. and Rees M., 1977. *Nature*, **267**, 211.

Blandford R. and Narayan R., 1986. *Astrophys. J.*, **310**, 568.

Blitz L., Binney J., Lo L. K., Bally J. and Ho P. T. P., 1993. *Nature*, **361**, 417.

Blitz L., Fich M. and Stark A. A., 1980. *IAU Symp.*, **87**, 213.

Boboltz D. A., Diamond P. J. and Kemball A. J., 1997. *Astrophys. J.*, **487**, L147.

Bogdanov S., Grindlay J. E., Heinke C. O. *et al.*, 2006. *Astrophys. J.*, **646**, 1104.

Boggess N. W., Mather J. C., Weiss R. *et al.*, 1992. *Astrophys. J.*, **397**, 420.

Bolton J. G. and Stanley G. J., 1948. *Nature*, **161**, 312.

Bolton J. G., Stanley G. J. and Slee O. B., 1949. *Nature*, **164**, 101.

Bondi H. and Gold T., 1948. *MNRAS*, **108**, 252.

Booth R. S., Kus A. J., Norris R. P. and Porter N. D., 1981. *Nature*, **290**, 382.

Bowers F. K., Whyte D. A., Landaker T. L. and Klinger R. J., 1973. *Proc. IEEE*, **61**, 1339.

Bracewell R. N., 1961. *IRE Trans. Antennas Propagation*, **9**, 59.

Bracewell R. N. and Roberts J. A., 1954. *Aust. J. Phys.*, **71**, 615.

Breton R. P., Kaspi V. M., Kramer M. *et al.*, 2008. *Science*, **321**, 104.

Bridle A. H., Hough D. H., Lonsdale C. J., Burns J. O. and Laing R. A., 1994. *Astron. J.*, **108**, 766.

Briggs F. H. and Sackett P. D., 1989. *Icarus*, **80**, 77.

Brisken W. F., Benson J. M., Goss W. M. and Thorsett S. E., 2002. *Astrophys. J.*, **571**, 906.

Brisken W. F., Fruchter A. S., Goss W. M., Herrnstein R. M. and Thorsett S. E., 2003. *Astron. J.*, **126**, 3090.

Britton M. C., 2000. *Astrophys. J.*, **532**, 1244.

Brouw W. N. and Spoelstra T. A. Th., 1976. *Astron. Astrophys. Suppl.*, **26**, 129.

Brown J. C., Haverkorn M., Gaensler B. M. *et al.*, 2007. *Astrophys. J.*, **663**, 258.

Burn B. J., 1966. *MNRAS*, **133**, 67.

Burns W. R. and Roberts M. S., 1971. *Astrophys. J.*, **166**, 265.

Burrows A., 2000. *Nature*, **403**, 727.

Burton W. B., 1988. In *Galactic and Extragalactic Radio Astronomy*, ed. Verschuur G. L. and Kellermann K. I., Berlin, Springer-Verlag, p. 295.

Burton W. B. and Gordon M. A., 1978. *Astron. Astrophys.*, **63**, 7.

Camilo F., Lorimer D. R., Friere P., Lyne A. G. and Manchester R. N., 2000. *Astrophys. J.*, **535**, 975.

Camilo F., Ransom S. M., Halpern J. P. *et al.*, 2006. *Nature*, **442**, 892.

Cane H. V., 1978. *Aust. J. Phys.*, **31**, 561.
   1979. *MNRAS*, **189**, 465.

Chang K. and Refsdal S., 1979. *Nature*, **282**, 561.

Chapman J. M. and Cohen R. J., 1986. *MNRAS*, **220**, 513.

Chatterjee S., Vlemmings W. H. T., Brisken W. F. *et al.*, 2005. *Astrophys. J.*, **630**, L61.

Chen G. H., Kochanek C. S. and Hewitt J. N., 1995. *Astrophys. J.*, **447**, 62.

Chen K. and Ruderman M., 1993. *Astrophys. J.*, **402**, 264.

Chi X. and Wolfendale A. W., 1993. *Nature*, **362**, 610.

Chluba J. and Sunyaev R. A., 2006. *Astron. Astrophys.*, **458**, L29.

Clark B. J., 1968. *IEEE Trans. Antennas Propagation*, **16**, 143.

Clemens D. P., 1985. *Astrophys. J.*, **295**, 422.

Clifton T. R., Frail D. A., Kulkarni S. R. and Weisberg J. M., 1988. *Astrophys. J.*, **333**, 332.

Cohen M., Witteborn F. C., Carbon D. F. *et al.*, 1992. *Astron. J.*, **104**, 1650.

Cohen M. H., Linfield R. P., Moffet A. T. *et al.*, 1977. *Nature*, **268**, 405.

Cohen R. J., 1989. *Rep. Prog. Phys.*, **52**, 881.

Cohen R. J., Brebner G. C. and Potter M. M., 1990. *MNRAS*, **246**, 3P.

Coles P. and Ellis G., 1994. *Nature*, **370**, 609.

Combes F., 1991. *Ann. Rev. Astron. Astrophys.*, **29**, 195.

Condon J. J., 1984. *Astrophys. J.*, **287**, 461.

1992. *Ann. Rev. Astron. Astrophys.*, **30**, 575.

Condon J. J., Anderson M. L. and Helou G., 1991. *Astrophys. J.*, **376**, 95.

Condon J. J., Cotton W. D., Greisen E. W. *et al.*, 1998. *Astron. J.*, **115**, 1693.

Cooper B. F. C., 1970. *Aust. J. Phys.*, **23**, 521.

Cordes J. M. and Lazio T. J. W., 2002. astro-ph 0207156.

Cordes J. M., Downs G. S. and Kraus-Polstroff J., 1988. *Astrophys. J.*, **330**, 847.

Cordes J. M., Romani R. W. and Lundgren S. C., 1993. *Nature*, **362**, 133.

Cotton W. D., Counselman C. C., Geller R. B. *et al.*, 1979. *Astrophys. J.*, **229**, L115.

Cox D. P., 2005. *Ann. Rev. Astron. Astrophys.*, **43**, 337.

Cox D. P. and Reynolds R. J., 1987. *Ann. Rev. Astron. Astrophys.*, **25**, 303.

Damashek M., Backus P. R., Taylor J. H. and Burkhardt R. K., 1982. *Astrophys. J.*, **253**, L57.

Dame T. M. and Thaddeus P., 1985. *Astrophys. J.*, **297**, 751.

Dame T. M., Ungerechts H., Cohen R. S. *et al.*, 1987. *Astrophys. J.*, **322**, 706.

Davies R. D., Dickinson C., Banday A. J. *et al.*, 2006. *MNRAS*, **370**, 1125.

Davis R. J., Muxlow T. W. B. and Conway R. J., 1985. *Nature*, **318**, 343.

de Pater I., 1990. *Ann. Rev. Astron. Astrophys.*, **28**, 347.

de Pater I. and Dickel J. R., 1991. *Icarus*, **94**, 474.

de Ruiter H. R., Parma P., Stirpe G. M. *et al.*, 1998. *Astron. Astrophys.*, **339**, 34.

Dennett-Thorpe J. and de Bruyn A. G., 2003. *Astron. Astrophys.*, **404**, 113.

Deshpande A. A., 2000. *ASP Conf.*, **202**, 149.

Dewi J. D. M. and van den Heuvel E. P. J., 2004. *MNRAS*, **349**, 169.

Diamond P. J., Goss W. M., Romney J. D. *et al.*, 1989. *Astrophys. J.*, **347**, 302.

Dicke R. H., Peebles P. J. E., Roll P. G. and Wilkinson D. T., 1965. *Astrophys. J.*, **142**, 414.

Dickel J. R. and Willis A. G., 1980. *Astron. Astrophys.*, **85**, 55.

Dickey J.M and Lockman F. J., 1990. *Ann. Rev. Astron. Astrophys.*, **28**, 215.

Dickinson C., Battye R. A., Carreira P. *et al.*, 2004. *MNRAS*, **353**, 732.

Dodson R., Lewis D. and McCulloch P., 2007. *Astrophys. Space Sci.*, **308**, 585.

Doeleman S. S., Weintroub J., Rogers A. E. E. *et al.*, 2008. *Nature*, **455**, 78.

Dougherty S. M., Bode M. F., Lloyd H. M., Davis R. J. and Eyres S. P. S., 1995. *MNRAS*, **272**, 843.

Dressel L. L. and Condon J. J., 1978. *Astrophys. J. Suppl.*, **36**, 53.

Drossart P., Maillard J.-P., Caldwell J. *et al.*, 1989. *Nature*, **340**, 539.

Duin R. M. and Strom R. G., 1975. *Astron. Astrophys.*, **39**, 33.

Duin R. M. and van der Laan H., 1975. *Astron. Astrophys.*, **40**, 111.

Dulk G. A., 1985. *Ann. Rev. Astron. Astrophys.*, **23**, 169.

Dunkley J., Komatsu E., Nolta M. R. *et al.*, 2008. arXiv 0803.0586.

Duyvendak J. J. L., 1942. *Publ. Astron. Soc. Pacific*, **54**, 91.

Edwards R. T., Hobbs, G. B. and Manchester R. N., 2006. *MNRAS*, **372**, 1549.

Elitzur M., 1992. *Ann. Rev. Astron. Astrophys.*, **30**, 75.

Emonts B. H. C., Morganti R., Oosterloo T. A. *et al.*, 2007. *Astron. Astrophys.*, **464**, L1.

Ewen H. I. and Purcell E. M., 1951. *Nature*, **168**, 356.

Eyres S. P. S., Davis R. J. and Bode M. F., 1996. *MNRAS*, **279**, 249.

Falco E. E., Gorenstein M. V. and Shapiro I. I., 1991. *Astrophys. J.*, **372**, 364.

Fanaroff B. L. and Riley J. M., 1974. *MNRAS*, **167**, 31P.

Fassnacht C. D., Blandford R. D., Cohen J. G. *et al.*, 1999. *Astron. J.*, **117**, 658.

Fender R. P., Garrington S. T., McKay D. J. *et al.*, 1999. *MNRAS*, **304**, 865.

Fermi E., 1949a. *Phys. Rev.*, **75**, 161.

   1949b. *Phys. Rev.*, **75**, 1169.

Fich M. and Tremaine S., 1991. *Ann. Rev. Astron. Astrophys.*, **29**, 409.

Fischer P., McKay T. A., Sheldron E. *et al.*, 2000. *Astron. J.*, **120**, 1198.

Fixsen D. J., Cheng E. S., Cottingham D. A. *et al.*, 1994. *Astrophys. J.*, **420**, 457.

   1996. *Astrophys. J.*, **470**, 63.

Flanagan C. S., 1990. *Nature*, **345**, 416.

Frail D. A., Kulkarni S. R., Nicastro L., Feroci M. and Taylor G. B., 1997. *Nature*, **389**, 263.

Frenk C. S., 1986. *Phil. Trans. RAS*, **A320**, 517.

Gamow G., 1946. *Phys. Rev.*, **55**, 718.

Gapper G. R., Hewish A., Purvis A. and Duffett-Smith P. J., 1982. *Nature*, **296**, 633.

Garn T., Green D. A., Hales S. E. G., Riley J. M. and Alexander P., 2007. *MNRAS*, **376**, 1251.

Garrett M. A., Calder R. J., Porcas R. W. *et al.*, 1994. *MNRAS*, **270**, 457.

Garrington S. T., Leahy J. P., Conway R. G. and Laing R. A., 1988. *Nature*, **331**, 147.

Gehrels N. and Wan Chen, 1993. *Nature*, **361**, 706.

Gentile G., Salucci P., Klein U. and Granato G. L., 2007. *MNRAS*, **375**, 199.

Genzel R., Pichon C., Eckart A., Gerhard O. E. and Ott T., 2000. *MNRAS*, **317**, 348.

Georgelin Y. M. and Georgelin Y. P., 1976. *Astron. Astrophys.*, **49**, 57.

Ghez A. M., Salim S., Hornstein S. D. *et al.*, 2005. *Astrophys. J.*, **620**, 744.

Giaconni R., Gursky H., Kellogg E., Schreier E. and Tananbaum H., 1971. *Astrophys. J.*, **167**, L67.

Ginzburg V. L. and Syrovatsky S. I., 1965. *Ann. Rev. Astron. Astrophys.*, **3**, 297.

   1969. *Ann. Rev. Astron. Astrophys.*, **7**, 375.

Gizani N. A. B. and Leahy J. P., 1999. *New Astron. Rev.*, **43**, 639.

Gold B., Bennett C. L., Hill R. S. *et al.*, 2008. arXiv 0803.0715.

Golden R. L., Manger B. G., Badhwar G. D. *et al.*, 1984. *Astrophys. J.*, **287**, 622.

Goldreich P. and Julian W. H., 1969. *Astrophys. J.*, **157**, 869.

Golla G. and Hummel E., 1994. *Astron. Astrophys.*, **284**, 777.

Gotthelf E. V., Halpern J. P. and Seward F. D., 2005. *Astrophys. J.*, **627**, 390.

Gotthelf E. V. and Vashisht G., 1997. *Astrophys. J.*, **486**, L133.

Gray R. O., 1998. *Astron. J.*, **116**, 482.

Gregory P. C. and Condon J. J., 1991. *Astrophys. J. Suppl.*, **75**, 1011.

Gregory P. C., Scott W. K., Douglas K. and Condon J. J., 1996. *Astrophys. J. Suppl.*, **103**, 427.

Griffith M. R. and Wright A. E., 1993. *Astron. J.*, **105**, 1666.

Grimani C., Stephens S. A., Cafagna F. S. *et al.*, 2002. *Astron. Astrophys.*, **392**, 287.

Gurvits L. I., Kellermann K. I. and Frey S., 1999. *Astron. Astrophys.*, **342**, 378.

Güsten R., Genzel R., Wright M. C. H. *et al.*, 1987. *Astrophys. J.*, **318**, 124.

Guth A. H., 1981. *Phys. Rev.*, **23**, 347.

Gwinn C. R., Moran J. M. and Reid M. J., 1992. *Astrophys. J.*, **393**, 149.

Hamaker J. P., 2000. *Astron. Astrophys. Suppl.*, **117**, 161.

Hamaker J. P., Bregman J. D. and Sault R. J., 1996. *Astron. Astrophys. Suppl.*, **117**, 137.

Han J. L., 2007. *IAU Symp. Proc.*, **242**, 55.

Han J. L., Manchester R. N. and Qiao G. J., 1999. *MNRAS*, **306**, 371.

Handa T., Sofue Y., Nakai N., Hirabayashi H. and Inoue M., 1987. *Publ. Astron. Soc. Japan*, **39**, 709.

Hankins T. H., Kern J. S., Weatherall J. C. and Eilek J. A., 2003. *Nature*, **422**, 141.

Hankins T. H. and Rickett B. J., 1975. In *Methods in Computational Physics*, Vol. 14, *Radio Astronomy*, New York, Academic Press, p. 55.

Harris S. and Wynn-Williams G. C., 1976. *MNRAS*, **174**, 649.

Harrison E. R., 1970. *Phys. Rev. D*, **1**, 2726.

Hartmann D. and Burton W. B., 1995. *Atlas of Galactic Hydrogen*, Cambridge, Cambridge University Press.

Harvanek M., Stocke J. T., Morse J. A. and Rhee G., 1997. *Astron. J.*, **114**, 2240.

Haslam C. G. T., Salter C. J., Stoffel H. and Wilson W. E., 1982. *Astron. Astrophys. Suppl.*, **47**, 1.

Heiles C., 1995. *PASP Conf. Series*, **80**, 507.

Heiles C., Chu Y.-H., Reynolds R. J., Yegingil I. and Troland T. H., 1980. *Astrophys. J.*, **242**, 533.

Helou G., Soifer B. T. and Rowan-Robinson M., 1985. *Astrophys. J.*, **298**, 7.

Henyey L. G. and Keenan P. C., 1940. *Astrophys. J.*, **91**, 625.

Herrnstein J. R., Moran J. M., Greenhill L. J. *et al.*, 1999. *Nature*, **400**, 539.

Hessels J. W. T., Ransom, S. M., Stairs I. H. *et al.*, 2006. *Science*, **311**, 1901.

Hewish A., Bell S. J., Pilkington J. D. H., Scott P. F. and Collins R. A., 1968. *Nature*, **217**, 709.

Hewitt J. N., Turner E. L., Schneider D. P. *et al.*, 1988. *Nature*, **333**, 537.

Hey J. S., Parsons S. J. and Phillips J. W., 1946a. *Proc. Roy. Soc. A*, **192**, 425.
    1946b. *Nature*, **158**, 234.

Hill R. J. and Clifford S. F., 1981. *Radio Sci.*, **16**, 77.

Hiltner W. A., 1949. *Astrophys. J.*, **109**, 471.

Hinshaw G., Spergel D. N., Verde L. *et al.*, 2003. *Astrophys. J. Suppl.*, **148**, 135.

Hinshaw G., Weiland J. L., Hill R. S. *et al.*, 2008. arXiv 0803.0732.

Hirabayashi H. and Hirosawa H., 2000. *Adv. Space Res.*, **26**, 589.

Hirabayashi H., Hirosawa H., Kobayashi H. *et al.*, 2000. *PASJ*, **52**, 955.

Hjellming R. M., Wade C. M., Vandenberg N. R. and Newell R. T., 1979. *Astron. J.*, **84**, 1619.

Hobbs G., Lorimer D. R., Lyne A. G. and Kramer M., 2005. *MNRAS*, **360**, 974.

Hobbs G. B. Edwards R. T. and Manchester R. N., 2006. *MNRAS*, **369**, 655.

Högbom J. H., 1974. *Astron. Astrophys. Suppl.*, **15**, 417.

Höglund B. and Mezger P. G., 1965. *Science*, **150**, 339.

Hoyle F., 1948. *MNRAS*, **108**, 372.
    1959. *Radio Astronomy*, ed. Bracewell R. N., Stanford, CA, Stanford University Press, p. 529.

Hu W. and Dodelson S., 2002. *Ann. Rev. Astron. Astrophys.*, **40**, 171.

Hu W., Scott D., Sugiyama N. and White M., 1995. *Phys. Rev. D*, **52**, 5498.

Jacoby G. H., Branch D., Ciardullo R. *et al.*, 1992. *PASP*, **104**, 599.

Jahoda K., Lockman F. J. and McCammon D., 1990. *Astrophys. J.*, **354**, 184.

Jansky K. G., 1933a. *Proc. IRE*, **21**, 1387.
    1933b. *Nature*, **132**, 66.
    1935. *Proc. IRE*, **23**, 1158.

Jennison R. C. and das Gupta M. K., 1953. *Nature*, **172**, 996.

Johnston K. J. and de Vegt Chr., 1999. *Ann. Rev. Astron. Astrophys.*, **37**, 97.

Johnston S., 2002. *Publ. Astron. Soc. Aust.*, **19**, 277.

Johnston S., Lorimer D. R., Harrison P. A. *et al.*, 1993. *Nature*, **361**, 613.

Johnston S., Manchester R. N., Lyne A. G. *et al.*, 1992. *Astrophys. J.*, **387**, L37.

Jones W. C., Ade P. A. R., Bock J. J. *et al.*, 2006. *Astrophys. J.*, **647**, 823.

Junor W., Biretta J. A. and Livio M., 1999. *Nature*, **401**, 891.

Kantharia N. G., Anupama G. C., Prabhu T. P. *et al.*, 2007. *Astrophys. J.*, **667**, L171.

Keeton C. R., Mao S. and Witt H. J., 2000. *Astrophys. J.*, **537**, 697.

Kellermann K. I. and Owen F. N., 1988. *Galactic and Extragalactic Radio Astronomy*, 2nd edn, Berlin, Springer-Verlag, p. 570.

Kellermann K. I. and Wall J. V., 1987. *IAU Symp.*, **124**, 545.

Kiepenheuer K. O., 1950. *Phys. Rev.*, **79**, 738.

Kochanek C. and Narayan R., 1992. *Astrophys. J.*, **401**, 461.

Komatsu E., Dunkley J., Nolta M. R. *et al.*, 2008. arXiv 0803.0547.

Komissarov S. S. and Gubanov A. G., 1994. *Astron. Astrophys.*, **285**, 27.

Kormendy J. and Norman C., 1979. *Astrophys. J.*, **233**, 539.

Kramer M., 1998. *Astrophys. J.*, **509**, 856.

   2005. *Advances in Astronomy*, ed. J. M. T. Thompson, London, Royal Society, p. 143.

Kramer M., Lange C., Lorimer D. R. *et al.*, 1999. *Astrophys. J.*, **526**, 957.

Kramer M., Stairs I. H., Manchester R. N. *et al.*, 2006. *Science*, **314**, 97.

Krichbaum T. P., Graham D. A., Witzel A. *et al.*, 1998. *Astron. Astrophys.*, **335**, L106.

Kronberg P. P., 1994. *Rep. Prog. Phys.*, **57**, 325.

Kulkarni S. R., 1989. *Astron. J.*, **98**, 1112.

Kuo C. L., Ade P. A. R., Bock J. J. *et al.*, 2004. *Astrophys. J.*, **600**, 32.

Kwee K. K., Muller A. C. and Westerhout G., 1954. *Bull. Astron. Inst. Netherlands*, **12**, 117.

Ladd E. F., Deane J. R., Sanders D. B. and Wynn-Williams C. G., 1993. *Astrophys. J.*, **419**, 186.

Lah P., Chengalur J. N., Briggs F. H. *et al.*, 2007. *MNRAS*, **376**, 1357.

Laing R. A., Longair M. S., Riley J. M., Kibblewhite E. J. and Gunn J. E., 1978. *MNRAS*, **184**, 149.

Lambert H. C. and Rickett B. J., 1999. *Astrophys. J.*, **517**, 299.

   2000. *Astrophys. J.*, **531**, 883.

Landecker T. L., Reid R. I., Wolleben M. *et al.*, 2006. *Bull. Am. Astron. Soc.*, **208**, 4909.

Lane B. F. and Colavita M. M., 2003. *Astron. J.*, **125**, 1623.

Lane B. F., Kuchner M. J., Boden A. F., Creech-Eakman M. and Kulkarni S. R., 2000. *Nature*, **407**, 485.

Langston G. I., Conner S. R., Lehar J., Burke B. F. and Weiler K. W., 1990. *Nature*, **344**, 43.

Larsson B., Liseau R., Pagani L. *et al.*, 2007. *Astron. Astrophys.*, **466**, 999.

Lattimer J. M. and Prakash M., 2001. *Astrophys. J.*, **550**, 426.

Leahy J. P. and Perley R. A., 1995. *MNRAS*, **277**, 1047.

Levy G. S., Linfield R. P., Ulvestad J. S. *et al.*, 1986. *Science*, **234**, 187.

Lifshitz E., 1946. *J. Phys. Acad. Sci. USSR*, **10**, 116.

Lin C. C. and Shu F. H., 1964. *Astrophys. J.*, **140**, 646.

Linde A., 1984. *Rep. Prog. Phys.*, **47**, 925.

Linfield R. P., Levy G. S., Ulvestad J. S. *et al.*, 1989. *Astrophys. J.*, **336**, 1105.

Lorimer D. R., Lyne A. G. and Camilo F., 1998. *Astron. Astrophys.*, **331**, 1002.

Lorimer D. R., Faulhner A. J., Lyne A. G. *et al.*, 2006. *MNRAS*, **372**, 777.

Lovell J. E. J., Jauncey D. L., Bignall H. E. *et al.*, 2003. *Astron. J.*, **126**, 1699.

Lynden-Bell D. and Rees M., 1971. *MNRAS*, **152**, 461.
Lyne A. G. and Manchester R. N., 1988. *MNRAS*, **234**, 477.
Lyne A. G. and McKenna J., 1989. *Nature*, **340**, 367.
Lyne A. G., Manchester R. N. and Taylor J. H., 1985. *MNRAS*, **213**, 613.
Lyne A. G., Manchester R. N., Lorimer D. R. *et al.*, 1998. *MNRAS*, **295**, 743.
Lyne A. G., Pritchard R. S., Graham-Smith F. and Camilo F., 1996. *Nature*, **381**, 497.
Lyne A. G., Shemar S. L. and Smith F. Graham, 2000. *MNRAS*, **315**, 534.
Maede Y. and Koyama K., 1996. *PASP*, **102**, 423.
Malov I. F., Malofeev V. M. and Sen'e D. S., 1994. *Astron. Zh.*, **71**, 762.
Manchester R. N., Lyne A. G., Camilo F. *et al.*, 2001. *MNRAS*, **328**, 17.
Marshall F. E., Gotthelf E. V., Zhang W., Middleditch J. and Wang Q. D., 1998. *Astrophys. J.*, **499**, L179.
Marti J., Parades J. M. and Estabella R., 1992. *Astron. Astrophys.*, **258**, 309.
Marti J., Rodriguez L. F. and Reipurth B., 1995, *Astrophys. J.*, **449** 184.
Massardi M., Ekers R. D., Murphy T. *et al.*, 2008. *MNRAS*, **384**, 755.
Mather J. C., Cheng E. S., Cottingham D. A. *et al.*, 1994. *Astrophys. J.*, **420**, 439.
Mathewson D. S., Ford D. L., Schwarz M. P. and Murray J. D., 1979. In *The Large-Scale Characteristics of the Galaxy*, ed. Burton W. B., Dordrecht, Reidel, p. 547.
McCready L. L., Pawsey J. L. and Payne-Scott R., 1947. *Proc. Roy. Soc. A*, **190**, 357.
McIntosh G. C., Predmore C. R., Moran J. M. *et al.*, 1989. *Astrophys. J.*, **337**, 934.
McKellar A., 1941. *Publ. Dominion Astrophys. Observatory*, **7**, 251.
McLaughlin M. A., Lyne A. G., Lorimer *et al.*, 2006. *Nature*, **439**, 817.
McMillan D. S. and Ma C., 1994. *J. Geophys. Res.*, **99**, 637.
Melatos A., Peralta C. and Wyithe J. S. B., 2008. *Astrophys. J.*, **672**, 1103.
Mellier Y., 1999. *Ann. Rev. Astron. Astrophys.*, **37**, 127.
Melrose D. B., 1992 *Proc. IAU Colloquium 128*, ed. Hankins T. H., Rankin J. M. and Gil J. A., Pedagogical University Press, Poland, p. 306.
Menten K. M. and Young K., 1995. *Astrophys. J.*, **450**, L67.
Mezger P. G. and Henderson Λ. P., 1967. *Astrophys. J.*, **147**, 471.
Migdal A. B., 1959. *Nucl. Phys.*, **13**, 655.
Mignani R. P., de Luca A. and Caraveo P. A., 2004. *IAU Symp.*, **218**, 391.
Miller-Jones J. C. A., Blundell K. M., Rupen M. P. *et al.*, 2004. *Astrophys. J.*, **600**, 368.
Mills B. Y., 1952. *Aust. J. Sci. A*, **5**, 456.
Minier V., Booth R. S. and Conway J. E., 1999. *New Astron. Rev.*, **43**, 569.
Mirabel I. F. and Rodriguez L. F., 1994. *Nature*, **371**, 46.
Miyoshi M., Moran J., Herrnstein J. *et al.*, 1995. *Nature*, **373**, 127.
Morris M. and Serabyn G., 1996. *Ann. Rev. Astron. Astrophys.*, **34**, 645.
Moskalenko I. V. and Strong A. W., 1998. *Astrophys. J.*, **493**, 694.
Mueller H., 1948. *J. Opt. Soc. Am.*, **338**, 661.
Muller C. A. and Oort J. H., 1951. *Nature*, **168**, 357.
Muller C. A., Raimond E., Schwarz U. J. and Tolbert C. R., 1966. *Bull. Astron. Inst. Netherlands Suppl.*, **1**, 213.
Muxlow T. W. B. and Wilkinson P. N., 1991. *MNRAS*, **251**, 54.
Muxlow T. W. B., Pedlar A., Wilkinson P. N. *et al.*, 1994. *MNRAS*, **266**, 455.
Neininger N., 1992. *Astron. Astrophys.*, **263**, 30.
Newell R. T. and Hjellming R. M., 1982. *Astrophys. J.*, **263**, L85.
Nolta N. R., Dunkley J., Hill R. S. *et al.*, 2008. arXiv 0803.0593.
O'Dea C. P., 1998. *Publ. Astron. Soc. Pacific*, **110**, 493.
O'Dea C. P. and Owen F. N., 1986. *Astrophys. J.*, **301**, 841.

Ojha R., Fey A. L., Jauncey D. L., Lovell J. E. J. and Johnston K. J., 2004. *Astrophys. J.*, **614**, 607.

Oort J. H., Kerr F. J. and Westerhout G., 1958. *MNRAS*, **118**, 379.

Oswianik I., Conway J. E. and Polatidis A. G., 1999. *New Astron. Rev.*, **43**, 669.

Owen F. N., Odea C. P., Inoue M. and Eilek J. A., 1985. *Astrophys. J.*, **294**, L85.

Özel F., Güver T. and Göğüs E., 2008. *AIP Conf. Proc.*, **983**, 254.

Page L., 1997. *Critical Dialogues in Cosmology*, ed. Turok N., Singapore, World Scientific, pp. 343–362.

Page L., Hinshaw G., Komatsu E. *et al.*, 2007. *Astrophys. J. Suppl.*, **170**, 335.

Pan X., Shao M. and Kulkarni S. R., 2004. *Nature*, **427**, 396.

Paresce F., 1984. *Astron. J.*, **89**, 1022.

Partridge R. B., 1995. *3 K: The Cosmic Microwave Background*, Cambridge, Cambridge University Press.

Patnaik A. R., Browne I. W. A., King L. J. *et al.*, 1993. *MNRAS*, **261**, 435.

Patnaik A. R., Browne I. W. A., Walsh D., Chaffee F. H. and Foltz C. B., 1992. *MNRAS*, **259**, 1P.

Pavelin P. E., Davis R. J., Morrison L. V., Bode M. F. and Ivison R. J., 1993. *Nature*, **363**, 499.

Pawsey J. L., 1951. *Nature*, **168**, 335.

Pearson T. J., Unwin S. C., Cohen M. H. *et al.*, 1981. *Nature*, **290**, 365.

Pedlar A., Muxlow T. W. B., Garrett M. A. *et al.*, 1999. *New Astron. Rev.*, **43**, 535.

Peebles P. J. E. and Yu J. T., 1970. *Astrophys. J.*, **162**, 815.

Penzias A. A. and Wilson R. W., 1965. *Astrophys. J.*, **142**, 419.

Perley R. A., Dreher J. W. and Cowan C., 1984. *Astrophys. J.*, **285**, L35.

Perley R. A., Schwab F. R. and Bridle A. H., 1989. *ASP Conf. Series*, **6**, 528.

Perley R. A., Willis A. G. and Scott J. S., 1979. *Nature*, **281**, 437.

Perlmutter S., Gabi S., Goldhaber G. *et al.*, 1997. *Astrophys. J.*, **483**, 565.

Petrova S. A., 2000. *Astron. Astrophys.*, **360**, 592.

Podsiadlowski Ph., Langer N., Poelarends A. J. T. *et al.*, 2004. *Astrophys. J.*, **612**, 1044.

Polatidis A. G., Wilkinson P. N., Xu W. *et al.*, 1999. *New Astron. Rev.*, **43**, 657.

Quireza C., Rood R. T., Balser D. S. and Bania T. M., 2006. *Astrophys. J. Suppl.*, **165**, 338.

Quirrenbach A., 2001. *Ann. Rev. Astron. Astrophys.*, **39**, 353.

Radhakrishnan V. and Cooke D. J., 1969. *Astrophys. Lett.*, **3**, 225.

Radhakrishnan V., Murray J. D., Lockhart P. and Whittle R. P. J., 1972. *Astrophys. J. Suppl.*, **24**, 15.

Rand R. J. and Kulkarni S. R., 1990. *Astrophys. J.*, **349**, L43.

Rand R. J. and Lyne A. G., 1994. *MNRAS*, **268**, 497.

Raymond J. C., 1984. *Ann. Rev. Astron. Astrophys.*, **22**, 75.

Readhead A. C. S. and Wilkinson P. N., 1978. *Astrophys. J.*, **223**, 25.

Readhead A. C. S., Mason B. S., Contaldi C. R. *et al.*, 2004. *Astrophys. J.*, **609**, 498.

Reber G., 1940. *Astrophys. J.*, **91**, 621.

1944. *Astrophys. J.*, **100**, 279.

1958. *Proc. IRE*, **46**, 15.

Rees M., 1966. *Nature*, **211**, 486.

Refsdal S., 1964. *MNRAS*, **128**, 295.

Reich P. and Reich W., 1986. *Astron. Astrophys. Suppl.*, **63**, 205.

1988. *Astron. Astrophys. Suppl.*, **74**, 7.

Reid M. J., 1993. *Ann. Rev. Astron. Astrophys.*, **31**, 345.

Reid M. J., Readhead A. C. S., Vermeulen R. C. and Treuhaft R. N., 1999. *Astrophys. J.*, **524**, 816.

Reid M. J., Schneps M. H., Moran J. M. *et al.*, 1988. *Astrophys. J.*, **330**, 809.

Reynolds S. P. and Chevalier R. A., 1984. *Astrophys. J.*, **281**, L33.

Reynolds S. P. and Gilmore D. M., 1986. *Astron. J.*, **92**, 1138.

Richards E. A., 2000. *Astrophys. J.*, **533**, 611.

Richards E. A., Kellermann K. I., Fomalont E. B., Windhorst R. A. and Partridge R. B., 1998. *Astron. J.*, **116**, 1039.

Rickett B. J., 1990. *Ann. Rev. Astron. Astrophys.*, **28**, 561.

Riess A. G., Filippenko A. V., Challis P. *et al.*, 1998. *Astron. J.*, **116**, 1009.

Riess A. G., Strolger L.-G., Casertano S. *et al.*, 2007. *Astrophys. J.*, **659**, 98.

Robinson B., 1999. *Ann. Rev. Astron. Astrophys.*, **37**, 65.

Rockstroh J. M. and Webber W. R., 1978. *Astrophys. J.*, **224**, 677.

Rogers A. E. E., 1976, in *Methods of Experimental Physics* Vol. 12c, ed. M. L. Meeks, p. 139.

Rogers A. E. E., 1993. In *Advances in Geodetic VLBI*, Washington, American Geophysical Union.

Rogers A. E. E., Doelman S. S. and Moran J. M., 1995. *Astron. J.*, **109**, 1391.

Rogers A. E. E., Dudevoir K. A. and Bania T. M., 2007. *Astron. J.*, **133**, 1625.

Rogers R. S., Costain C. H. and Landecker T. L., 1999. *Astron. Astrophys. Suppl.*, **137**, 7.

Rood R. T., Bania T. M. and Wilson T. L., 1984. *Astrophys. J.*, **280**, 629.

Rots A. H. and Shane W. W., 1975. *Astron. Astrophys.*, **45**, 25.

Rowan-Robinson M., 1968. *MNRAS*, **141**, 445.

Rubin V. C., Burstein D., Ford W. K. Jr and Thonnard N., 1985. *Astrophys. J.*, **289**, 81.

Rubin V. C. and Ford W. K., 1970. *Astrophys. J.*, **159**, 379.

Ruderman M., 1991. *Astrophys. J.*, **366**, 261.

Ruze J., 1966. *Proc. IEEE*, **54**, 633.

Ryle M., 1950. *Rep. Prog. Phys.*, **13**, 184.

1958. *Proc. Roy. Soc. A*, **248**, 289.

1962. *Nature*, **194**, 517.

Ryle M. and Neville A. C., 1962. *MNRAS*, **125**, 39.

Ryle M. and Smith F. G., 1948. *Nature*, **162**, 462.

Ryle M., Smith F. G. and Elsmore B., 1950. *MNRAS*, **110**, 508.

Ryle M. and Vonberg D. D., 1946. *Nature*, **158**, 339.

Sachs R. K. and Wolfe A. M., 1967. *Astrophys. J.*, **147**, 73.

Safouris V., Subrahmanyan R., Bicknell G. and Saripalli L., 2008. *MNRAS*, **385**, 2117.

Sandage A. and Hardy E., 1973. *Astrophys. J.*, **183**, 743.

Sanders R. H. and Huntley J. M., 1976. *Astrophys. J.*, **209**, 53.

Saripalli L., Subrahmanyan R. and Shankar N. U., 2002. *Astrophys. J.*, **565**, 256.

2003. *Astrophys. J.*, **590**, 181.

Schechter P., 1996. *IAU Symp.*, **169**, 633.

Scheuer P. A. G., 1957. *Proc. Camb. Phil. Soc.*, **53**, 764.

1968. *Nature*, **218**, 920.

1974. *MNRAS*, **166**, 329.

Schilizzi R. T., Burke B. F., Booth R. S. *et al.*, 1984. *IAU Symp.*, **110**, 407.

Schmidt M., 1968. *Astrophys. J.*, **151**, 393.

Schödel R., Ott T., Genzel R. *et al.*, 2003. *Astrophys. J.*, **596**, 1015.

Schwab F. R., 1980. *Soc. Photo-Opt. Inst. Eng.*, **231**, 18.

Seaquist E. R., 1989. *Classical Novae*, ed. M. F. Bode and A. Evans, New York, Wiley, pp. 143–161.
Seiradakis J. H. and Wielebinski R., 2004. *Astron. Astrophys. Rev.*, **12**, 239.
Shao M., Colavita M., Staelin D. H. *et al.*, 1987. *Astron. J.*, **93**, 1280.
Shemar S. L. and Lyne A. G., 1996. *MNRAS*, **282**, 677.
Sheridan K. V. and McLean D. J., 1985, in *Solar Radiophysics*, p. 443.
Shklovsky I. S., 1949. *Astron. Zh.*, **25**, 237.
    1953. *Dokl. Akad. Nauk. SSSR.*, **90**, 983.
    1960. *Cosmic Radio Waves*, Harvard, MA, Harvard University Press.
Silk J., 1968. *Astrophys. J.*, **151**, 459.
Sironi G., Limon M., Marcellino G. *et al.*, 1990. *Astrophys. J.*, **357**, 301.
Smail I., Dressler A., Kneib J.-P. *et al.*, 1996. *Astrophys. J.*, **469**, 508.
Smart W. M., 1977. *Spherical Astronomy*, 6th edn, Cambridge, Cambridge University Press.
Smith E. K., 1982. *Radio Sci.*, **17**, 455.
Smith E. K. Jr. and Weintraub S., 1953. *Proc IRE*, **41**, 1035.
Smith F. G., 1951. *Nature*, **168**, 962.
Smoot G. F., Gorenstein M. V. and Muller R. A., 1977. *Phys. Rev. Lett.*, **39**, 898.
Snellen I. A. G., Schilizzi R. T., Miley G. K. *et al.*, 1999. *New Astron. Rev.*, **43**, 675.
Sokoloff D. D., Bykov A. A., Shukurov A. *et al.*, 1998. *MNRAS*, **299**, 189.
Solomon P. M. and Vanden Bout P. A., 2005. *Ann. Rev. Astron. Astrophys.*, **43**, 677.
Southworth G. C., 1945. *J. Franklin Inst.*, **239**, 285.
Sparks W. B., Biretta J. A. and Machetto F., 1996. *Astrophys. J.*, **473**, 254.
Spencer R. E., 1996. *ASP Conf. Series*, **93**, 252.
Spencer R. E., Swinney R. W., Johnston K. J. and Hjellming R. M., 1986. *Astrophys. J.*, **309**, 694.
Spencer R. E., Vermeulen R. C. and Schilizzi R. T., 1993. In *Stellar Jets and Bipolar Outflows*, p. 203.
Spergel D. N., Bean R., Doré O. *et al.*, 2007. *Astrophys. J. Suppl.*, **170**, 377.
Stanghellini C., O'Dea C. P., Baum S. A., Dallacasa D., Fanti R. and Fanti C., 1997. *Astron. Astrophys.*, **325**, 943.
Stark A. A., Gammie C. F., Wilson R. W. *et al.*, 1992. *Astrophys. J. Suppl.*, **79**, 77.
Strom R., de Pater I. and van der Tak F. (eds.), 1996. *Proc. Conf. High Sensitivity Radioastronomy, Jodrell Bank*, Cambridge, Cambridge University Press, 1996.
Sturrock P. A., 1971. *Astrophys. J.*, **164**, 529.
Sun X. H., Reich W., Waelkens A. and Enβlin T. A., 2008. *Astron. Astrophys.*, **477**, 573.
Sunyaev R. A. and Zel'dovich Ya. B., 1980. *Ann. Rev. Astron. Astrophys.*, **18**, 537.
Swarup G., Ananthakrishnan S., Kapahi V. K. *et al.*, 1991. *Current Sci.*, **60**, No 2.
Tang K.-K., 1984. *Astrophys. J.*, **278**, 881.
Taylor J. H. and Cordes J. M., 1993. *Astrophys. J.*, **411**, 674.
Taylor J. H. and Huguenin G. R., 1971. *Astrophys. J.*, **167**, 273.
Taylor J. H. and Weisberg J. M., 1989. *Astrophys. J.*, **345**, 434.
Terzian Y. and Parrish A., 1970. *Astrophys. Lett.*, **5**, 261.
Toomre A., 1981, in *Structure and Evolution of Normal Galaxies*, ed. Fall and Lynden-Bell, p. 111.
Toomre A. and Toomre J., 1972. *Astrophys. J.*, **178**, 623.
Toscano M., Sandhu J. S., Bailes M. *et al.*, 1999. *MNRAS*, **307**, 925.
Townes C. H., 1947. *Astrophys. J.*, **105**, 235.

1957. *IAU Symposium* **4**, ed. van der Hulst H. C., Cambridge, Cambridge University Press, p. 92.

Townes C. H. and Schawlow A. L., 1955. *Microwave Spectroscopy*, New York, McGraw-Hill.

Tully R. B. and Fisher J. R., 1977. *Astron. Astrophys.*, **54**, 661.

Turner E. L., Ostriker J. P. and Gott J. R., 1984. *Astrophys. J.*, **284**, 1.

Urry C. and Padovani P., 1995. *Publ. Astron. Soc. Pacific*, **107**, 803.

Uyaniker B., Fürst E., Reich W., Reich P. and Wielebinski R., 1999. *Astron. Astrophys. Suppl.*, **138**, 31.

van Dishoek E. F., Jansen D. J. and Phillips T. G., 1993. *Astron. Astrophys.*, **279**, 541.

van Dyk S. D., Weiler K. W., Sramek R. A., Rupen M. P. and Panagia N., 1994. *Astrophys. J.*, **432**, L115.

van Paradijs J., Taam R. E. and van den Heuvel E. P. J. 1995. *Astron. Astrophys.*, **299**, L41.

van de Hulst H., 1945. *Ned. Tijd. Natuurkunde*, **11**, 210.

van den Heuvel E. P. J., 2007. *AIP Conf. Proc.*, **924**, 598.

Verschuur G. L., 1989. *Astrophys. J.*, **339**, 163.

Vlemmings W. H. T., Harvey-Smith L. and Cohen R. J., 2006. *MNRAS*, **371**, 26.

von Hoerner H., 1967. *Astron. J.*, **72**, 35.

Wakker B. P. and van Woerden H., 1997. *Ann. Rev. Astron. Astrophys.*, **35**, 217.

Waldram E. M., Pooley G. G., Grainge K. J. B. *et al.*, 2003. *MNRAS*, **342**, 915.

Walker R. C., Benson J. M. and Unwin S. C., 1987. *Astrophys. J.*, **316**, 546.

Wall J. V., Jackson C. A., Shaver P. A., Hook I. M. and Kellermann K. I., 2005. *Astron. Astrophys.*, **434**, 133.

Walsh D., 1993. In *Proc. Conf. on Subarcsec Astronomy*, ed. Davis R. J. and Booth R. S., Cambridge, Cambridge University Press, p. 111.

Walsh D., Carswell R. F. and Weymann R. J., 1979. *Nature*, **279**, 381.

Walter F., Carilli C., Bertoldi F. *et al.*, 2004. *Astrophys. J.*, **615**, 17.

Wambsganss J., 1993. *Gravitational Lenses in the Universe; Proceedings of the 31st Liège International Astrophysical Colloquium*, ed. Surdej J., Fraipont-Caro D., Gosset E., Refsdal S. and Remy M., Liège, Institut d'Astrophysique, Université de Liège, p. 369.

Warwick J. W., Pearce J. B., Evans D. R. *et al.*, 1981. *Science*, **212**, 239.

Warwick J. W., Pearce J. B., Riddle A. C. *et al.*, 1979. *Science*, **206**, 991.

Weinreb S., 1963. *MIT Technical Report* 412.

Weinreb S., Barrett A. H., Meeks M. L. and Henry J. C., 1963. *Nature*, **200**, 829.

Weltevrede P., Edwards R. T. and Stappers B. W., 2006. *Astron. Astrophys.*, **445**, 243.

Werner M. W., 2004. *Astrophys. J. Suppl.*, **154**, 1.

Wevers B. M. H., van der Kruit P. C. and Allen R. J., 1986. *Astron. Astrophys. Suppl.*, **66**, 505.

White G. J. and Padman R., 1991. *Nature*, **354**, 511.

White R. L., Becker R. H., Gregg M. D. *et al.*, 2000. *Astrophys. J. Suppl.*, **126**, 133.

Wielebinski R. and Krause F., 1993. *Astron. Astrophys. Rev.*, **4**, 449.

Wild J. P., Roberts J. A. and Murray J. D., 1954. *Nature*, **173**, 532.

Wild J. P., Sheridan K. V. and Neylan A. A., 1959. *Aust. J. Phys.*, **12**, 369.

Wilkinson P. N., 1991. In *IAU/URSI Symposium* 131, ed. T. J. Cornwell, Washington, Astronomical Society of the Pacific, p. 381.

Wilkinson P. N. *et al.*, 1997. In *Observational Cosmology with the New Radio Surveys*, ed. Bremer M. N., Jackson N. and Nirez-Fournon I., Dordrecht, Kluwer Academic Publishers, p. 221.

Wilkinson P. N., Polatidis A., Readhead A. C. S., Xu W. and Pearson T., 1994. *Astrophys. J.*, **432**, L87.

Will C. M., 2006. *Prog. Theor. Phys. Suppl.*, **163**, 146.

Wilson R. W., Jefferts K. B. and Penzias A. A., 1970. *Astrophys. J.*, **161**, L43.

Wilson T. L., Walmsley C. M. and Baudry A., 1990. *Astron. Astrophys.*, **231**, 159.

Wolleben M., 2007. *Astrophys. J.*, **664**, 349.

Wolleben M., Landecker T. L., Reich W. and Wielebinski R., 2006. *Astron. Astrophys.*, **448**, 411.

Wolszczan A., 1994. *Science*, **264**, 538.

Wolszczan A. and Frail D. A., 1992. *Nature*, **355**, 145.

Wood-Vasey W. M., Miknaitis G., Stubbs C. W. *et al.*, 2007. *Astrophys. J.*, **666**, 694.

Woods P. M. and Thompson C., 2006. *Compact Stellar X-Ray sources*, ed. Lewin W. and van der Klis M., Cambridge, Cambridge University Press, p. 547.

Wright A. E. and Barlow M. J., 1975. *MNRAS*, **170**, 41.

Wright E. L., Chen X., Odegard N. *et al.*, 2008. arXiv 0803.0577.

Wynn-Williams G. C., Becklin E. E. and Neugebauer G., 1972. *MNRAS*, **160**, 1.

Yakovlev D. G. and Pethick C. J., 2004. *Ann. Rev. Astron. Astrophys.*, **42**, 169.

You X. P., Hobbs G., Coles W. A. *et al.*, 2007. *MNRAS*, **378**, 493.

Young M. D., Manchester R. N. and Johnston S., 1999. *Nature*, **400**, 898.

Yun M. S., Ho P. T. P. and Lo K. Y., 1994. *Nature*, **372**, 530.

Yusef-Zadeh F., Morris M. and Chance D., 1984. *Nature*, **310**, 557.

Zel'dovich Ya. B., 1970. *Astron. Astrophys.*, **5**, 84.

1972. *MNRAS*, **160**, 1P.

Zirin H., Baumert B. M. and Hurford G. J., 1991. *Astrophys. J.*, **370**, 779.

Zwicky F., 1933. *Helvetica Phys. Acta*, **6**, 110.

Index